D. Crookall · K. Arai (Eds.)

Global Interdependence

Simulation and Gaming Perspectives

Proceedings of the 22nd International Conference
of the International Simulation and Gaming Association (ISAGA)
Kyoto, Japan: 15–19 July 1991

Editorial Assistants
J.C. Hilbun · B.F. Harris

With 72 Illustrations

Springer-Verlag
Tokyo Berlin Heidelberg
New York London Paris
Hong Kong Barcelona

Dr. *David Crookall*
Department of English, University of Alabama, Tuscaloosa, AL 35487-0244, USA

Dr. *Kiyoshi Arai*
Department of Industrial Engineering and Management, Kinki University, Kyushu, Iizuka-shi, Fukuoka 820, Japan

Editorial Assistants
Joy C. Hilbun and *Brenda F. Harris*
University of Alabama, Tuscaloosa, AL 35487–0244, USA

ISBN-13:978-4-431-68191-5   e-ISBN-13:978-4-431-68189-2
DOI: 10.1007/978-4-431-68189-2

Library of Congress Cataloging-in-Publication Data. International Simulation and Gaming Association. International Conference (22nd : 1991 : Kyoto, Japan) Global interdependence : simulation, gaming, and perspectives : proceedings of the 22nd International Conference of the International Simulation and Gaming Association (ISAGA), Kyoto, Japan, 15–19 July 1991 / D. Crookall, K. Arai, eds.; editorial assistants, J.C. Hilbun, B.F. Harris. p.   cm.   Includes bibliographical references. ISBN-13:978-4-431-68191-5
1. Game theory — Congresses.   2. Simulation methods—Congresses. 3. Intercultural communication—Congresses. 4. Economic development—Mathematical models—Congresses. 5. Economics—Mathematical models—Congresses. 6. Environmental policy—Mathematical models—Congresses. I. Crookall, David. II. Arai, K. (Kiyoshi), 1950–   . HB144.I58 1991. 519.3—dc20. 92-5641

© Springer-Verlag Tokyo 1992
Softcover reprint of the hardcover 1st edition 1992

Typesetting, printing and binding: Best-set Typesetter Ltd., Hong Kong

Welcome Messages

I wish to express a most heartfelt welcome to our distinguished visitors from all over the world, who have come to attend this 22nd Annual International Conference of the International Simulation and Gaming Association, co-sponsored by the Science Council of Japan and the Japanese Association of Simulation and Gaming. I am sure the conference will be a tremendous success.

Toshiki Kaifu
Prime Minister

本日ここに、日本学術会議およひ日本シミュレーション＆ゲーミング学会の共催により第22回国際シミュレーション＆ゲーミング学会が開催されるにあたり、世界各国からの参加者の皆様を心から歓迎申し上げるとともに、貴国際会議が多大な成果を修められることを祈念いたします。

海部俊樹
内閣総理大臣

Congratulations to the ISAGA'91 organizers for your outstanding international leadership on hosting the conference on the theme of Global Modeling for Solving Global Problems. Please extend my best wishes to all ISAGA participants now in Kyoto.

Harold Guetzkow
Northwestern University

Post-Conference Reflections

The Science Council of Japan was established in 1949 as a governmental organization, representing 530,000 qualified Japanese scientists, both domestically and internationally, and covering all scientific fields from the cultural and social to the natural sciences. The aim of the Council is to promote scientific development and to help improve administration, industry, and the well-being of the general public through science.

As President of the Council, I am pleased that it was able to sponsor the 22nd Annual International Conference of the International Simulation and Gaming Association, in cooperation with the Japanese Association of Simulation and Gaming (which is registered with the council). It was an extraordinary pleasure for me to have had the opportunity of being with such distinguished scientists from around the world, and to have listened to their lectures and presentations on their many achievements.

The field simulation and gaming has been developing in step with advances in computer technology, and has become a valid scientific method for making predictions about the future using sophisticated computerized models to simulate complex natural and social phenomena. Today, this methodology is becoming increasingly important in many areas, such as the prediction of global environmental problems in the 21st century, the future prospects of the world economy and international politics, and the evaluation of business strategies.

The science of simulation and gaming in Japan has been gaining international reputation in some areas, such as pioneering studies for predicting the lessening of international political tensions and predictive research concerning the world economy. Global simulation models, such as the FUGI Model, are based on quantitative descriptions of dependence among many nations.

The applications of simulation/gaming methodologies, however, are spread across a wide range of scientific fields; this makes the recently recognized need for cooperation among the researchers even more urgent. Fortunately, in 1988, the Japanese Association of Simulation and Gaming was established through the efforts of the scientists concerned. I am very honored to have been elected as first president of the association.

It was indeed a great honor and privilege for me to attend the Conference, with distinguished delegates active in this scientific field, and it is gratifying to witness the wide ranging discussions in these proceedings.

Jiro Kondo
Science Council of Japan

Preface

This volume records the proceedings of the 22nd Annual International Conference of the International Simulation and Gaming Association (ISAGA), 15–19 July, 1991, Kyoto, Japan, sponsored by the Science Council of Japan and the Japanese Association of Simulation and Gaming (JASAG). The conference theme was Global Modeling for Solving Global Problems.

The first 2 days of the conference were held in the magnificent Kyoto International Conference Hall; the 3rd day was spent admiring the floats of the famous Gion Festival in the exquisite city of Kyoto and the Daibutsu (or Great Buddha) of the Todaiji Temple in Nara and visiting one of the Sharp factories. During the last 2 days of the conference we were made most welcome in the Faculty of International Relations of Ritsumeikan University. The day after the conference, a number of delegates went to Hiroshima (the Peace Memorial Hall, Museum and Park) and also to one of Japan's "Scenic Trio," the island of Miyajima with its breathtaking views and the Itsukushima Shrine.

The conference was attended by some 400 delegates from over 30 different countries. Over 100 sessions, both theoretical and practical, were given: keynote speeches, round-table discussions, workshops, papers. This volume reflects most of those sessions, in the form of either a full paper or a short abstract.

Simulation/gaming is fast gaining recognition as an extraordinary powerful methodology in many areas. Few other methodologies appear to have as much immediate relevance to global interdependence and to have as great a potential in dealing with global problems and in making links between the local and the global.

Increasing numbers of people are becoming aware of the need to think and act in a perspective of global interdependence, whether this be couched in political, environmental, or cultural terms. The world is becoming smaller as contact among peoples of the planet grows; vital concerns, such as the environment or development, are becoming ever more tightly woven into a transnational web. Events and issues, mundane or momentous, from the local to the global, increasingly mesh with a wider planetary context. The need, therefore, is greater than ever before, not only to encourage more effective communication between, but above all to enable deeper understanding and tolerance of, the various cultural groups around the world. Because of its dynamic, interactive, learner-centered nature, and because it is grounded in emerging, concrete realities, simulation/gaming is admirably equipped to help us gain greater understanding of these complex international phenomena and thus to contribute to the enhancement of international understanding [Crookall D (1990) Editorial: Future perfect? Simulation & Gaming: An International Journal of Theory, Design, and Research 21(1)3–11].

The papers in this book are grouped into clusters, which—as usual when dealing with simulation material—overlap in many ways. The eclectic nature of simulation/gaming as well as its immediate application to real-world problems mean that conventionally defined disciplinary boundaries are inadequate for many publications on simulation/gaming. The clusters chosen here serve to indicate a general emphasis or theme, rather than a speciality; some papers could easily have appeared in more than one category. Each cluster is contained in a section, as follows:

1. Professional and Methodological Issues
2. Communication and Culture
3. Environmental and Developmental Issues
4. Economics and Business
5. Abstracts

The first section contains an important chapter (Meadows) on the need to make simulation/gaming more professional. The other chapters look at various aspects of simulation/gaming as a methodology, for example, design, evaluation, and modeling. The second section tackles a broad array of topics and concerns related to communication and culture, which are both interdisciplinary and interdependent. The first chapter (Lederman) provides a valuable overview and analysis of key concepts, while the other chapters deal with certain facets of this double theme: intercultural communication, global awareness, language games, international negotiation. Section 3 also deals with two themes, the environment and development, which are closely related and which tend to involve policy exercises. The first chapter (Greenblat) illustrates the complexity and breadth of the issues involved, while the other chapters focus on specific areas, such as agriculture, town planning, and ecology. (The paper by Toth is good example of an account which, involving both culture and the environment, could have easily been located in the previous section.) The next section, 4, looks at some of the topics from the angle of business and economics. A brief predictive overview is provided in an intriguing first chapter (Teach), while the other chapters dwell on several topics, including global modeling, post-communist economies, decision making, and stock markets. Finally, Section 5 contains the abstracts of many of the remaining conference sessions. Taken together, these papers provide a glimpse of the state of simulation/gaming and its contribution to global interdependence.

Acknowledgments

In addition to sponsorship by the Science Council of Japan and the Japanese Association of Simulation and Gaming, a great many people and organizations were involved in organizing and supporting the conference. They are all to be thanked heartily; without each of their contributions, the success of the conference would have been diminished. I wish here to record many people's deep gratitude to Kiyoshi Arai for his tremendous work, and to express my own warm appreciation for his co-editing of these proceedings. The following people and organizations all deserve our gratitude; those omitted in error are asked to forgive the oversight.

Welcome and Keynote Speakers

First thanks go to Toshiki Kaifu, then Prime Minister of Japan, for his welcome message. Harold Guetzkow, Professor Emeritus, Northwestern University and one of the world's most respected simulation scholars, is thanked for his good wishes, read at the ceremony. Jiro Kondo, President of the Science Council of Japan, has very kindly formulated some after-conference thoughts. (These three texts appear before this Preface.)

Several people gave opening keynote presentations; they deserve special thanks for the insight they provided and the outlook which they encouraged. Their words (some of which appear in this volume) set the scene and tone for the rest of the conference week.

Jiro Kondo, Science Council of Japan, Japan
Linda C. Lederman, Rutgers University, USA
Dennis L. Meadows, University of New Hampshire, USA
Anatol Rapoport, University of Toronto, Canada
Hiroharu Seki, Ritsumeikan University, Japan
Diana Shannon, Rutgers International Simulation and Gaming Center, USA

This volume is dedicated to Hiroharu Seki for his enthusiasm in applying the methodology of simulation/gaming to global problems and for his initiative in establishing the Japanese Association of Simulation and Gaming (JASAG).

Committees

Ten different committees were involved in one way or another with the conceptualization, planning, funding, and organization of the conference. Each committee is to be warmly thanked for the work it accomplished to make the conference the success that it was.

Honorary Committee

Teiichi Aramaki, Governor of Kyoto Prefecture
Koji Fushimi, Post-President of the Science Council of Japan
Masateru Ohnami, President of Ritsumeikan University
Yutaka Ohno, Director of the Institute of Science and Engineering, Ritsumeikan University
Naoshi Sanada, Vice-president of Ritsumeikan University
Tomoyuki Tanabe, Mayor of Kyoto City

Organizing Committee

Chair: Hiroharu Seki, Ritsumeikan University
Vice-chair: Akira Onishi, Soka University
General Secretary: Kanji Sato, Soka University
Members:

Hirotsugu Akaike, Institute of Statistical Mathematics
Ikuro Anzai, Ritsumeikan University
Yoichi Erikawa, KOEI Co.
Tadashi Hidano, University of the Air
Kazuo Inoue, Ritsumeikan University
Tadashi Kawata, Sophia University
Sakutaro Kyozuka, Chuo University
Kinhide Mushakoji, Meiji Gakuin University
Yozo Shimizu, Japan Personal Computer Software Association
Mitsuo Takahashi, Tsukuba University
Fukuji Taguchi, Nagoya University
Tatsuo Urabe, Nagoya University
Kenji Yamashita, Ritsumeikan University

Kiyoshi Arai, Kinki University
Takeshi Arai, Ritsumeikan University
Norio Baba, Osaka Kyoiku University
Ikuhiko Hata, Takushoku University
Arata Ichikawa, Ryutsukeizai University
Chiyoko Ishida, Niigata University
Tosiyasu Kunii, University of Tokyo
Takehiko Matsuda, Sanno University
Sogo Okamura, Tokyo Denki University
Toshiyuki Sakai, Ryukoku University
Rei Shiratori, Tokai University
Paul Smoker, Antioch College
Sakio Takayanagi, Chuo University
Kei Takeuchi, University of Tokyo
Masaya Yamaguchi, Ryukoku University

Program Committee

Chair: Akira Onishi, Soka University
Vice-chair: Kiyoshi Arai, Kinki University
Members:

Norio Baba, Osaka Kyoiku University
Ikuhiko Hata, Takushoku University
Kazuo Inoue, Ritsumeikan University
Tomio Kinoshita, Kyoto University
Jan Klabbers, Erasmus University Rotterdam
Kinhide Mushakoji, Meiji Gakuin University
Hiroharu Seki, Ritsumeikan University
Paul Smoker, Antioch College

David Crookall, University of Alabama
Arata Ichikawa, Ryutsukeizai University
Toshio Iwasaki, Japan Economics Foundation
Tosiyasu Kunii, University of Tokyo
Masatsugu Matsuo, Hiroshima University
Hideki Ohata, Waseda University
Yukio Sato, Hiroshima University
Yoshihisa Shinagawa, University of Tokyo
Sakio Takayanagi, Chuo University

Public Relations Committee

Chair: Ikuro Anzai, Ritsumeikan University
Vice-chair: Yoichi Erikawa, KOEI Co.

Local Arrangements Committee

Chair: Kazuo Inoue, Ritsumeikan University
Vice-chair: Ikuro Anzai, Ritsumeikan University

Finance Committee

Chair: Arata Ichikawa, Ryutsukeizai University
Vice-chair: Minoru Asahi, Ritsumeikan University

Planning Committee

Chair: Kanji Sato, Soka University
Vice-chair: Chiyoko Ishida, Niigata University

Funding Committee

Chair: Kinhide Mushakoji, Meiji Gakuin University
Vice-chair: Arata Ichikawa, Ryutsukeizai University

Conference Consultant

David Crookall, University of Alabama

Special thanks must be expressed to a number of individuals for their very special help on various aspects of the conference. By last-minute, popular request, several people—Cathy Greenblat and Richard Powers, for example—conducted evening workshops. Ikuro Anzai, Kazuo Inoue, and Katsuari Kamei made everyone most welcome (unfailing help, friendly smiles, buffet suppers, going-away presents, and even real magic) in Ritsumeikan University. Kanji Sato and Arata Ichikawa held the purse strings with great expertise, knowing when to tighten and untie the knots—we are most grateful to them. Two wonderful secretaries provided unfailing service over many weeks and their work is greatly appreciated: Yuriko Shimano and Chikako Nakahara. Atsushi Takamizawa and Akira Watanabe helped out on all sorts of things with un-flinching good humor. Yoshiro Tabuchi, JASAG secretary, was a tremendous help in many ways. Joy Hilbun and Brenda Harris, with no uncertain skill, much good humor, and great willingness, put in many hours helping to put together this book. Thanks also go to Alexandra Bernstein (National Academy of Sciences) and Barbara Jacobsen (Rutgers) for their editorial help and to the employees of Springer-Verlag Tokyo. Last (but by no means least), we thank Springer-Verlag (Tokyo). All these deserve gratitude from all those who attended the conference and who contributed to this book.

Support

A large number of professional associations and societies very kindly lent their name in support of the conference. This brought the conference to the attention of people working in many different fields and did much to attract delegates. The organizations to be thanked include:

Architectural Institute of Japan
Behaviormetric Society of Japan
Ecological Society of Japan
Information Processing Society of Japan
Institute of Electrical Engineers of Japan
Institute of Statistical Research
Japan Association of Economics and
  Econometrics
Japan Association for Philosophy of
  Science
Association of Japanese Geographers
Center for Environmental Information
  Science
Information Science and Technology
  Association
Institute of Systems, Control, and
  Information Engineers
Japan Association of International
  Relations
Japan Association for Planning

Japan Association for Social and
  Economic Systems Studies
Japan Election Studies Association
Japan Industrial Management Association
Japan Social Studies Research Association
Japan Society for Industrial and Applied
  Mathematics
Japan Society for the Promotion of
  Science
Japan Society of Civil Engineers
Japan Society of Educational Technology
Japanese Association of International Law
Japanese Group Dynamics Association
Japanese Society of Social Psychology
Japanese Society for Artificial Intelligence
Law and Computers Association of Japan
Mathematical Society of Japan
Nippon Toshi Joho Gakkai
Peace Studies Association of Japan
Robotics Society of Japan
Society of Environmental Science, Japan
United Nations University

  Administration
Japan Association for Urban Sociology
Japan Ergonomics Research Society
Japan Personal Computer Software
  Association
Japan Society for Comparative
  Civilizations
Japan Society for Science Education
Japan Society for the Study of Business
  Administration
Japan Society of Educational Information
Japanese Association for Mathematical
  Sociology
Japanese Psychological Association
Japanese Society of Soil Mechanics and
  Foundation Engineering
Japanese Society for Science of Design
Linguistic Society of Japan
National Defense Society
Operations Research Society of Japan
Remote Sensing Society of Japan
Society of CAI in Japan
Society of Information Theory and its
  Applications

Generous financial aid was received from several public bodies, and they are warmly thanked. Their aid enabled the conference to achieve its objectives with great success.

Kyoto Municipal Government   Kyoto Prefectural Government

The conference also benefitted immensely from a grant from the Commemorative Association for the Japan World Exhibition (1970).

In addition, donations were given by a large number of organizations, corporations, or companies. Each donor is thanked; they all contributed greatly to the success of the conference.

ArgoTechnos21
Association of Tokyo Stock Exchange
Canon
CREO
Dempa Publications
DEC Japan
Fujitsu
Imagineer
Itoman
Japan Automobile Manufacturers
  Association
Japan Foreign Trade Council
Japan Systems Engineering Corporation
Kanda Tsushinki
Kankaku Securities
KOEI
Matrix

Artdink
C.Itoh
Chori
Daiwa Securities
Dexter
ENIX
Hitachi
Intercom
Iwatani International Corporation
Japan Federation of Construction
  Contractors
Justsystem Corporation
Kanematsu
Kawasho
Marubeni
Matsushita Electric Industrial Company
Milky Way

Microsoft
Mitsubishi
Mitsui
NEC
Nichimen
Nippon Telenet
Nissho Iwai
Obayashi
PFU
Regional Banks Association of Japan
Softbank
Softwing
Sumitomo
Federation of Electric Power Companies
Japan Iron & Steel Federation
Marine and Fire Insurance Association of
    Japan
Tomen Corporation
Toshiba
Toyota Motor

Mitsubishi Heavy Industries
Nagase
NHK Joho Network
Nippon Express
Nissei Sangyo
Nomura Research Institute
Okura
Professional Computer Automation
Shimizu
Software Japan International
Sogo Toshikaihatsu
Suzuyo
Japan Gas Association
Life Insurance Association of Japan
Nomura Securities
Tokyo Bankers Association
Toppan Printing
Toshoku
Toyota Tsusho

In addition, we thank several private organizations for their greatly appreciated services and cooperation.

Apple Computer Japan
Fujitsu
Hi-Vision Promotion Center
Japan Airlines
Konami
NEC

Artdink
Hitachi
Imagineer
KOEI
Maruzen
Sharp

About ISAGA

The main aims of the International Simulation and Gaming Association (ISAGA) are to serve as a forum for the exchange of ideas and information on all aspects of simulation/gaming and to further the development and use of simulation/gaming around the world. Previous ISAGA conferences have taken place in such locations as:

Aberdeen, Scotland
Caracas, Venezuela
Elsinor, Denmark
Geneva, Switzerland
Toulon, France
Venice, Italy

Alma Ata, Kazakhstan
Durham, NH, USA
Haifa, Israel
Sofia, Bulgaria
Utrecht, Netherlands
Weimar, Germany

The official periodical of ISAGA is *Simulation & Gaming* (see below). Proceedings of some of the previous conferences have been published; they are as follows:

Bruin K, de Haan J, Teijken C, Veeman W (eds) (1979) How to build a simulation/game. (Two volumes) Centrale Reproductiendienst der Rijksuniversiteit, Gronigen

Crookall D, Greenblat CS, Coote A, Klabbers JHG, Watson DR (eds) (1987) Simulation-gaming in the late 1980s. Pergamon Press, Oxford

Crookall D, Klabbers JHG, Coote A, Saunders D, Cecchini A, Delle Piane A (eds) (1988) Simulation-gaming in education and training. Pergamon Press, Oxford

Klabbers JHG, Scheper WJ, Takkenberg CAT, Crookall D (eds) (1989) Simulation-gaming: On the improvement of competence in dealing with complexity, uncertainty and value conflicts. Pergamon Press, Oxford

The most recent ISAGA Steering Committee is composed of the following members:

Officers

President	Hiroharu Seki, Ritsumeikan University, Kyoto, Japan
Past President	Dennis Meadows, University of New Hampshire, USA
President-Elect	Fred Percival, Napier Polytechnic, Scotland
General Secretary	Jan Hisok G. Klabbers, Erasmus University, Rotterdam, Netherlands
Treasurer	Alan Cudworth, Nottingham Polytechnic, England
Journal Editor	David Crookall, University of Alabama, USA
SAGSET Representative	Alan Coote, p.i. Associates, Wales
ABSEL Representative	Richard Teach, Georgia Institute of Technology, USA
NASAGA Representative	Steven Underwood, University of Michigan, USA

Regional Secretaries

Africa	Kunle Akinyemi, University of Ilorin, Nigeria
Australia (AUSSAGA)	Elizabeth Christopher, Charles Sturt University, Australia
Central/Eastern Europe	Victor I. Rybalski, Kiew Institute of Technology, Ukraine
Japan (JASAG)	Hiroharu Seki, Ritsumeikan University, Japan
Italy (SIGIS)	Arnaldo Cecchini, Venice University, Italy
Latin America	Leopoldo Schapira, Universidad Nacional de Cordoba, Argentina
Western Europe	Jan H. G. Klabbers, Erasmus University, Rotterdam, Netherlands
USA	Steven Underwood, University of Michigan, USA
Philippines (PHILSAG)	Florosita Q. Pimental, Quezon City, Philippines

Members at Large

Kiyoshi Arai, Kinki University in Kyushu, Japan
Douglas W. Coleman, University of Toledo, USA
Richard D. Duke, University of Michigan, USA
Adriana Frisenna, Venice University, Italy
Hans Gernert, Humbold University, Berlin, Germany
Cathy Stein Greenblat, Rutgers University, USA
Dmitri Kavtaradze, Moscow State University, Russia
Ludmilla Kraova, Academy of Sciences, Moscow, Russia

Hubert Law-Jones, Technion, Israel
Linda Langerman, Rutgers University, USA
Laurent Mermet, Paris, France
Elzbieta Naumienko, Warsaw, Poland
Krzysztof Nowak, Katowice, Poland
E. Radaceanu, Institual Roman de Management, Bucarest, Romania
R. Siebecke, Friedrich-Schiller Universitat, Jena, Germany
Wladyslaw R. Switalski, Warsaw University, Poland
Ivo Wenzler, Multilogue International, Ann Arbor, USA

To obtain information on ISAGA and its conferences, either write to JHG Klabbers, Secretary General, ISAGA, Gostervelden 59-6681, WR Bemmel, Netherlands, or consult a recent issue of the journal.

Simulation & Gaming

Simulation and Gaming: An International Journal of Theory, Design and Research is the world's foremost journal devoted to academic and applied issues in the fields of simulation, computerized simulation, gaming, modeling, play, role-play, and active, experiential learning, and related methodologies in education, training, and research. The broad scope and interdisciplinary nature of *Simulation & Gaming* is demonstrated by the variety of its readers and contributors, such as sociologists, political scientists, economists, psychologists, and educators, as well as experts in environmental issues, international studies, management and business, policy and planning, decision making and conflict resolution, cognition, learning theory, communication, language learning, media, educational technologies, and computing. The journal carries scholarly research articles, reports, read-to-use simulation/games, reviews of books and simulation/games, news of ISAGA activities, and announcements about events and resources.

Before submitting a manuscript, authors should write (enclosing a self-addressed, self-adhesive label and, in the USA, $2 in stamps) for a copy of the Guide for Authors to David Crookall, Editor, *Simulation & Gaming*, English/Morgan, University of Alabama, Tuscaloosa, AL 35487, USA, or c/o Sage Publications, 2455 Teller Road, Newbury Park, CA 91320, USA or 6 Bonhill Street, London EC2A 4PU, UK. To subscribe, contact Sage Publications at one of the above addresses.

Finally . . .

Each ISAGA conference is a wonderful event, and the conference in Kyoto was a truly enriching experience—one which all delegates will cherish for years to come. Thank you, Kiyoshi, for pulling it all together.

David Crookall
Tuscaloosa and Washington, USA

Contents

Section 1   Professional and Methodological Issues

Section 2   *Communication and Culture*

Section 3   *Environmental and Developmental Issues*

Section 4   Economics and Business

Section 5   *Assembled Abstracts*

List of Contributors

Section 1
Professional and Methodological Issues

Preparing the Gaming Profession to Deal with Problems of the Twenty-first Century

Dennis L. Meadows[1]

Abstract. Solving global problems will require great foresight, concerted action across national boundaries, and profound changes in technology, culture, and economics. This conclusion is fully recognized by New Earth 21, a recent Japanese initiative to deal with climate change. I propose here a partnership between the global gaming community and the scientific and political leaders who have begun to plan the New Earth 21 initiative. Collaboration would pay dividends for both sides. Those concerned with global problems will gain access to a set of powerful tools for social research, communication, and training. Gamers would gain political and financial support for their work. They would also be confronted with the need to make long overdue changes in the conduct of their profession: standardizing and extending professional training programs, raising the quality of documentation, and developing better means for accessing the growing body of important educational games.

Key words. climate change; gaming profession; global models; global problems; Limits to Growth; New Earth 21; simulation profession

For two days this past June I participated in a meeting organized by the Japanese to focus on a fundamental problem of the twenty-first century. About one hundred of Japan's most senior government, corporate, and scientific leaders came to Atlanta, Georgia in the United States to meet with a similar number of their American counterparts. The purpose of the seminar was to find partners who could help undertake a monumental new project, "New Earth 21," a 100-year effort to counteract threats of global climate change [Fukukawa (1991) The challenge of Japan and the United States: Recovery of the global environment (unpublished). Available from The Global Industrial and Social Progress Research Institute, 3-8-21 Toranomon, Minato-ku, Tokyo, 105 Japan].

My purpose today is to summarize why and how the international community of gamers should offer to become a partner in the New Earth 21 initiative. I have four reasons for my proposal:

[1] Institute for Policy and Social Science Research, Hood House, University of New Hampshire, Durham, NH 03824, US; phones 603/862-2186 (w) 603/868-1942 (h); facsimile 603/862-1488; telex 493 0372; e-mail d_meadows@unhh.unh.edu

1. This Japanese initiative is unique in the breadth of its vision and in the level of national support. It is the only plan I have seen that could possibly avert what I believe will otherwise be the catastrophic consequences of global warming. It deserves support by all the globe's citizens.
2. I know absolutely that efforts to avoid climate change cannot succeed without the forms of social innovation and learning that emerge from gaming and simulation experiments.
3. Creating tools useful to the New Earth 21 program will enable our profession to build important foundations for dealing with other global problems as well.
4. Efforts to become an effective partner in the New Earth 21 program will confront our discipline clearly with the need to make some long overdue improvements in its own administration and practice.

Outline of My Remarks

My comments today will elaborate on each of those four points. First, I will define the difference between global problems and other challenges with which society has had more experience and more success. I will summarize several relevant conclusions that have emerged from my own 20 years' professional experience using global models in games and simulations. Included in my summary will be some new results that are reported publicly here for the first time. I will briefly mention the six main goals of the New Earth 21 initiative and show how achieving them will require new advances from the gaming and simulation community. Finally, I will point to several changes our profession should bring to its own practice, if it is to be an effective and a credible partner in humanity's efforts to cope with global problems.

The goals I have outlined for this speech constitute an ambitious agenda. Obviously I cannot deal conclusively with any of these topics. But my purpose is not to lay out a final plan. I want to raise your interest in this proposal. I want to give you some basic information related to it. I want to suggest a list of topics that we might fruitfully discuss in more informal settings during and after this conference. Throughout this discussion, I will use the terms "we" and "our" specifically to designate the global community of those who use educational games and simulations in their profession.

Universal Versus Global Problems

As we begin to confront issues like climate change, it is crucial to recognize the fundamental differences between universal and global problems. For dealing with the first, society can draw on many decades of experience, numerous institutions, and countless technologies. However, mankind is in a very preliminary stage of preparing to cope with the second.

Universal problems are those which affect the peoples of many countries, but they can and must be dealt with internally to a region or a nation. Soil erosion, ground water pollution, deforestation, municipal waste accumulation, and hazardous waste spills in the soil are examples. While international collaboration can be helpful, for example through the transfer of appropriate technologies, the solution to these problems lies intrinsically with those who live and produce within the region.

Unilateral initiatives can be effective with universal problems. For example, West Germany was able to reduce contamination of the air in its own cities even while this problem grew to catastrophic proportions in East Germany.

Global problems are fundamentally different. While they also affect peoples of many countries, global issues can be dealt with only through concerted action across regional, national, and even continental boundaries. Global problems generally do not respond quickly. They typically require actions that must be sustained in a consistent fashion for many decades in order to achieve fundamental solutions. They generally do not have purely technical solutions. Initiative cannot be unilateral. For example, Japan cannot avoid the consequences of climate change by reducing its own CO_2 emissions, so long as other major nations ignore the problem of global warming.

Global problems require unprecedented foresight. In some ways, dealing with global problems is like steering a mighty ship to avoid icebergs in the fog. Because of the inertia in a large ship, it can take a great distance and substantial time to change its course. As a consequence, when the ship is passing through a fog, its navigators use radar to identify obstacles early enough so that they still have time to pick a safer course. Global modelling first began to develop in the early 1970s as a form of radar for use with social systems. Those early models let policy makers begin to peer several decades ahead in the evolution of their social and economic systems to anticipate the nature and the magnitude of emerging global problems.

I directed one of the earliest global modelling efforts, and the conclusions that emerged from our research are still relevant today. Let us look at some of those conclusions.

Perspectives on Global Problems from *The Limits to Growth*

In the period 1970–1972, I carried out with colleagues at MIT a program of data research, mathematical modelling, and simulation to understand the long-term causes and consequences of growth in the globe's population and of exponential expansion in the physical aspects of the world's economic output. Our analyses lead to construction of WORLD3, a computer simulation model, and our results were reported in four texts including *The Limits to Growth* (Meadows et al. 1972, 1973, 1974, 1977). The basic conclusion was simply stated:

If the present growth trends in world population, industrialization, pollution, food production, and resource depletion continue unchanged, the limits to growth on this planet will be reached sometime within the next 100 years. The most probable result will be a rather sudden and uncontrollable decline in both population and industrial capacity.

With this conclusion came two insights. First, changes in technology alone will not solve global problems. Technological advance generally serves only to shift the burden of the problem from one sector to another. For example, increasing fertilizer use will increase food production and thereby decrease hunger, but it also raises pollution of ground and surface waters while hastening the depletion of fossil fuels. New technologies can be helpful, but changes in social values and institutions are required as well. For example, the ultimate solution to hunger lies in stabilizing the population—a goal that implies many changes including reductions in desired family size and improvements in the status of women.

The second insight was to recognize the difference between problem and symptom. The underlying cause of most global problems is the continuation of exponential growth in population, resource use, and energy consumption on a finite planet. Hunger, ozone depletion, and climate change are symptoms of that problem. You will never get rid of all the symptoms as long as the real problem is not addressed. For example, attempting to deal with climate change while ignoring population growth and material economic expansion is like prescribing aspirin for a chronic headache while ignoring the growing brain tumor that is causing it.

In April 1991, we began a comprehensive reanalysis of the data, references, and computer simulation runs used in our original analysis. The results will lead to the publication of a new book in the spring of 1992, *Beyond the Limits* (Meadows et al. 1992).

This recent work already makes one conclusion absolutely clear. The principal facts that led to our original results have not been changed during the past 20 years. The growth trends in world population, industrialization, pollution, food production, and resource depletion still continue essentially unchanged. Figures 1 and 2 show two of the graphs that were published originally in *The Limits to Growth*. Here they have been extended with data for the past 20 years. The two figures are characteristic of most data plots you could draw to show behavior of the globe's demographic and economic parameters over the past 2 decades. Most physical indices of mankind's activities have nearly doubled since our first study, and population has grown even faster than we projected—rising from 3.7 in 1970 to 5.3 billion in 1990 (UNEP 1991).

WORLD3 was not designed for the task of making precise predictions, but anyone who cared to interpolate on our graphs could see that our standard reference projection produced turning points in population and economy around the year 2030. I still believe that our standard run is a useful forecast. Since *The Limits to Growth* was published in 1972, humanity has come

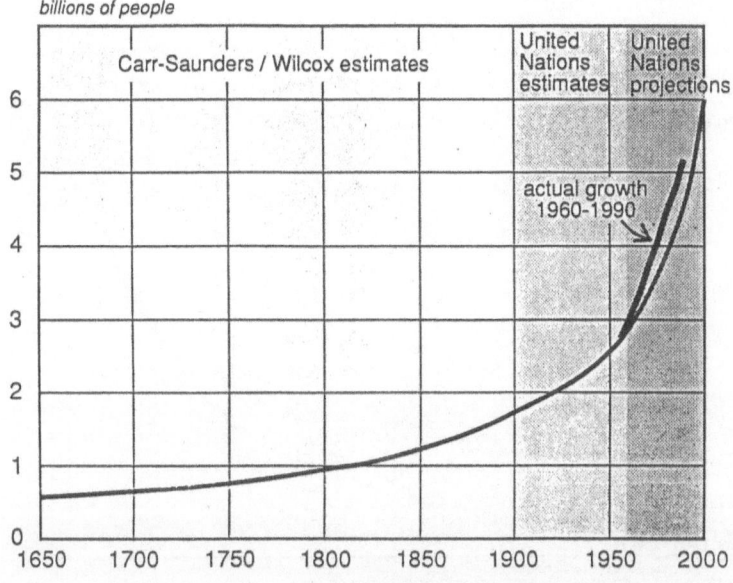

Source: DJ Bogue
UN, *World Population Prospects: estimates
& projections as assessed in 1984.*

Fig. 1. World population projections. From Meadows et al. 1992 with permission

approximately one-third of the way towards the "Global Turning Point" in its physical expansion.

Although global population, energy use, and resource usage are now above the planet's long-term carrying capacity, our new simulations still show that there is time to change course and move along an orderly path towards a sustainable and equitable global society which can satisfy the basic needs of all its citizens. But this can be done only with major new initiatives, enacted urgently, that confront realistically the nature of our problems, our resources, and our options.

Social Versus Technical Solutions

Up to now most international initiatives dealing with a global problem have been fundamentally unrealistic; they have treated some individual symptom of the limits to growth as if it were an independent problem. They have assumed that this problem could be addressed and solved in isolation from the underlying growth trends. They have assumed that technological solutions would be possible and that social change was neither possible nor necessary. New Earth 21 does not make those mistakes.

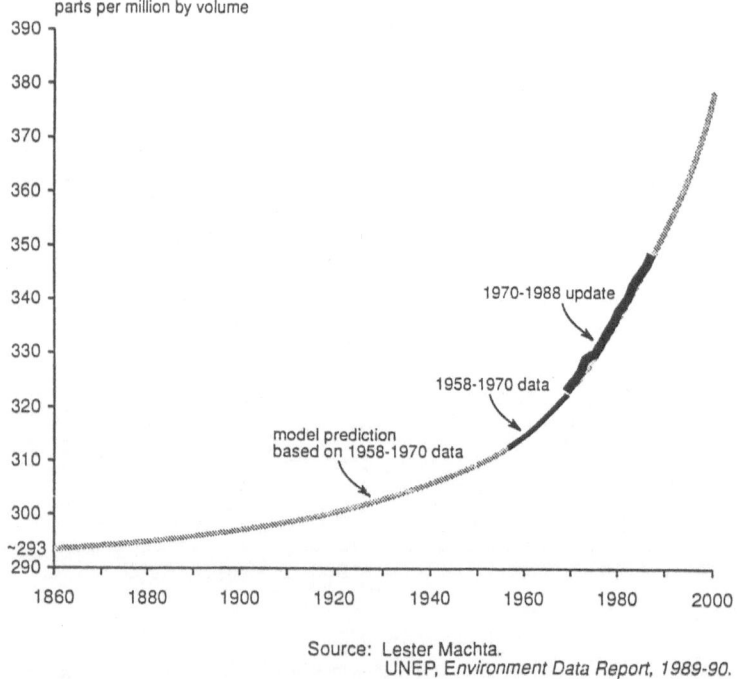

parts per million by volume

Source: Lester Machta.
UNEP, *Environment Data Report, 1989-90.*

Fig. 2. Carbon dioxide concentration in the atmosphere. From Meadows et al. 1992 with permission

There is explicit recognition in the Japanese proposal that there must also be great advances in social innovation—changes in personal lifestyles, international ethics, and even in the paradigms governing present policies of industrialization and energy use.

The New Earth 21 project will produce fundamental solutions, but only if these other aspects of the problem can receive the same attention that is given to the technical and economic issues. It is here that I see that major opportunities and challenges for the simulation/gaming profession—or gaming profession for short. We can provide tools to assist in reaching each of the project's six objectives. The precise nature of our work would have to be designed through discussions with other disciplines engaged in the New Earth 21 project. But I can suggest possible goals here and cite examples of simulations and games that are already available. I know that each of you could add many suggestions to the short list I will provide. I hope you will take the time to do so.

Tools Available from Gaming

The recommendations presented in Atlanta at the meeting on the New Earth 21 project included six sweeping changes:

1. *Build international systems that will serve nations' short-term and long-term interests*. This requires games which show the long-term dynamics of change in demography, economy, and environment. STRATEGEM is such a game. I designed it originally for the US Agency for International Development to serve as a training tool for development specialists in Central America. The game represents the major forces governing development of a region over 50 years. Volunteers have disseminated the game widely; it has been translated into at least eight languages. Over 2,000 sets have been manufactured for use in about 30 countries.

2. *Improve the science for predicting causes and consequences of environmental deterioration*. During the 1980s the International Geophysical and Biosphere Program (IGBP) supported extensive efforts to develop models of the natural systems involved in climate change. But their simulations take demography and economy as exogenous factors. We could reintegrate those essential causes of climate change back into the modelling framework through games that represent international negotiation processes.

3. *Accelerate environmental assistance and technology transfer to the poorer nations*. Technology transfer programs have been notoriously ineffective. Games like Duke's ADVANTIG could be used to clarify the institutional and psychological factors that determine one country's use of tools developed by another nation.

4. *Develop the required new technologies and create new businesses to support and disseminate them*. New devices will have little effect if there are not trained managers to sell and service the technologies as well as to train clients in their effective use. Of course business schools around the world use games for training managers. There is an enormous inventory of simulations and games that could be useful for training those who will build the new industries that "sell" protection against climate change. For example, Elgood (1988) presents a wonderfully useful catalogue of several hundred commercially available business games that are used in training business managers.

5. *Establish international codes of environmental conduct for corporations*. Ethics are not conveyed by lectures. They come out of perceptions, habits, and norms that survive the test of practical utility within a community. We need to help corporate officials see operationally what norms prevail at present, and get them to understand that a different code of conduct with more emphasis on sustaining the environment's resources will serve their underlying interests much better. Games like Shirts' WHERE DO YOU DRAW THE LINE? can be useful in understanding what ethics currently prevail. I worked with colleagues in Budapest to adapt Shirts' exercise for use with Hungarian industry; the resulting game has given managers surprising insights about their collective ethic related to disposal of industrial wastes.

6. *Educate the public to increase their wisdom and spirit of cooperation related to these problems*. Role-playing simulations are an effective way for individuals to experiment with alternative modes of behavior and to see the

long-term problems that can be caused by short-sighted policies. New games-based curricula, which teach creative modes of cooperation and negotiation, are especially important for children in school; they will inherit the problems that have accumulated through our actions.

To this list we could add other tools. Reaching consensus on solutions to global problems will place dramatic new demands on leaders' capacities to negotiate and compromise. This is already a major focus of gaming activity (Christopher and Smith 1987).

Each summer in Salzburg, Austria, MIT Professor Lawrence Susskind (see Dolin and Susskind 1992) of the Harvard/MIT Negotiation Project conducts a 2-week training program on environmental negotiation. The basic tools for his training program are case studies and games. The enormous demand by countries for participation in this program shows the effectiveness of games for this purpose.

The Ministry of the Environment in the Netherlands has just committed major funding to a new project that will develop games useful for training Ministry staff involved in climate change negotiations. That effort could be enlarged and enriched through more widespread participation by the gaming community.

Games such as BAFA BAFA have been created to inform individuals about the nature of, and the implications resulting from, differences in cultural norms. It would be useful to extend this type of simulation to incorporate factors that produce differing national responses to the threat of climate change.

Our work can theoretically contribute these and many other important new tools for social research, negotiation, education, and social system design. But our capacity for this work is limited by several practical considerations. If our discipline is to be an equal partner with many others in projects like New World 21, we need to improve our practice until it meets the standards observed by most of the other professions.

New Standards for the Profession

For almost a decade I served as Associate Editor for four of the principal English language journals that report on work involving simulations and games related to social systems (*System Dynamics Review*; *The Journal of the System Dynamics Society*; *Simulation & Gaming: An International Journal of Theory, Design, and Research*; *Technological Forecasting and Social Change*). From that vantage point, I have seen three areas in which our discipline falls short of the standards observed by many of our peers.

The first area involves our educational programs. There are few formal training programs in game design and use. There is no generally accepted list of requirements for those who wish to work in our field. Thus, there is no process of accreditation that outsiders can use to gauge the professional qualifications of those involved in gaming.

It would be impossible and undesirable to expect that all formal training programs in gaming would have an identical curriculum. But it would be very helpful to compile a list of the various topics which provide expertise in gaming, and then collect course curricula related to each topic. This would facilitate the design of game-related curricula and courses, and permit new training programs to be initiated with less work and with more deliberation. Making widely available the curricular materials from the Certificate Program in Gaming administered by Professor Richard Duke at the University of Michigan, Ann Arbor, Michigan would give us a foundation for starting this effort.

Gaming draws on many different areas of expertise—business management, organizational psychology, economics, environmental sciences, to name just a few. Nonetheless, there must be some core curriculum that we would expect most professionals to master. We should identify the members of this list and decide how they can best be incorporated in the formal education of those who join our field.

Until there is some basis for evaluating different academic programs and certifying professionals at different levels of competence, the academic community is not likely to accord us nearly as much respect as we think we deserve. More importantly, there will be few new training programs created in our area, and our capacity to play a major role in any program of social change will be severely limited by personnel constraints.

The second area where we need work is in documentation standards. The past work in global models nicely illustrates the problem. During the 1970s, approximately 20 major global modelling projects were carried out in at least a dozen nations. Published papers result from all of these efforts, and many of them produced major books (Meadows et al. 1982).

Unfortunately, the standards of documentation observed by the field during this frenzy of work was so poor that the reader of these publications has almost no possibility to reproduce the published computer simulations or to test the sensitivity of published results to changes in assumptions. Of the 20 models that were published, only one gave readers all the information necessary for them to reproduce all published results independently. Indeed, I believe that scientists in at least 15 of the projects could no longer reproduce the results that they themselves published only a decade ago. If the professional literature in the physical science disciplines were so lax, progress within the discipline would be slow and there would justifiably be little respect for practitioners in the field.

I think that we can and must do better. We could easily start by adopting, as a norm, the requirement that any reader of a published report should have access to all information required independently to conduct the game or simulate the model referenced in the text. If the information necessary to achieve this cannot be published as part of the article, it could be provided to the central archives of the journal with permission for it to be reproduced and distributed on request.

Even when we have achieved some uniformity in documentation standards, we will need to work towards making our products more permanently available. I know that many of you in this audience have experienced the frustration of learning that some "wonderful" game you have heard about is not available in a form that would let you take it over for use in your own programs.

For every 100 simulations and games that are created by someone for use in their own teaching or to provide the basis for a professional publication, probably less than five are documented sufficiently that others can actually use them. And most of those five rapidly go out of print.

Many times I have turned to that wonderful catalog of games and simulations compiled by Horn and Cleaves (1980) to find exercises that would be useful to me. Invariably I find several entries of potential value, but almost never have I managed to get full documentation of the kit. The author has moved and left no forwarding address, the computer software is now obsolete and cannot be operated on current machines, or the documentation is no longer reproduced. For these and many other reasons, our best work is quickly lost to the general community. If our work is to produce a steadily accumulating library of tools for dealing with global problems, we will need to invest more effort in archiving our products. It would cost comparatively little to achieve this.

A third deficiency of our field is the low interest and the sparse knowledge related to the real impact of the games and simulations we develop. All of us have repeated the Chinese proverb "When I hear I forget, when I see I remember, when I do I understand." I use the phrase myself, and I am sure it is true. But how much objective evidence do we have for that statement? I recently went through several dozen back issues of *Simulation & Games: An International Journal of Theory, Design, and Research* in search of theoretical or empirical articles on the effectiveness of games. I found practically nothing—certainly not a single article that could be used to select which of several alternative games might be most effective in a given situation; nothing also that differentiated between short-term and long-term learning effects.

Here again the problem remains unsolved, not because it is inherently difficult, but because no one in the profession has given it much attention. I suggest that we might turn to the example of the National Diffusion Network (NDN) which has been operated by the US Department of Education for the past 15 years (Ralph and Dwyer 1988). NDN has constituted a Program Effectiveness Panel, a group of 25 experts who use rigorous statistics and sound didactic theory to test and certify teaching materials that have proven effectiveness. Their approach could be modified for our use.

How Can this Be Done?

Three facts are relevant, as we think how a program of this sort could be implemented. First, there is an enormous depth and diversity of competence in the international gaming community. Second, there is no coordination and not even much communication among gaming professionals within different branches and in different countries. Third, there has not been any attempt to secure funding that would support the profession as a whole.

The last time I checked, I identified eight professional societies created around the interests of professionals involved in gaming: ISAGA, NASAGA, ABSEL, SAGSET, MORS, JASAG, AEE,[1] and an association of East European gamers. I am sure this list is incomplete, and I know that the majority of those with expertise in this field do not affiliate with any society. The United States government alone spends many tens of millions of dollars annually on game-related research and training programs. Expenditures by industry are not tabulated, but they certainly exceed those by the federal government. This is an enormous resource.

I have asked myself why there is no fairly coherent international community for the gaming discipline as there is for many of the other disciplines. For example, Japanese physicists regularly meet their colleagues from other countries at international meetings. They all read the same journals, and the leaders of the field are well known to each other and in frequent contact with one another. The difference is that basic science is an end in itself; gaming is a means to an end—actually many different ends. So the gaming community has been pulled into a large number of groups, each pursuing radically different goals, working with different clients, and gauging their success according to different criteria—high school education, corporate training, social science research, and military strategy.

Until now gaming has attracted little program support. There are no endowed chairs for gamers in universities; no foundation has created a new institute for the advancement of our field. But that is easy to explain. There is no critical mass of professionals seeking support in a common area of concern, the senior professionals do not constitute a coherent community of interest, and there is no way for an outsider to judge who is doing the best work.

I know that if gaming is going to make a major contribution to the way we solve global problems we will need more funding. That will require of us more coordination. Let me cite an example from my own work to show how this might be done without encroaching on individual and group perogatives.

---

[1]ISAGA: International Simulation and Gaming Association, Jan Klabbers, General Secretary, Oostervelden 59, 6681 WR Bemmel, Netherlands

NASAGA: North American Simulation and Gaming Association, John del Regato, President, Pentathlon Institute, P.O. Box 20590, Indianapolis, IN 46220 US

ABSEL: Association for Business Simulations and Experiential Learning, Robert Wells, Center for Business Simulation, LB 8127, Georgia Southern College, Statesboro, GA 30460-8127 US

SAGSET: The Society for the Advancement of Games and Simulation in Education and Training, Secretariat, Center for Extension Studies, University of Technology, Loughborough, Leicestershire, LE11 3TU UK

MORS: Military Operations Research Society

JASAG: Japan Association for Simulation and Gaming

AEE: The Association for Experiential Education, CU—Box 249, Boulder, CO 80309 US

Building a New Professional Network

Twelve years ago Donella Meadows and I observed that around the world there were many centers of teaching, research, and consulting on issues related to sustainable development. But the staff members of these institutes did not communicate with each other, and there was no central forum for them to meet, exchange information, and provide mutual support. None of these people were about to accept someone else's directives about the focus of their work. But they were united by three common features: the desire and the capacity to do work that addressed important problems, intense interest in global problems, and a proven record of achieving influence on local decision makers. Most importantly, each of the centers clearly saw that they could be more effective in achieving their own goals within their own region if there was an exchange of people and ideas between centers engaged in related work.

To realize the potential for synergism, we called leaders of this group together for a week-long conference in the fall of 1982. The meeting was an opportunity to become personally and professionally acquainted, to identify areas of common interest, and to sort out a set of norms and administrative procedures that would let all members of the group realize the benefits of association without intruding too much on their own programs and goals.

We decided to call the association INRIC—The International Network of Resource Information Centers. It was incorporated as a non-profit education and research organization in the United States. We have found it comparatively easy to raise about US$50,000/year to pay for the central activities of the group.

The money pays for a quarterly newsletter, *The Balaton Bulletin*, that runs about 50 pages an issue and that is comprised mainly of news from the members about their work, new programs, goals, and the opportunities they see for joint work. Another major use of funds is to cover expenses of an annual meeting, which brings together once a year about 50 members from the 25 centers.

The third category of expense is a venture fund that pays for travel and meeting expenses whenever personnel from two or more of our member centers wish to meet for discussions about a joint project. This model would work as well for those in the gaming community who wish to come together around projects that address the longer-term global issues.

These annual meetings have been an incredible stimulus to the work of INRIC's members. They attract potential funders and they let us reinforce and refine our special goals, tools, and concerns. They are a way of bringing younger staff members rapidly to the mastery of technique and the acquaintance of the world's leading scientists in their field. There is no permanent staff, just an administrator who works about 2 months a year to coordinate the distribution of the newsletter, the submission of reports required to maintain our tax-exempt status, and the organization of our annual meeting.

This approach could easily be employed to bring coordination and significant financial support to the work of those in the gaming community who wish to collaborate with the New Earth 21 initiative and other projects that address global issues.

The Next Steps

ISAGA is a network, not a hierarchy. That means that work does not get done by one person assigning it to another. Instead we move ahead as a profession when one individual or group volunteers to take on some responsibility. I will close my formal remarks by listing some of the concrete steps we could undertake now—especially if you find the idea appealing of a partnership in efforts to avoid climate change.

Incorporate ISAGA as a non-profit research and teaching organization so that it qualifies to receive foundation support. Design a first teaching workshop that will demonstrate the use of games that address aspects of the New Earth 21's six point program. Propose to the Dutch Ministry of the Environment that ISAGA convene a meeting of gamers who could evaluate the Ministry's proposals for a project to develop climate change negotiation games. Develop recommendations on documentation standards, so that they could be considered for adoption by the leading journals in which we publish our work. Prepare a large, multi-year proposal for an American or a Japanese foundation requesting support to survey gaming curricula and gaming professionals around the world and to develop a list of central requirements, and a proposal for accrediting academic programs. Assemble a small group who will study the example of INRIC and then adapt the design of that network for the purposes of those who would like to promote the application of gaming to global issues.

The list could go on, but I believe I have made my point. There is much to be done, and it will be easy to do it. These steps sound insignificant, but they will lay foundations for us to tap the enormous resources of our discipline. As Daly (1973) once observed, "The path to a sustainable society is unclear, not because it is hard to see, but because so little effort has been spent in finding the way."

References

ADVANTIG. Duke R (1988) Multilogue (Multilogue, 321 Parklake Ave., Ann Arbor, MI 48103, US)

BAFA BAFA. Shirts G (1973) Simile II, Del (PO Box 910, Del Mar, CA 92014, USA)

Christopher E, Smith L (1987) Leadership training through gaming. Kogan Page, London

Daly H (1973) Toward a steady-state economy. Freeman, San Francisco

Dolin EJ, Susskind LE (1992) A role for simulations in public policy disputes: The case of national energy policy. Simulation and Gaming: An International Journal of Theory, Design, and Research (23)1:20–44

Elgood C (1988) Handbook of management games. Gower, Hants, UK

Horn RE, Cleaves A (1980) The guide to simulations/games for education and training, 4th edn. Sage, Newbury Park, CA

Meadows DH et al. (1972) The limits to growth. Universe Books, New York

Meadows DL et al. (1973) Toward global equilibrium. Productivity Press, Cambridge, MA

Meadows DL et al. (1974) Dynamics of growth in a finite world. Productivity Press, Cambridge, MA

Meadows DL et al. (1977) (ed) Alternatives to growth I: A search for sustainable futures. Heronbrook Publications, Box 844, Durham, NH03829, USA

Meadows DH et al. (1982) Groping in the dark: The first decade of global modelling. Wiley, New York

Meadows DH et al. (to be published) Beyond the limits. Chelsey Greene, Vermont

Ralph J, Dwyer MC (1988) Making the case: Evidence of program effectiveness in schools and classrooms. US Government Printing Office, Washington, DC, USA Publication #1989-248-893/00755

STRATEGEM. Meadows D (1984) IPSSR (Hood House, UNH, Durham, NH 03824, US)

UNEP (1991) United Nations Environment Programme Environmental Data Report. Basil Blackwell, Cambridge Center, MA

WHERE DO YOU DRAW THE LINE? Shirts G (1977) Simile II, Del (PO Box 910, Del Mar, CA 92014, US)

Applying Principles of Graphic Design to Game Design

Diana E. Shannon[1]

Abstract. Historically, graphic design considerations have failed to be fully realized during the early stages of game design and construction, entering in only during the final production stage to facilitate mass production or to meet certain publishing criteria. Consequently, critical communication and visual decisions were left to artists most often removed from the experience and dynamics of the gaming/simulation session. The increasing sophistication and availability of a variety of graphics and publishing software for PCs offer game designers the means to create and modify game materials during the critical early stages of game construction. Appreciation, knowledge, and consideration of the principles of graphic design is crucial, however, in order to use any graphics or publishing software effectively, as well as to critique the work of graphic designers hired for a game design project. This paper will recommend how best to utilize graphic design expertise for game design projects. Through illustrated examples, this paper will introduce principles of layout, typography, and color in the creation of game materials. This paper will also consider the implications and appropriateness of format, intended audience, and budget on the design process.

Key words. communication; desktop publishing; graphic design; printing

Does gaming/simulation have an image problem? As both a game as well as a graphic designer, I believe it does. Many games I have examined and played suffer as a result of last-minute, retrofit graphic design, if there is any graphic design at all. Poorly designed game materials result in inefficient communication dynamics which both hamper the game process and diminish the educational impact. Furthermore, such games appear dull and uninviting to play. Consequently gaming simulations remain ineffectively used and unexploited.

What impact do your games have on current and potential users? How do the visual and graphic components of your games affect game play? In other words, do they keep game play running, or do they impede play by forcing participants to scan confusing or poorly printed materials? Are your game materials' directives self-evident, or do they require extensive explanation by the game facilitator? Are your games designed so that they attract and hold the attention and interest of the participants, or do they rest unexplored, perhaps

---

[1] Shannon Associates, RR1 Box 221A, Strafford, VT 05072, US

even abandoned, before or during play? Do your game materials enliven play? Do they encourage creativity or spontaneity? Are your game materials internally consistent, and does the game as a whole convey an identity which is sustained after the experience, for both educational and promotional benefits?

If you have concerns raised by these questions, then you need to seek out graphic design expertise at the start of your next game design endeavor. Graphic design concerns must be seriously considered during the entire game design process. In this paper, I will describe what graphic design is and how it best serves game design. I will then discuss a few major principles of graphic design and show how I have applied them in some of my own game design projects.

At this point it is interesting to examine where game design fits into the larger perspective of a communications development timeline. As you can see from Table 1, game design as we practice it today is very much a product of a long and rich history of technological advance and social development. It has been influenced by and owes much of its present form to cultures and trades undergoing development and improvement for thousands of years. As game designers, we have yet to truly appreciate and learn from the lessons and practices of other older disciplines, among them graphic design.

What Is Graphic Design?

In his masterfully written, designed, and printed *Envisioning Information*, Edward Tufte states:

> We thrive in information-thick worlds because of our marvelous and everyday capacities to select, edit, single out, structure, highlight, group, pair, merge, harmonize, synthesize, focus, organize, condense, reduce, boil down, choose, categorize, catalog, classify, refine, abstract, scan, look into, idealize, isolate, discriminate, distinguish, screen, sort, pick over, group, pigeonhole, integrate, blend, average, filter, lump, skip, smooth, chunk, inspect, approximate, cluster, aggregate, outline, summarize, itemize, review, dip into, flip through, browse, glance into, leaf through, skim, list, glean, synopsize, winnow wheat from chaff, and separate the sheep from the goats (Tufte, 1990).

Gaming/simulations are "information-thick worlds" of their own. A typical gaming session requires participants to utilize many of the above-mentioned capacities under the pressure of practical time constraints. It is crucial, therefore, that a game's visual format aid and even encourage a game participant's efforts under such conditions. In *Graphic Design for the Electronic Age*, Jan White states that "information is neutral material. It is a data base, lying fallow. It has to be accessed and used. Only when it is transformed into knowledge does it have value" (White, 1988). Graphic design enables that transformation.

Table 1. Communications timeline. Adapted from: White 1988

32,000 BC	Earliest known cave decorations	1622	First graphite pencil
1600 BC	First real alphabet developed in Mideast; brought to Greece by Phoenician traders around 1100 BC	1709	First modern copyright act law in England
		1714	Typewriter patent issued
1500 BC	First book: Egyptian book of the Dead, a long papyrus scroll	1718	First banknotes printed in England
1500 BC	Chinese develop ideographs	1791	First Amendment to the US Constitution which guarantees freedom of the press
105	Ts'ai Lun invents paper made from tree bark, cloth, hemp waste, and fish net		
150	Books of folded parchment begin replacing scrolls	1798	Invention of the paper-making machine
600	Paper making spread from China to Korea, Japan, and Persia	1839	First use of daguerreotypes (photographs) in European journals.
748	First newspaper printed in Beijing		
770	Japanese Empress Shotoku sanctions first printing on paper; a million prayers to ward off smallpox epidemic	1855	Vegetable parchment paper (tracing) introduced
		1866	Lithographic printing on metal produces decorated tin cans
1120	Playing cards invented in China.		
1221	Chinese developed movable type made of wood blocks	1880	Halftone perfected for use in newspapers
1450	Gutenberg invents printing press with movable type matrices in Mainz	1904	Offset printing developed
		1930	Four-color offset press
		1940	Simulation gaming begins
1465	Gutenberg's 42-line Bible is first printed book with movable type	1962	Xerox, electrostatic printing introduced
1465	First printed music	1972	Color xerography introduced
1470	Nicolas Jenson produces Jenson, the first roman typeface, in Venice	1972	Personal computer concept and name comes into use
1525	Newsletter developed as an early form of newspapers and are widely use to keep trading houses informed		

I have found that the role of graphic design in game design is often misperceived by game designers. To begin with, graphic design is not decoration, icing on the cake, or frills. Rather, it is functional. Graphic design organizes, clarifies, and sharpens the communication process, thereby increasing the rate at which game players understand their role and responsibilities during a gaming session. Good graphic design is not a luxury only rarely affordable with large project budgets. Instead, graphic design should be considered an investment. Successful graphic design sells itself.

Graphic design is not an afterthought, but is planned. Design success results not from luck, innate artistic talent, intuition, or flashes of brilliance, but from experience and from trial and error. The addition of graphic design cannot cover for or "mask" a poor game. At its best, graphic design strengthens the impact of a well-designed game.

Graphic design is not an ends in itself—beauty, prettiness, fashion—but a means to an end—a more effective game. And, contrary to what advertising might suggest, graphic design is not desktop publishing but a profession with its roots in the early printing establishments, a pursuit much older than game design. (Desktop publishing is a new technology available to a mass market of users, including graphic designers. However, just as an ability to use a paintbrush does not make one a fine arts painter, using a desktop publishing software package does not qualify one to be a graphic designer, either.)

How Does Graphic Design Fit into the Game Design Process?

Game design projects present enormous challenges for graphic designers. Games must be designed for multiple audiences—for players as well as for facilitators. Game projects often involve the design or use of a myriad of related pieces, parts, and paraphernalia. Complex game topics (complex because game designers often grapple with issues difficult to teach in traditional ways using books and lectures) may require a graphic designer to spend an extensive amount of time researching appropriate symbology and styles. And game design often includes many different specialty areas of graphic design known formally as publication, presentation, package, and industrial design. Altogether, not a simple job assignment.

To clarify the task facing the graphic designer, I will examine each stage of what Greenblat (1988) has defined as the five stages of game design and suggest how a graphic designer might best contribute to a game design team. Stage one, *Setting Objectives and Parameters*, includes deciding subject matter, purpose, potential operator, potential players, context of use, resources (for design and play), time, and conditions of a proposed gaming simulation. If you lack access to a graphic designer at this stage—perhaps you are writing a proposal—then these parameters should be handed to a graphic designer at your very first meeting. This is extremely critical information; all of these parameters will shape the majority of graphic design decisions, particularly the game's format (size, shape and appearance). Having these parameters early on in a game design schedule will enable the graphic designer to think about the game, to begin a period of creative distillation, even if the designer is not yet actively participating in the project's work. From these parameters, a graphic designer can provide some idea of the estimated cost of a projected production run, always useful information when writing a proposal budget.

A graphic designer can aid greatly in stage two, *Model Development*. A graphic designer's expertise in creating visual aids and schematics might produce a sophisticated visual model most representative of the game model. Many of my early game projects, including STRATEGEM-1, and FISH BANKS, LTD., made repeated use of causal loop and system dynamics modeling diagrams as the visual basis for the game model and subsequent game materials, particularly the game boards. A recent game design consultation for the United Nations Development Programme at the National Development Planning Commission in Ghana with Cathy Greenblat involved a great amount of graphic design time that yielded the game model schematic shown in Fig. 1. I strongly believe that a well-designed model graphic during this stage has an enormous chance of becoming a significant game piece. Time well-invested here will accelerate the design process at later stages.

During stages three and four, *Representation and Construction*, game model elements are assigned a particular game form and created in draft form. Here a graphic designer may bring fresh ideas into feasible, new ways in which model elements can be represented in the game. Game models are "communicated" through a final physical set of game materials. Graphic designers have a great experience determining the most relevant and efficient printed forms for different communication needs. Constructing draft materials is also very much a part of the graphic design routine. Graphic designers are particularly skilled at creating "comps," which are mock-ups of a graphic piece prepared for final client approval before being sent to the printer. Creating draft game sets with graphic designers will teach you many new representational tricks and save time and money in later stages, especially production. Too many games have been created with no consideration of print production, resulting in draft game sets so complicated that they have to be entirely redesigned at great expense. Or even worse, they remain on the designer's shelf, in draft form, unavailable to potential users.

If the previous stages were done well, stage five or *Preparation for Use by Others*, should require little creative effort. In fact, most of the decisions regarding mass production should have been planned in advance as a result of work completed during the earlier stages, particularly stages one and four.

A Few Principles of Graphic Design

I have referred in preceding sections to the graphic design format of games. By format, I mean a particular identity conveyed to users through a consistent visual structure of all game materials. Achieving a consistent graphic design format for a game produces a familiarity which results in more efficient game play, as well as an identity which is remembered after the session, enhancing the impact of lessons learned as well as advertising the game to new users. Format is the results of numerous factors, among them layout, typography, and color.

Layout Principles

Layouts are specific solutions to the communication needs of each game piece, be it a booklet, a game board, or a form. The foundation of any layout is its grid. Grids can be composed of several columns or of a single column, of equal or varying sizes. In some cases, as with forms and game boards, they may also

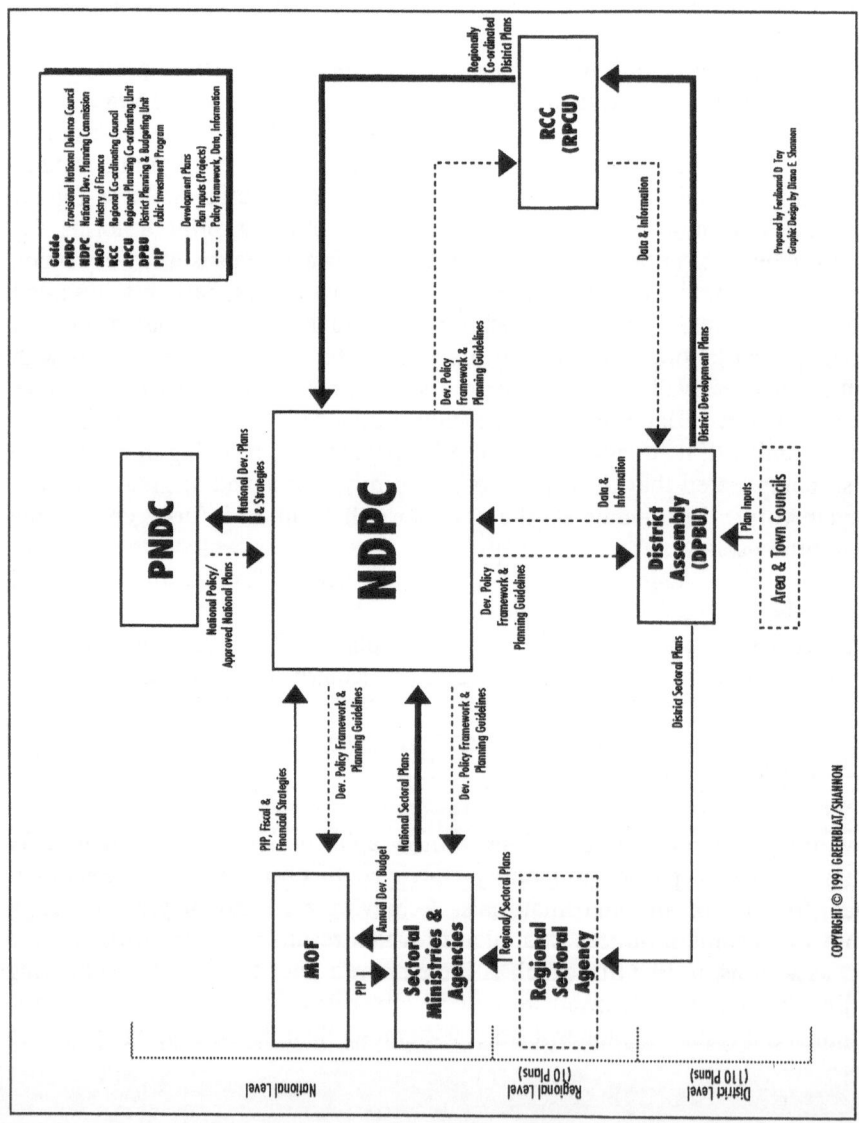

Fig. 1. Graphic design principles greatly influenced the development of this game model schematic of the new planning system in Ghana

consist of rows. The function of grids is to provide consistency, in spite of the fact that text and graphic elements may be changing on each page. A grid sets the boundaries for the "live matter area," that is, the area outside of the margins where text and graphics are placed. Do not assume that live matter area consists of "white space" to be filled up by text and graphics. White space on its own is a critical element of any layout, serving to contrast any black elements and to provide visual "breathing room" to get ideas off the page and transformed into working knowledge. Be generous in its use. Remember the ying and yang principle, where black cannot exist without white (White 1974).

Within the layout itself, care should be taken when arranging objects relative to each other. For example, on a game board, be aware how various borders and area outlines impact on white space and on each other. Tufte (1983) discusses the problem of graphical clutter whose resulting visual noise impedes the flow of information. Desktop publishing and the plethora of accompanying drawing programs are major culprits, providing would-be graphic designers with exotic patterns and borders to use when simple black lines would be more effective. Figure 2 shows a desktop-published chart in contrast to a graphically-designed chart. The lower chart maximized what Tufte defines as the "data-ink ratio," the proportion of ink devoted to conveying data information over the total ink used to print the graphic (Tufte 1983). Viewers can grasp the message of the lower chart in a shorter, less visually fatiguing amount of time, freeing them to proceed with other information gathering.

Layouts should entice the game players to read and to use the materials. This can be achieved by designing interesting opening pages, employing signals such as drop caps to announce beginning sections, or adding occasional dashes of color. For the opening page of the HOSTAGE CRISIS player booklet, shown in Fig. 3, I allowed much white space; the window graphic symbolically invites players to look through or step into the game. Figure 4 shows how a distant sun and the suggestion of a journey have been utilized to intrigue players at the start of a CAPJEFOS session.

Layouts should consist of easy transitions from section to section, preventing the game player from getting lost. This is why I have repeatedly chosen to design games almost entirely of participants' booklets that contain all or a majority of the critical materials [ENCOUNTERS WITH AIDS (1988), HOSTAGE CRISIS (1988), DEATH OF A DISSIDENT (1990), FIRE IN THE FOREST (1990)]. A participant booklet with clear sections, directives, and smooth transitions requires far less work for the facilitator as well. Draft game sets, whether consisting of booklets of players' materials or a collection of related printed pieces and paraphernalia, should not be critiqued statically like snapshots, but collectively like a movie, as a continuous stream of images, much as they are presented to the player during the game. I often accomplish this by tacking all of the pages and materials to the wall in a room, in the sequence they are viewed by players in a game. This enables me to view the game in its entirely, as well as evaluate its collective visual and psychological impact.

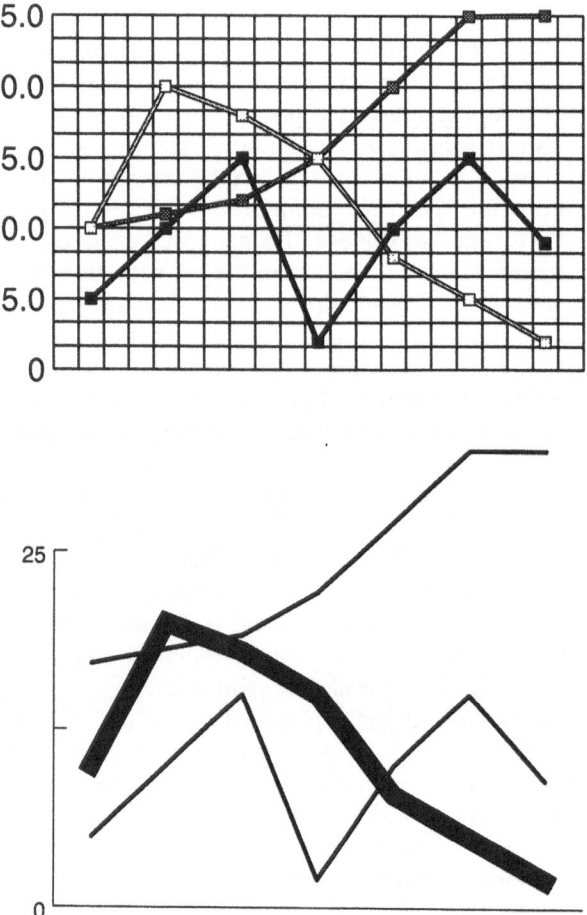

Fig. 2. Comparison of a desktop-published chart (*upper*) and a graphically-designed chart (*lower*)

Typographic Principles

Roman alphabets boast over 10,000 different typefaces. Typography alone is capable of vast expression; some examples are presented in Fig. 5. Lacking graphics or photographs, you might find simple typographic variations to be a neat solution for a game format. Figure 6 depicts the opening pages of SACRILEGE IN TALBOTSVILLE, a game about the First Amendment in which proponents of artistic freedom of expression clash with defenders of public decency laws in a small community. Besides two typefaces, I used no other graphical elements for the game's format, choosing to counterplay a gothic typeface, Fetter Fraktur, with a modern face, Bodoni, and in a subtle manner extend the game's inherent tensions (both typefaces represent conflicting philosophical and artistic *esprit du temps*).

you are about to embark on an adventure, involving the hijacking by Middle East terrorists of an international airliner, and the holding of hostages. You will be learning much about the Middle East, Islam, and terrorism. You will also find yourself thinking and talking about justice and injustice, ethics, power, violence, human rights, national self-determination and aspirations, and why people from other cultures often think very differently from us.

Participants who have played this simulation have told us that they found most challenging the complexity of the issues raised. They said that they had learned things that they could not have learned from books, movies or television. The reason is that you will be playing roles, of terrorists, the president and his advisors, hostages, or television correspondents. One of you may be asked

HOSTAGE CRISIS   1

Fig. 3. A visually inviting opening page of HOSTAGE CRISIS

The intended audience represents the major factor influencing typographic choice. In Fig. 7, two formats of the ENCOUNTERS WITH AIDS series were adopted to address the needs of two different audiences: adults and adolescents. Note how the format, both the typography and layout, vary based on the intended game player.

CAPJEFOS
A simulation of
village development

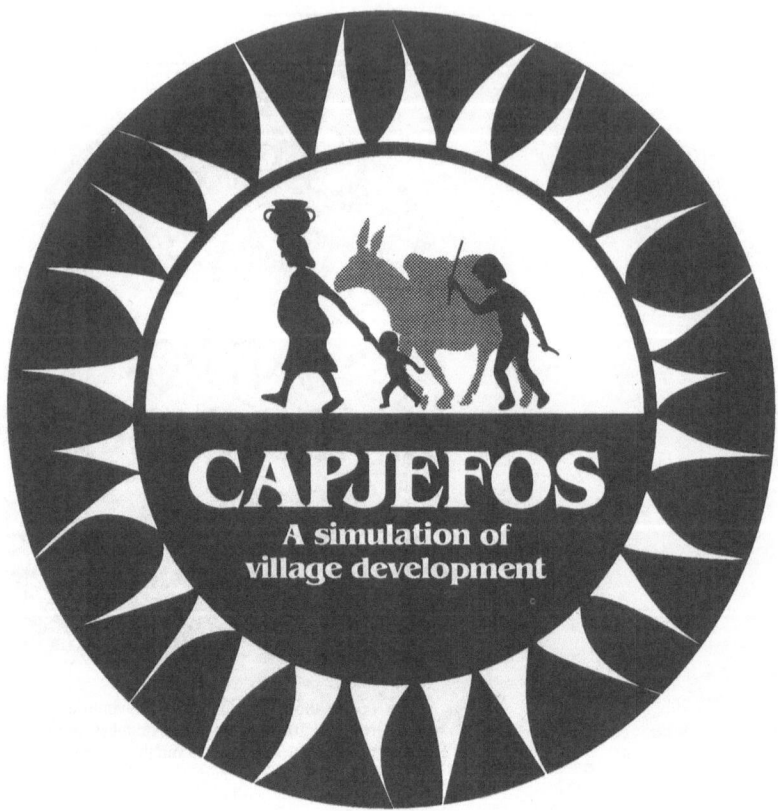

Fig. 4. A distant sun and the suggestion of a journey intrigue players at the start of CAPJEFOS

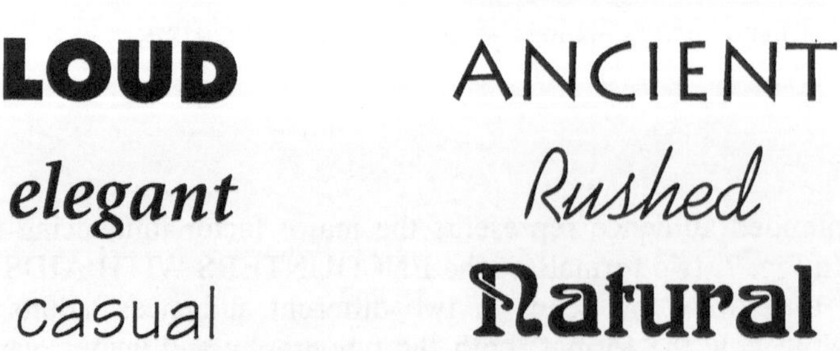

timid
ORNATE
LOUD
ANCIENT
elegant
Rushed
casual
Natural

Fig. 5. Typography and expression

Fig. 6. Simple typographic design gives SACRILEGE IN TALBOTSVILLE its distinctive character

Other key design considerations related to typographic decisions include legibility. Legibility is determined by many factors, including size of type, line length, space between lines, page size, paper type, toner/ink type, difficulty of language, and organization. If your materials appear illegible, it is important to consider not one, but all of these factors.

ENCOUNTER 3

HIV & Your Child

ROLE

You are the Group Leader for this encounter. Speaker A, the mother of a 12-year-old HIV-positive hemophiliac, will discuss with Speaker B, the father, how to talk to their son about his health condition.

GOALS

Your goals are to help the participants keep the encounter going smoothly, to maintain the time limits, and to lead a discussion afterwards.

RESOURCES

You have some information about the issues that arise in this encounter. You also have a timer to keep track of the time available for each step of the round. When you set it, push the number of minutes you want. Then push "Start." The timer will immediately show one minute less than you indicated, as it shows the time remaining in whole minutes until it gets to 1 minute; then it shows seconds remaining.

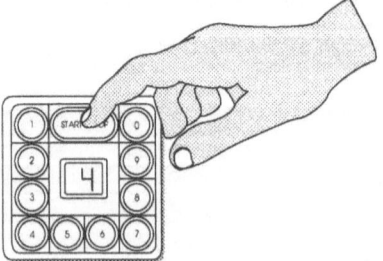

ENCOUNTERS WITH AIDS—GROUP LEADER   27

a

Fig. 7a,b Two formats of the ENCOUNTERS WITH AIDS series used to address adults (**a**) and adolescents (**b**)

Who are the Speakers?

Speaker A is Joyce, a sixteen year old. Keisha, one of her best friends since third grade, was in a serious car accident in 1984. It seems that the U.S. blood supply wasn't being screened for HIV at that time, and Keisha had a lot of blood transfusions. Last month, Keisha's doctor told her to get tested for HIV. The test showed that she is HIV positive—that she is infected with HIV. She has no symptoms of AIDS.

Speaker B is Darryl, a 16 year old. Several years ago, Keisha, a friend from his neighborhood, was in a car accident. Because she had a lot of blood transfusions then, her doctor had her tested for AIDS. The test showed that she is infected with HIV.

Darryl was shocked at the news because he never thought someone he knew could get AIDS. He and other people have been staying away from her lately. It's sad, but Darryl can't blame them—nobody wants to get AIDS and die. He likes Keisha, but he's not taking any chances.

In this encounter, you will

1 help the encounter run smoothly
2 keep the **Speakers** with the time limits, and
3 lead a discussion afterwards.

b

Color Principles

Like typography, color is capable of enormous expression (White 1990). Color has many functions in graphic design, among them to label, emphasis, locate, analyze, identify, unify, and pace. Color attracts attention, and anything on a page in color will be noticed first. While it is critical that important page elements be in color so they are noticed, too much color highlighting decreases the impact of all color use. Too much color, particularly in text sections, is not readable. Bright colors are effective in small spaces, dull or pale colors in large ones.

When choosing colors, keep in mind the intended audience. While some colors have universal meanings (red means "stop" or "warning" in most parts of the world), color has inherent cultural significance. Remember that 8% of men and 1% of women are color-impaired, with red and green being most problematic. In consideration of such audiences as well as in anticipation of less than ideal photocopying conditions, try to introduce shape and texture, as well as color, into critical game symbology. Some examples are shown in Fig. 8.

Conclusion

What has been lacking to date in many game design endeavors is the contribution of a sufficiently trained and experienced visual perspective. I have noted just a few graphic design principles and examples which will aid in the

Fig. 8. Examples of shape and texture that can be used for game symbology

creation of more effective game materials. The nature of gaming/simulation has attracted many users and designers who appreciate multi-disciplinary approaches to problem solving and education. Let us then find more opportunities to enlist the expertise and experience of graphic designers in our work.

References

Capjefos. Greenblat CS, et al. (1987) CSG Enterprises (40 East 19th St., New York, NY 10003)

Death of a Dissident. Kennedy M, Keys M, Shannon DE (1990) The Moorhead Kennedy Institute of the American Forum for Global Education (45 John St., Suite 908, New York, NY 10038)

Encounters with Aids. Greenblat CS, Gagnon, Shannon DE (1988) CSG Enterprises (40 East 19th St., New York, NY 10003)

Fire in the Forest. Kennedy M, Keys M, Shannon DE (1990) The Moorhead Kennedy Institute of the American Forum for Global Education (45 John St., Suite 908, New York, NY 10038)

Fish Banks, Ltd. Meadows DL, Shannon DE, Fiddaman T (1987) IPSSR (University of New Hampshire, Durham, NH, 03824)

Greenblat CS (1988) Designing Games and Simulations. Sage, Newbury Park

Hostage Crisis. Kennedy M, Keys M, Shannon DE (1988) The Moorhead Kennedy Institute of the American Forum for Global Education (45 John St., Suite 908, New York, NY 10038)

Sacrilege in Talbotsville Kennedy M, Keys M, Shannon DE (1990) The Moorhead Kennedy Institute of the American Forum for Global Education (45 John St., Suite 908, New York, NY 10038)

Strategem-1. Meadows DL, Toth FL, Naumienko E, Shannon DE (1986) IPSSR (University of New Hampshire, Durham, NH, 03824)

Tufte ER (1983) The Visual Display of Quantitative Information. Graphics, Chesire

Tufte ER (1990) Envisioning Information. Graphics, Chesire

White JV (1974) Editing by Design. Bowker, New York

White JV (1988) Graphic Design for the Electronic Age. Watson-Guptill, New York

White JV (1990) Color for the Electronic Age. Watson-Guptill, New York

Diana Shannon is a freelance graphic and game designer based in Strafford, Vermont. She has collaborated on numerous game design projects in Africa, Mexico, Europe, and the CIS. When not designing games, she enjoys cross-country skiing, biking, and gardening.

A Global Model of Simulation and Game Evaluation

Dany Laveault[1], Michel St-Germain[2], and Pierre Corbeil[3]

Abstract. A global model of simulation and game evaluation is necessary if one is to demonstrate the anticipated benefits of games for learning or to develop more useful and efficient games. Many criteria and aspects of evaluation have already been documented. The relationship among these criteria and the functions and goals of evaluation are not, however, always clear. Important new criteria and interactions need to be emphasized. We have developed an evaluation model that blends new and well-established criteria. We have applied them to different aspects of evaluation of simulation and games and tried to integrate them in a global and systemic model. An application of the model to the evaluation of educational software is also considered.

Key words. design; education; educational software; evaluation; global model; simulation/gaming

Confusion in determining proper evaluation procedures is a direct consequence of confusion in defining the function of evaluation at different phases of the game development. Each function of evaluation requires specific instruments of measurement. The goals of evaluation are different for game designers and game consumers. The first aspire to formative evaluation that will help build a better game. The second prefer a more summative form of evaluation that provides the criteria for the selection of a game among its competitors.

The model we have developed builds on some of the criteria used in research methodology, in education and in the social sciences. It applies them to different aspects of evaluation of simulation and games. The selection of an appropriate methodology of evaluation is a function of the goals of evaluation, of the moment the evaluation occurs, and of the person doing the evaluation. Some of these methodologies concern the evaluation of a game while it is still

---

[1] Faculté d'éducation, 145 Jean-Jacques-Lussier, Ottawa, Ontario, Canada K1N 6N5; phone (613)564-7728; facsimile (613)564-9098; e-mail ladta@uottawa.bitnet
[2] As[1]; phone (613)564-6561
[3] 960, rue Saint-Georges, Drummondville, Québec, Canada J2C 6A2; phone (819)478-4671; facsimile (819)474-6859

in its conceptual phase. Some others are more appropriate to test a prototype of the game and the success of its application in different contexts.

Two important phases of evaluation are described in general terms in Fig. 1. They are:

1. *Ex ante* evaluation. At this stage, the goals of evaluation are mainly formative. The game designer wants to know how the prototype can be improved. This phase of evaluation includes two different aspects. First, in terms of internal criticism, the evaluation deals with absolute criteria of evaluation that determine the value of the game in itself. Second, in terms of external criticism, evaluation deals with relative criteria. These criteria help to compare the value of the game to that of similar games. Both evaluations involve the game designer.
2. *Ex post* evaluation. At this stage, the designer has produced a working version of the game to be field-tested. The goals of this evaluation are both formative and summative. They subsume three main aspects. The first, evaluation of learning, deals with several facets of learning, whether it is *learning the game* or *learning from the game*. The second, game calibration, considers how well adapted the game is to the needs of the end users. The third, evaluation of the debriefing session, assesses the merits of the debriefing on the learning process. These aspects involve both the game users and the game participants. They are useful in helping the designer to improve the game and its documentation.

For conceptual simplicity, we have defined separately each of these phases. A systemic model of game evaluation, however, implies there are interactions among the aspects. They are addressed in the section dealing with the evaluation of important interactions.

Ex Ante Evaluation

Ex ante evaluation occurs at the conceptual phase of game construction. Figure 2 illustrates the criteria associated with the two main aspects of *ex ante* evaluation: internal and external criticism. Following is a description of the criteria associated with each aspect of this evaluation.

Internal Criticism

The criteria of internal criticism correspond to validity, reliability, and utility. These criteria and their connection with simulation and game evaluation have been described at length in a paper by Ruben and Lederman (1982). Our purpose here is not to describe them again but to relate them to other criteria in a global model of evaluation.

Validity seeks to determine whether the game model is a proper reconstruction of the phenomenon it describes. It includes face validity and construct validity (for more information, see Ruben and Lederman 1982). Norris and

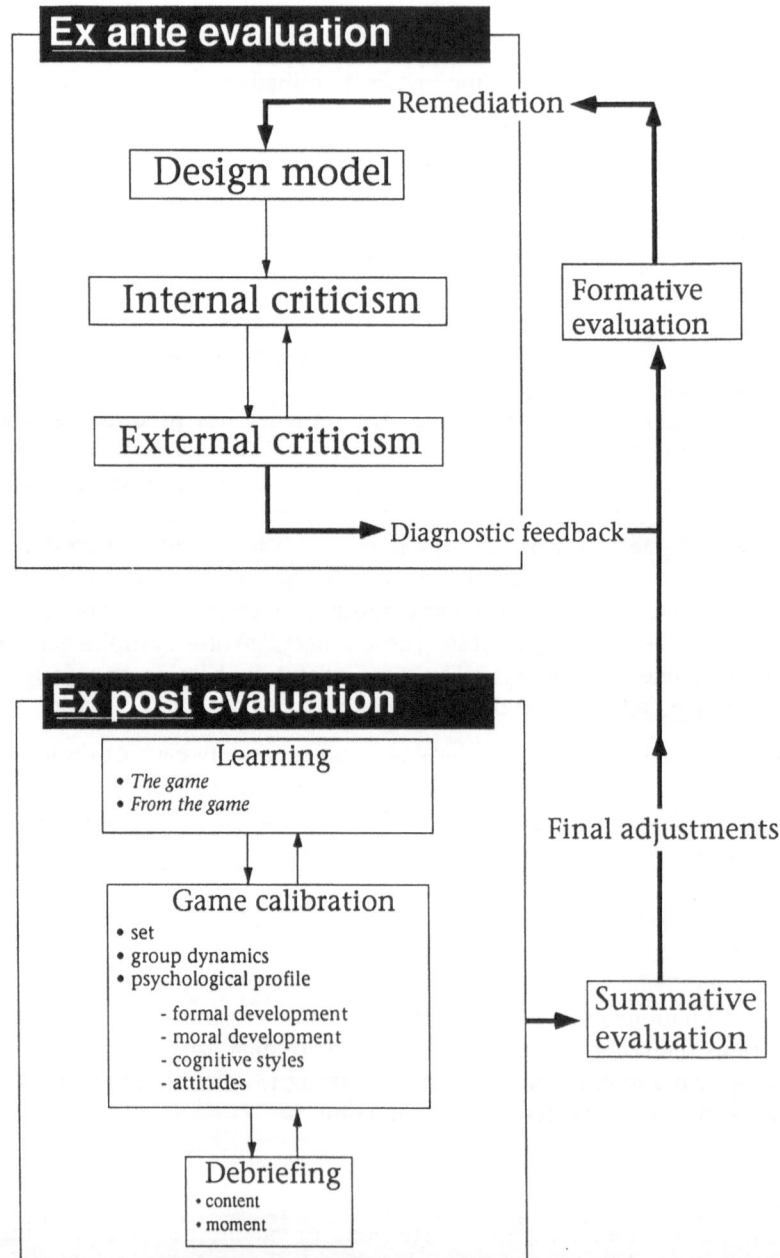

Fig. 1. Outline of a global model of simulation and games evaluation

Internal criticism		
Validity	Reliability	Utility
- Internal	- Product stability	- Cost-benefit ratio
- External	- Process stability	
- Face		
- Construct		

External criticism		
Ergonomics of the material	Suitability of activities	Compatibility with other games
- Manageability	- Credibility	- Concomitant validity
- Ease of administration	- Synchronicity	- Predictive validity
- Continuity of feedback	- Symmetry	

Fig. 2. Criteria associated with *ex ante* evaluation

Snyder (1982) offer an additional consideration on validity. They distinguish between internal and external validity.

Reliability refers to the reoccurrence of learning outcomes across several plays. Depending on whether the game is internally or externally parametered, the assessment of reliability will focus more on learning processes or on products.

Utility involves a cost and benefits analysis of the game. This analysis checks that the goals of the game are achieved at a reasonable cost. Costs include not only financial units, but also time on task, efforts, and so on.

External Criticism

Elder (1975) lists six characteristics that can help to rate a game in comparison to other games. We have added two more characteristics to his initial list and grouped all eight as follows: suitability, ergonomics, and compatibility.

Suitability seeks to determine whether a game is appropriate for a certain clientele. It includes credibility, symmetry, and synchronicity. Ergonomics considers whether the game is user friendly and allows for an easy and fast flow of information among players. Criteria about ergonomics include manage-ability, ease of administration, and continuity of feedback.

Game designers can build on the frame of previous games. Standard frames are specially useful since they allow users to try out new games without spending too much time on learning the game. The use of computers increases the need for common interfaces. Corbeil et al. (1989) distinguish between two kinds of compatibility: concomitant (or frame) validity and predictive (or chain) validity.

A game may be more effective than another because it has a better framework or more familiar user interface: this is concomitant validity. Furthermore, simple, interacting games are preferable to a single, complex game. When the approach and activities are compatible throughout the series of games and when a game is a pre-condition to other activities in the chain, the game has predictive (or chain) validity.

Summary

Internal criticism is useful in assessing whether the conceptual model of the game is appropriate. External criticism aims at assessing whether a game is user friendly. The coordination of these two aspects is important: a bad conceptual model of the game may reinforce some partial knowledge or simplistic perceptions of reality. The conceptual model is of utmost importance. External criticism is not relevant if the conceptual model is flawed. However good the model is, it is not going to be efficient if the ergonomy and the suitability of the game are not appropriate.

Ex Post Evaluation

Ex post evaluation may proceed once a prototype of the game is available. It is then possible to experiment with the different modalities of playing the game. This is necessary to determine how successful the game is in reaching its goals. *Ex post* evaluation deals with the fine-tuning of the game model, as well as the game materials, including the user's and the game leader's guide. After the game has been played several times or for a period of time, some insight may emerge on the best way to use the game and on what the debriefing approach should be. This phase of game evaluation includes three different aspects: (1) evaluation of learning, (2) evaluation of game calibration, and (3) evaluation of the debriefing session. Here is a description of each of these aspects.

Evaluation of Learning

This evaluation is two-fold: it deals as much with the processes involved in *learning the game* as with the processes involved in *learning from the game*. Laveault and Corbeil (1990) have used Kolb's model of experiential learning (Kolb 1974) to describe the processes involved in both kinds of learning. Figure 3 illustrates the two learning cycles involved.

Each cycle follows this sequence: concrete experience, reflexive observation, abstract conceptualization, and active experimentation. The first cycle starts with the concrete experience of game rules and ends with the active experi-mentation of game strategies. The second cycle starts with the concrete experience of the conceptual model and ends with the active experimentation of the concepts to be learned through the conceptual model.

The first cycle is about *learning the game*. No *ex post* evaluation is possible if the participants do not have some expertise with the game rules. This means that the participants are not only aware of the game rules but have developed some fluency in game strategy through the complex coordination of game rules. The easier it is to learn the game, the sooner learning will arise from the concepts modeled by the game.

The second cycle is about *learning from the game*. There is little learning *from* the game without some learning *of* the game. Evaluation of learning from the game has to be related to the game's learning objectives and to the game's type. For example, it may be easier to reach objectives of creativity with an internally parametered game. Other reasons that may explain why learning does not occur include (1) learning the game takes too much time, (2) the game design (e.g., degree of parametrization) is not appropriate for the learning objectives, and (3) the game model is invalid or unreliable.

Evaluation of Game Calibration

This aspect of evaluation determines how appropriate the game activities are in terms of the learning experience. Relevance of the activities depends on the

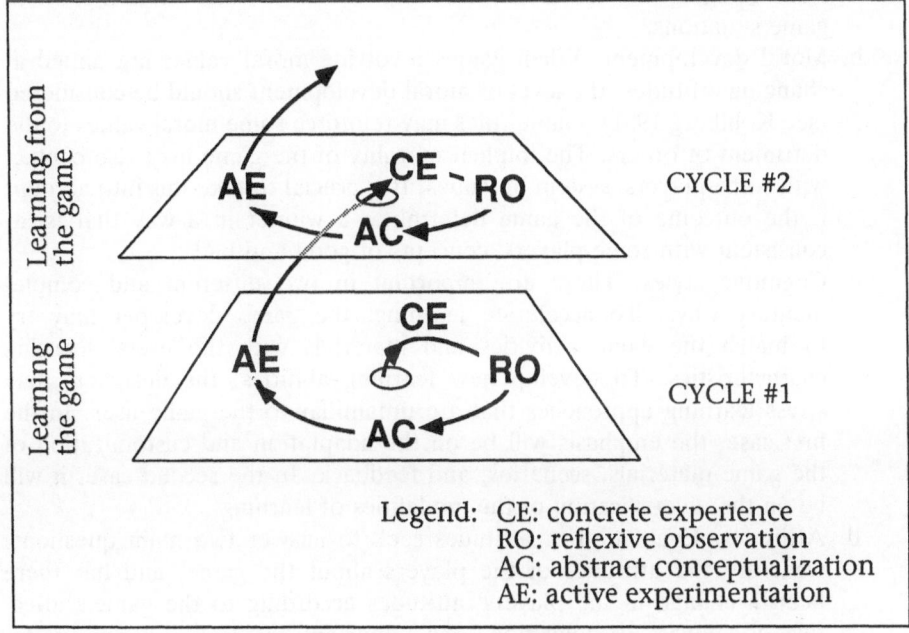

Fig. 3. Learning cycles in a simulation game

objectives of the game. The game may be intended for the discovery of new facts, new ideas, new concepts, new ways of dealing with a problem (exploratory learning). It may also be intended for the actualization of previously acquired knowledge, concepts, or ideas (confirmatory learning).

To check the game calibration, the game evaluator has to pay attention to the following factors:

1. *Set.* Participants are not always prepared to use games as a method for learning. They should be aware that their involvement in the game is important. They should also have a rough idea of the way they will learn from the experience. Evaluation should include consideration of what has been done to create the right learning set among the participants.
2. *Group dynamics.* Some games involve group dynamics to an important extent. The group dynamics and the game user's ability to capitalize on them may affect the prospect of reaching the game's objectives. Evaluation should determine how important the group dynamics are to the achievement of the game's objectives.
3. *Psychological profile of the participants.* To produce stable learning, the game developer must consider the personal characteristics of the game users. They include:
   a. Formal development. The ability to play the game depends partly on the participant's cognitive ability. Furthermore, the likelihood of using higher forms of reasoning is dependent on the subject's familiarity with the content of problem-solving situations. Game players may thus refrain from using higher levels of reasoning in non-familiar or non-credible game situations.
   b. Moral development. When games involving moral values are aimed at changing attitudes, the level of moral development should be considered (see Kohlberg 1981). Game rules may reinforce some moral values to the detriment of others. The implicit morality of the game may also conflict with some players' system of values. It is crucial to take this into account if the outcome of the game determines a winner in a way that is inconsistent with some players' concepts of good and bad.
   c. Cognitive styles. These are important in two different and complementary ways. To accelerate learning, the game developer may try to match the game activities and materials with the users' learning characteristics. To develop new learning abilities, the designers may stress learning approaches that are unfamiliar to the game user. In the first case, the emphasis will be on the adaptation and customization of the game materials, scenarios, and feedback. In the second case, it will be on the diversification of the modalities of learning.
   d. Attitudes. Evaluation of attitudes seek to answer two main questions: what are the attitudes of the players about the game, and has there been a change in the players' attitudes according to the game's affective objectives. In education, one important reference on this matter is the taxonomy of affective objectives of Krathwohl et al. (1964),

which describes five hierarchical levels of interiorization of affective learning.

Evaluation of the Debriefing Session

Debriefing is a basic part of game and simulation activities if participants are to benefit from their social or transactional aspect. A good debriefing session will help the participant to reflect objectively on the learning experience and gain new knowledge from this reflection. Debriefing may occur alone or in a group and may take place whenever the game has started. Evaluation of a debriefing session should pay special attention to the content of debriefing and to the moment it occurs.

Evaluation of debriefing sessions must determine if appropriate means have been taken to prepare the participants. Many factors may impinge on the usefulness of a group debriefing session. The discussion on the learning experience should not refer to a participant's learning capacity or capacity to change. Such matters may make participants defensive. It should be done independently of the evaluation of the game as a learning tool or as an entertaining activity. It must also reinforce the expression of opinions and the description of one's own learning experience.

Though comparisons among players are inevitable, the debriefing session should not emphasize them. Possible consequences of excessive interindividual comparisons may result in an increase in conformity, social desirability, or an "Emperor's new clothes" syndrome, where participants feel compelled to perceive what has been perceived by others.

To assess the merits of a debriefing session, learning from the game should be measured before and after the debriefing session. In formal testing procedures, one can use a typical measure of change, such as the sensitivity to teaching index (see Roid and Haladyna 1982). Such an index may help to discover the facets of learning where the debriefing session has been the most instrumental.

Evaluation of Important Interactions

The global evaluation of a game must consider the interactions that may occur among criteria. Some of these interactions may change the final judgement on the game and may also convey significant formative information. For illustrative purposes, five of these interactions are detailed below.

1. *Moral development and credibility.* Some simulations or games may lack credibility because they are in contradiction with some of the players' moral convictions or system of values.
2. *Formal development and manageability.* Some aspects of game management may be too much for people who have not reached a certain level of formal development. In this condition, the game has to be either more manageable or targeted more precisely to a specific population.

3. *Cognitive styles and ease of administration.* The ease of administration may depend on the perceptual and learning modes of the players. One solution would consist in multiplying the ways of playing with the game materials. This would bring out, however, some other problems of manageability and ease of administration.
4. *Set and attitudes.* Overstating the advantages of a game as a learning method may be as detrimental as understating the value of a game. Evaluation of attitudes towards the game must include the effect of set on the players' attitudes and learning expectancies.
5. *Compatibility and learning the game.* Less time will be spent on learning the game if the game frame is already familiar to the players. As a result, more time will be spent on learning from the game. The time on task is important in assessing the game's effectiveness.

Evaluation of Computer-Based Games and Simulations

The evaluation of computer-based games and simulations proceeds as in the global model of evaluation described above. Some criteria, however, have been added and receive a different emphasis. Among the evaluation criteria set up by Gillis for the Ministry of Education of Ontario (Gillis undated) one can identify the following as specially important:

1. *Compatibility.* There is no point in using a computer if the same results may be achieved by similer means, such as paper and pencil materials.
2. *Technical reliability.* The software must run correctly despite human mistakes. It must also be able to run on different machines of the same type, whatever their specific configuration.
3. *Documentation.* The evaluation of documentation is specially important. It must illustrate clearly how to use the software, whether this be on screen or on paper.
4. *Feedback.* Software can be specially good or bad at feedback. In some cases, it is possible for a player to use the feedback option as a tool to reach the right answer, instead of trying the different strategies that are intended as the path to learning. A good feedback procedure requires that the information received by the participant does not exceed his capacity to process it.
5. *Manageability.* Complex games are difficult to manage. They gain from being done on a computer. The ergonomics of the screen/page design and of the input and output formats influence how the player manages this information. There is a point, however, when one should consider splitting a game that is too complex. Very often, the players themselves will make a complex software more simple. They simply skip some of the important options of the game.
6. *Symmetry.* Computers have great potential to deal adequately with symmetry. The players may assume different roles. The computer also

makes it easier to replay a game since the presence of other players is not always essential.

7. *Ease of administration*. Some software may be difficult to install. Some also requires some form of help during the game or simulation session. Proper documentation, on-line help, and self-tutorial materials will ease administration. Evaluation must consider the interaction between ease of administration and documentation quality to make valid recommendations.

Conclusion

A global system of evaluation of game and simulation activities must consider a series of criteria as well as their interactions. Failure to do so may result in either rejecting potentially good games or accepting inadequate games. On the one hand, one faulty criterion interacting with many others may give the wrong impression that a game is irreparable. On the other hand, omission of some important criteria may lead to the acceptance of a weak game or simulation activity. This is why a global model of evaluation is so important.

The same global model may be used for the evaluation of educational computer software. The specific nature of the computer media requires, however, that some criteria receive special attention.

So far, we have considered double interactions in our model. The study of even higher interactions, though more difficult, may prove very useful. Such conditions of evaluation are necessary, not only to make better judgement but also to better understand what makes a good game or simulation activity.

Acknowledgments

The authors wish to thank Martine Clément, consultant in Educational Technology, for her help in creating the artwork for the figures.

References

Corbeil P, Laveault D, St-Germain M (1989) Games and Simulation Activities: Tools for International Development Education. Canadian International Development Agency, Ottawa

Elder CD (1975) Problems in the structure and use of educational simulations. In: CS Greenblat, RD Duke (eds) Gaming-Simulation: Rationale, Design and Applications. Wiley, New York

Gillis L (undated) Un plan d'évaluation formative des logiciels types. Centre d'informatique scolaire, Ministère de l'éducation de l'Ontario, Toronto

Greenblat CS (1975) Gaming-simulation for teaching and training: an overview. In: CS Greenblat, RD Duke (eds) Gaming-Simulation: Rationale, Design and Applications. Wiley, New York

Kohlberg L (1981) The philosophy of moral development: Moral stages and the idea of justice. Harper and Row, San Francisco

Kolb DA (1974) On management and the learning process. In: DA Kolb et al. (eds). Organizational Psychology. Jossey Bass, Englewood Cliffs, NJ

Krathwohl DR, Bloom BS, Masia BB (1964) Taxonomy of Educational Objectives, Handbook II: The Affective Domain. McKay, New York

Laveault D, Corbeil P (1990) Assessing the impact of simulation games on learning: A step-by-step approach. Simulation/Games for Learning 20(1):42–54

Norris DR, Snyder CA (1982) External validation of simulation games. Simulation & Games: An International Journal of Theory, Design, and Research 13(1):73–85

Roid GH, Haladyna TM (1982) A Technology for test-item writing. Academic, New York

Ruben BD, Lederman LC (1982) Instructional simulation gaming: Validity, reliability and utility. Simulation & Games: An International Journal of Theory, Design, and Research 13(2):233–244

Dany Laveault is full professor in Measurement and Evaluation, University of Ottawa. His main interest is the evaluation of simulation and games, specially those used as educational software for computer-assisted instruction.

Michel St-Germain, associate professor, is currently the director of the Formation à l'enseignement (Teacher Education Department) at the Faculty of Education of the University of Ottawa. He regularly uses the case method in his teaching activities in education administration. He has developed a computer management game and integrated decision-making assitance systems for the Education Policy and Planning Division of UNESCO.

Pierre Corbeil teaches history at CEGEP de Drummondville. He holds degrees from the University of Toronto and the Université de Montréal. He has designed several simulation games. His interests include war-gaming and writing science-fiction (in French). He is presently a member of the NASAGA Board.

Information Technology and Simulation Games

Gee Kin Yeo[1]

Abstract. Most simulation games have been designed with predetermined rules. When computerized, their game data and administration rules are locked in the program codes with changes permitted only to the extent allowed for by the program. A framework for developing simulation games using the present information technology has been proposed in order to decrease the difficulty in the transfer of the game from the game designer to the game administrator, to increase the flexibility of game administration, and to improve the learning process for the game participants. In turn, simulation games can be useful in the learning of information technology. Within the suggested framework, the roles of the game designer and the game administrator can be combined easily to enhance the educational perspectives of a game, and using it in courses such as decision–support systems or quantitative methods.

Key words. game administration; game design; information technology; learning; simulation games

Most simulation games have been designed with predetermined rules. When computerized, their game data and administration rules are locked in the program codes, which allow only certain changes. Many recent computerized simulation games also come with administration systems and certain support features for participants. They are known as *simulation game systems* (SG^2). A framework for developing and maintaining SG^2 has been proposed by Yeo (1991). It can decrease the difficulty of transfer of a game from the game designer to the game administrator, increase the flexibility of game administration, and improve the learning process for the game participants. The framework takes into consideration the different requirements, of the different groups of people involved with a simulation game, namely, the game designer, the game administrator, and the game participants. In the framework, a *Model Specification System* is used to describe the game model and a *Game Specification System* is used to describe how the game is to be played. The result is a set

[1]Department of Information Systems and Computer Science, National University of Singapore, Kent Ridge, Singapore 0511, Republic of Singapore; phones 065-772-2908 (w), 065-469-3238 (h); facsimile 065-779-4580; e-mail yeogk@nusvm.bitnet

of generated routines that can be combined with some development tools, as well as in the framework, to facilitate the building of the game administration system and the participants' support system. In Yeo and Ho (1990), it has been shown that a system of management games developed in such a framework allows business information to be flexibly disseminated and game performance of participants to be more adequately evaluated. Figure 1 shows the analogy of the SG^2 to the decision-support system (DSS) generator. Decision-support systems can be built with DSS tools, but are more easily generated with DSS generators; likewise, simulation game systems can be built with general programming tools but are more easily generated in SG^2.

In this paper, I shall discuss the information technology employed in the implementation of the framework. Relevant features of database management systems (DBMS), fourth generation languages (4GL), and DSS generators will be drawn. I shall also discuss how the roles of the game designer and the game administrator can easily be combined within the framework to enhance the educational perspectives of a game, and use of it in a course in DSS or in quantitative methods.

Data Management

Most existing simulation games use the traditional approach to data management, resulting in the well-known, associated problems such as difficulty in accessing data, redundancies in data, data security, and the lack of data independence. In view of these problems, a database approach is used in the

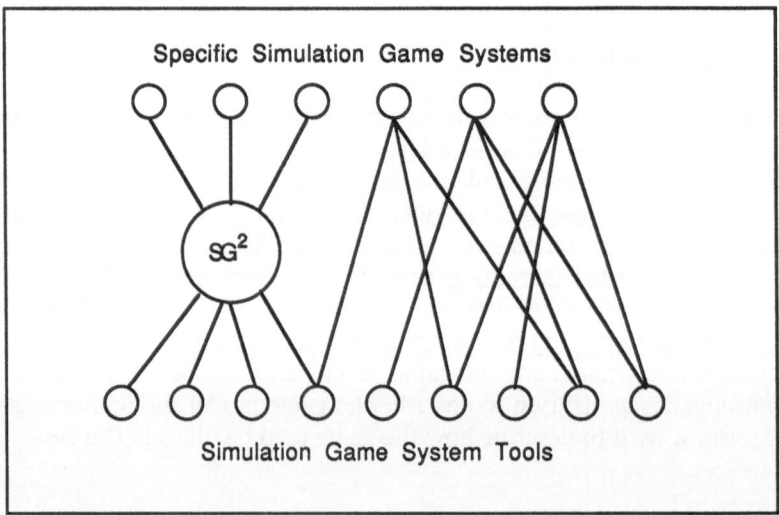

Fig. 1. Analogy of the simulation game systems (SG^2) to the decision-support system (DSS).

SG^2. The three-level architecture (the internal, conceptual, and external levels) provides physical and logical data independence. The data required by the entire simulation/game system are defined through the conceptual schemas. How the data is stored in the database, that is, the internal schema, is taken care of by the DBMS, thereby providing physical data independence. External schemas with the appropriate access authorizations are defined to give the simulation software, the game administrator, and the participants their "views" of the database. Any change in their data access rights are changed through their respective external schemas, thus providing logical data independence.

At the heart of the SG^2 is a data dictionary (Allen et al. 1982). The data dictionary is the knowledge base of the simulation game. It will hold all the information the SG^2 needs to know about a specific simulation/game in order to support the development and maintenance of its administration system and participants' support systems. Minimally, it is to contain the attributes and edit criteria for each data element, such as type, width, output format, range, input pattern, and screen and report labels. It can also be extended to maintain a centralized source of information on all program modules and their required resources. Thus the detailed function, data requirement, or side effects of a PRODUCTION subroutine in a management game, for example, can also be documented in the dictionary. Ideally, it is to be an active data dictionary, meaning that additions or modifications to the data elements, the data structures, and the way in which they are used must be made through additions or changes to it. Maintenance of the possible versions of a simulation/game can therefore be controlled.

When data elements have been defined and their relationship captured in the data dictionary and the DBMS, the simulation/game becomes only an application program manipulating them. With an existing simulation/game, a parser can be used to derive the data elements from the program. Otherwise, the semantics implied by the relation expression documented in the data dictionary can actually be used to provide the programming logic in routines that can eventually be combined to become the whole simulation program.

Obviously, queries on the database and even complex manipulation can be achieved with the data manipulation language provided with the DBMS.

Fourth Generation Languages and DSS Generators

A good user interface, consistent screen design, and quick report and program generation are the main features of the 4GL (Martin 1985) used in prototyping the SG^2. On-line contextual help is easily provided (i.e., the information given depends on the position of the cursor at the time the HELP key is pressed). The interface between the 4GL with the DBMS provides the facilities for maintaining data integrity. For example, when help is requested for a particular Participant Game Data or Game Parameter, the information given is taken from their description stored in the data dictionary. Because 4GLs are designed for on-line operations, the following features are possible:

— Errors can be detected as soon as they are made.
— Cross-validations dependent on other data values can be performed prior to the values being saved.
— Pop-up windows and form-filling for data entries can be used where possible.

Standard screens for entering, displaying, and checking data are usually generated as soon as data have been defined. Facilities are also available for controlling the layout, content, and information shown on the screen to suit user's needs. Designing reports in the 4GL environment is easy. The report generator will display the fields of the database tables and let the user choose those to be included in the report. Facilities such as sorting, record grouping, and data aggregation are commonly available. Thus, performance evaluation of the simulation game with simple evaluation rules can easily be built in with just the report generation facilities.

Continuing development in the 4GLs leads to the enhancement of decision-support facilities. These include functions for modeling and multidimensional presentation of data. With proper interfaces built on the data dictionary and other components of the simulation/game system, these facilities can be made available for game participants to build their own support systems for decision making in the game. Yeo and Nah (to be published) describe the construction of a participants' support system making use of a well-known DSS generator. The framework of SG^2 supports the extraction of relevant data from the game database to form a database just for the participants. Data access tables from the DBMS defined earlier by the game administrators are consulted during the extraction. Decision model building and maintenance, what-if, and scenario analysis capabilities are provided with the DSS language.

CAL for Information Technology

Management games have traditionally been used to support management courses in providing a simulated environment for exercising managerial skills in problem solving. Courses on information systems may also benefit from the use of simulation games. Building a DSS requires task context training where the user is shown how to produce output and how to use that output in the context of a given job function. I have employed a management game to support a course on DSS in the Department of Information Systems and Computer Science at the National University of Singapore (Yeo et al. 1990). The course is offered to the 3rd-year students in information systems who have studied core courses in computer science. A first objective of the course is to let the students see the advantages of the need to support decision making with information systems. The second objective is to apply the present information technology to fulfill such needs. The management game is used to provide an environment for simulated decision making for which information systems may be built. Another promising area in using the management game in connection with the course is in examining the group decision-making process.

Extensions or modifications can easily be made to the simulation/game to provide simulated data evolution to support a course in DBMS. For example, the periodically-generated data from the game provides an excellent example of a temporal database. In supporting a course in quantitative methods for management, different problem areas could be explored to provide exercises in applications of operations research and statistical analysis. For example, decisions on the production levels of consumers and industrial products can be formulated as a linear programming problem. The complex decision situation generally existing in simulation/games provides good examples for practice on multi-criteria decision making.

Software engineering economics can also be learnt using simulation/game systems. A simulated software house can be designed to incorporate decision making in areas such as hardware and software acquisition, project scheduling, control, and manpower planning.

Conclusion

The preceding paragraphs examine the support of existing information technology in building a new generation of simulation/game systems, and how simulation/game systems may, in reverse, support the teaching of existing information technology. New information technologies continue to emerge. To this date, concepts in multimedia and hypertext, neural networks, object-oriented design, and programming are being actively discussed in computer science and information systems literature. Applications of these concepts in building better simulation/game systems and the utilization of them in teaching should be included in the directions for researchers in simulation/games.

References

Allen FW, Loomis MES, Mannino MV (1982) The integrated dictionary/directory system. Computing Survey pp 245–286

Martin J (1985) Fourth generation languages (vol. 1). Prentice-Hall, Englewood Cliffs, NJ

Yeo GK (1991) A framework for developing simulation game systems. Simulation & Gaming: An International Journal of Theory, Design, and Research 3(22):308–327

Yeo GK, Ho CL (1990) Information dissemination and performance evaluation in simulation game systems, Technical Report TR21/90. Department of Information Systems and Computer Science, National University of Singapore

Yeo GK, Nah FH (to be published) A participants' DSS for a management game with a DSS generator. Simulation & Gaming: An International Journal of Theory, Design, and Research

Yeo GK, Nah FH, Chew BW (1990) Initial observations from learning DSS in a simulated decision environment, Technical Report TR10/90. Department of Information Systems and Computer Science, National University of Singapore

Gee Kin Yeo has been responsible for developing MAGNUS, a management game. She is a senior lecturer in the Department of Information Systems and Computer Science and has been teaching courses in operations research and decision support systems. A member of the ACM, the Computer Society of IEEE, and the International Association for Statistical Computing, her current research interests include model management in decision–support systems and computers in education. She is a member of the Editorial Board of *Simulation & Gaming: An International Journal of Theory, Design, and Research.*

Towards a Concept of Meta-Game: Some Applied Results

Yusaku Shibata[1], Hirofumi Kurihara[2], and Shinzo Takatsu[3]

Abstract. When the situation of a game is extremely uncertain, players should first define a general game framework. Such a game can be called a meta-game because the purpose of the game is to define the framework. Global problems are infinitely uncertain, and require a high degree of cooperation between international organizations that generally does not occur without some kind of catalytic coordination. The first objective of the meta-game is for all players to understand the process of global problem solving. The second objective is to understand team coordination. Sustaining such a massive coordination over a long period and retaining policy stability throughout the process are the basic requisites of complex problem solving. Sustained drive and energy, as well as a creative catalytic leadership style, are scarce skills within our society.

Key words. catalytic coordination; crisis management; decision making; general framework; global problem; meta-game; policy

Uncertainty and complexity are buzz words in the study of global problems. In order to solve these problems, international organizations must cooperate under focused strategies yet still have a flexible management style (Mushakoji 1990, Chadwick 1990). The key factor in this effort is the structural change of institutions based on a new way of thinking. Puri and Bhide (1981) insisted that the difficulties and risks in institutional change will be especially formidable for the Japanese culture:

> ... the ability to innovate and explore the uncharted is precisely what
> separates the leader from the industry follower. The Japanese system

[1] Coordination Bureau, SINPL-MEGANET, 7-22-14 Imaizumidai, Kamakura, 247 Japan; phone and facsimile 0467-45-5981
[2] Information and Systems Laboratory, Tonen Corporation, Shuwa Iidabashi Building, 2-3-19 Koraku, Bunkyo-ku, Tokyo, 112 Japan; phones 03-5684-1525 (w) 0474-48-1194 (h); facsimile 03-5684-1539
[3] Department of Business Administration, Senshu University, 2-1-1 Higashimita, Tama-ku, Kawasaki, 214 Japan; phones 044-911-7131 (w) 03-3488-0320 (h); facsimile 044-911-1241

has shown great strength in coming up from behind. But the kind of organizational skills that were needed for catching up are probably not appropriate for being at the cutting edge. In time, Japanese society and industrial organizations may transform themselves in order to fulfill these new tasks. Until then, non-Japanese competitors have real opportunities to hold their own by exploiting Japanese institutional rigidities.

Drucker (1981) has a completely different opinion:

> What is a fact is that the secret behind Japan's economic achievement is not a mysterious Japan, Inc., a creation that belongs, if anywhere, in some Hollywood grade B movie. Far more likely, it is that Japan—at present alone among the major industrial nations—has addressed herself to defining the rules for a complex, pluralist society of large organizations in a world of rapid change and increasing interdependence.

However, these contradictory statements indicate an important common point, which is that many organizations, regardless of whether they are Japanese or non-Japanese, are trying to adapt to a changing world. Complex innovations will come up against many obstacles. This makes a new meta-game not only desirable but also indispensable. This paper describes the features of complex problem solving, the obstacles, a conceptual design of a meta-game named SINPL, and experimentally applied results of this concept to the planning of future generation manufacturing systems.

An Example of a Complex Problem

Future Manufacturing Trends

According to a report by ATKearney (1988), manufacturing industries are faced with new challenges which are caused by mounting global competition and accelerated innovation. Decision making will become more complicated in terms of where to make or buy materials and parts across the world. Strategic alliances between companies and cooperative production agreements will be added to an increasingly complex manufacturing equation. External social influences will also alter the manufacturing environment. There will be more women and minorities at work. The work force will become older, and will have greater difficulty in adapting to changes. While some countries will continue to use innovative methods to curtail imports, aggressive companies will learn how to overcome trade obstacles. The political climate between the East and the West will continue to improve.

In a report to the EC Commission, Cooley (1989) pointed out that, although many people believe that we need better technology and more financial power to be competitive, what firms truly need is better human and organizational innovation and greater integration of technical, human, and organizational elements. However, our future will never belong to powerful bureaucracies, but to human organizations that will be able to set human goals shared by the largest fraction of people, and to achieve them by using knowledge and technology, in the public and general interest. Therefore, the main challenge for the manufacturing industry in the next 20 years is not competition between the richest areas of the world for control, but its contribution to the sustainable development of planet earth, and the needs of around 8 billion people.

ESPRIT of the EC, the Society of Manufacturing Engineers of the US, and IMS (Intelligent Manufacturing System) of Japan are concurrently developing human-centered manufacturing systems. However, the various worldwide practices in shaping manufacturing systems and the generally weak attempts to follow anthropocentric orientations in manufacturing seem to be the major hindrances for the realization of the objectives (see, e.g., Japan Electronics Industry Development Association 1991).

ATKearney (1988) has studied the issues and implications of the next 10 years for the Society of Manufacturing Engineers. Manufacturing planners should cover a wide variety of tasks and activities. Since no one individual can be omnipotent, manufacturing planning teams must be made up of strategic planners, systems integrators, and technical specialists. Each of these roles requires a different balance of skills. All players must understand the process of global problem solving, which requires sustained effort over a period of time, and a catalytic coordination team which is ultimately concerned with the process.

Although many organizations are trying to adapt themselves to the changing world, uncertainties and contradictions remain. Specialization and organizational segmentation hinder an integrated approach to the development of more human production systems. For example:

> Anthropocentric production systems offer an interesting avenue to enhance competitiveness of industry in Europe. They are also compatible with the social values of Europe. In spite of these advantages, the development of anthropocentric production systems is slow and uneven. Structural rigidities enhance development along traditional lines and hinder promising innovations. In order to support constructive break-throughs, systematic knowledge has to be developed that can be easily adjusted to different environments. Such a type of knowledge can only be developed within an institutionalized and permanent setting (Lehner 1991).

Planners are confronted with nothing less than a sweeping cultural change. During times of crisis, great leaders have also emerged. However, we now need managers with a very different leadership style, a style which has not been systematized and is still not taught in engineering or business schools.

The major challenge for manufacturing industries will not be the competition between rich countries, but the contribution they have to make to the sustainable development of the planet. To find this creative direction, the meta-game players must learn how to analyze the interaction of complex obstacles. They will become the catalytic forerunners of great leaders.

Usually, there is a gap between objectives and realistically possible actions. A new strategy is a bridge over this gap. According to Petrella (1989), what firms need is better human organization, and integration between technical, human, and organizational elements. This calls for a qualitative jump in the understanding of how humans interact with technology and, above all, in the design of increasingly complex techno-organizational systems. This will enable the setting up of human goals shared by the largest proportion of people throughout the world.

How can the new objectives of world manufacturing industries be attained? It is not only by making a good plan, or by establishing a new institution, but also by mobilizing existing organizations and motivating people.

Pascale (1978) found that a different approach is effective in uncertain and complex situations. Generally speaking, whereas Westerners regard ambiguity as a symptom of organizational ills, the notion of ambiguity is normal in Japanese business and it helps Japanese managers to justify their practice of making tentative decisions or no decisions. In their estimation, the solution always involves groups of persons at different levels with different mandates. The distribution of power is such that they lack full control. Perhaps these circumstances have made well-structured games of little benefit in Japan, but will render ill-structured meta-games of greater service, not only for the Japanese, but also for Westerners in a future where uncertainty and complexity are likely to prevail.

A Conceptual Framework SINPL

It is extremely difficult to play a meta-game without predetermined rules. The difficulty could be alleviated to some extent by preparing a conceptual framework (meta-rule) which consists of a planning process (meta-process model), a problem structure (meta-problem model), and a team structure (meta-role model) (see, e.g., Abt 1970, Duke 1974, Arai 1990, Ozbekahn 1971). One realization of these features is SINPL.

A special feature of SINPL is that it does not start from a given problem but from a desirable future norm, from which the name SINPL (SImplified Normative PLanning method) is derived. The game contains five steps, as follows (Fig. 1):

1. Forecasting desirable future (process) and vision scenario (structure)
2. Insight of contradictions (process) and problem structure (structure)
3. New idea proposal (process) and normative objectives (structure)
4. Strategic planning (process) and strategic patterns (structure)
5. Action proposal (process) and launching tactics (structure)

The chief advantage in using meta-games like SINPL is that they provide a mode of experimentation with alternative strategies and tactics in a constantly changing environment. The fluid nature of a meta-game approximates the uncertainties encountered in a real situation. The artificially controlled contexts

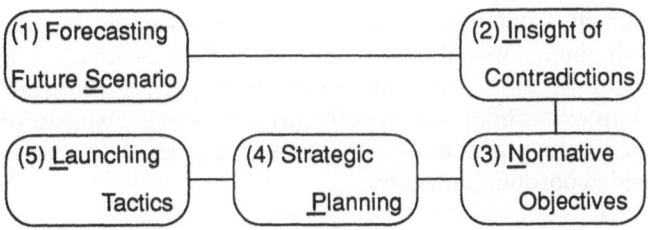

Fig. 1. Planning process model of SINPL

which the conceptual framework imposes has built into it a certain relationship between the decisions made by some of the players and the responses of other stakeholders. These relationships are initially unknown to the players and are only revealed as the meta-game proceeds. So the outcome of a particular decision or strategy has immediate effects, thereby providing data for analyzing and evaluating the selected course of action, with neither the time lag nor the potentially irreversible consequences of a decision in the real world.

Forecasting a Future Scenario (Scenario: S)

The purpose of this step is for participants to share a common future scenario.
   Teams brainstorm, which reveals the hidden thinking of the participants. The session leader, in collaboration with the participants, classifies and systematizes the idea cards. In one run, participants produced an attractive view of the future: multinationalization of the manufacturing company, emphasis on systems integration, diversification of business, humanization of the company, improvement of welfare facilities, longer paid vacation and overseas travel, new factories in provincial districts, and one computer terminal for each employee.

Insight into Contradictions (Insight: I)

An underlying contradiction is a set of obstacles held together by a common underlying factor. Obstacles are solid, real factors that actually exist and indicate contradictions. This was the time to examine the real situation. It was very painful, but it was the key to a creative approach to the future. Participants used their intuition to cluster related obstacles as follows: obstacles to communication within the company, lack of employee responsibility, sense of urgency, competition in the market, ambiguity in customer specification, ambiguity in company vision, and inadequate employee education and training.

Normative Objectives (Norm: N)

New idea proposals point to major new directions which will effectively break through the contradictions. Proposals should be unique and creative, and not simply improvements on existing programs and procedures. Also, proposals are independent from each other and are not sequential steps. Proposals require fresh, innovative thinking. The workshop resulted in the following recommendations: shift national government policy, reorganize industrial structure, improve school and in-company education, instigate organizational reform, re-educate top management, recruit top level people, and set up a company-wide coordination center.

Strategic Plan (Plan : P)

There was a gap between the objectives and the realistically possible actions. Based on brainstorming, the session leader classified and systematized the cards and made a strategic plan to bridge this gap as follows: organize a new informal network of concerned engineers, organize a new trade association, assign a training manager, support meta-game participation in the company, use outside consultants, and distribute suggestion boxes.

Launching Tactics (Launch : L)

At the last session of the SINPL workshop, the participants created the steps necessary to implement the first 3 months. They answered specific questions focusing on the their own actions rather than others'. Some of the commitments were: decide company policy at the board meeting and publicize it to all employees, have a periodical meeting between employers and employees, have company-wide SINPL workshops, consult with headhunting companies, conduct an opinion survey covering all employees, and start human relations education.

Concluding Remarks

We feel that SINPL as a meta-game is an insightful conceptualization of a strategic planning process. It highlights critical issues and provides a guideline for developing creative strategies where diverse organizations and individuals interact. It further suggests group processes and agenda formats which guide each step of strategic planning. From a number of applications of SINPL, the authors are convinced of its practicality and power, but also convinced that considerable training is required to (1) internalize cognitively the planning process as a discrete series of workable steps, (2) integrate diverse participants and idea cards in each step, and (3) transfer and apply the findings to their daily problems.

In addition, our experience suggests that timely support from an outside network of creative catalytic consultants is indispensable to sustain such a massive coordination drive over a long period of time (for further discussion, see, e.g., Friedman 1987, McCarthy 1984, Satou 1986, Shibata 1984, Takatsu 1986).

References

Abt CC (1970) Serious games. Viking, New York

Arai K (1990) "SIMPLE" as a Policy Formation Exercise (in Japanese). Proceedings of the 2nd JASAG Meeting. Japanese Association of Simulation and Gaming, Tokyo

ATKearney (1988) Countdown to the future: The manufacturing engineer in the 21st Century. Society of Manufacturing Engineers, New York

Chadwick RW (1990) Global modeling for global responsibility (in Japanese). Simulation and Gaming 1(1):45–49

Cooley M (1988) European competitiveness in the 21st Century. Commission of the European Communities, Strasbourg

Drucker PF (1981) Behind Japan's success. Harvard Business Review 59(1):83–90

Duke RD (1974) Gaming: The future's language. Sage, Beverly Hills

Friedman J (1987) Planning in the public domain—From knowledge to action. Princeton University Press, Princeton

Japan Electronics Industry Development Association (1991) Study Report on the New Factory System (in Japanese). Japan Electronics Industry Development Association, Tokyo

Lehner F (1991) The development of anthropocentric production systems in Europe: Challenges to science and technology. In: Proceedings of 2nd International Conference on Science and Technology Policy Research organized by the Science and Technology Agency of Japan, Jan 24–26, Kanagawa, Japan

McCarthy T (1984) The critical theory of Juergen Habermas. Polity Press, Cambridge

Mushakoji K (1990) The study of global problems: A new challenge for simulation modeling (in Japanese). Simulation and Gaming 1(1):2–6

Ozbekhan H (1971) Planning and human action. In: Weiss PA (ed) Hierarchically organized systems in theory and practice. Hafner, New York

Pascale RT (1978) Zen and the art of management. Harvard Business Review 56(2): 153–162

Petrella R (1989) Competitiveness for what? In: Cooley M (ed) European competitiveness in the 21st Century. Commission of the European Communities, Strasbourg

Puri T, Bhide A (1981) The crucial weaknesses of Japan. Wall Street Journal, June 8

Satou Y (1986) From Weber to Habermas—A horizon of association (in Japanese). Sekai Shoin, Tokyo

Shibata Y (1984) Toward a policy guidance system for complex innovation. In: Eto H, Matsui K (eds) R&D management systems in Japanese industy. North-Holland, Amsterdam

Takatsu S (1986) Experimental study of design factors for MIS/DSS (in Japanese). Senshu University Business Review 41:261–326

Yusaku Shibata has been hooked on complex problem solving for more than 30 years, mainly as an engineering-manager with Hitachi. Now, he is trying to apply his lifelong experience to global problems, such as future generation manufacturing systems planning in developed countries and sustainable regional development planning in developing countries. He is a member of the Operations Research Society of Japan and the Japanese Society of Organizational Science.

Hirofumi Kurihara is a member of the Operations Research Society of Japan, in which he is an active member of a study team for organizational intelligence. His special interest is the application of psychology to organizational intelligence.

Shinzo Takatsu, Dr. of Engineering, Tokyo Institute of Technology, is Professor of Systems Analysis in the Department of Business Administration, Senshu University. His current interests are general systems theory and organizations theory. He is a member of the Operations Research Society of Japan and the Japan Industrial Management Association.

Software to Communicate Global Models

Kenneth L. Simons[1] and Peter J. Poole[2]

Abstract. Global simulation models can be powerful tools (1) to develop academics' knowledge of the global socio-politico-environmental world and (2) to analyze possible policies. However, they require considerable knowledge about their workings and the types of questions they are designed to answer before one can use them. Hence it is useful for communication among academics, analysis, and education in schools and universities to develop descriptions of the models and accessible computerized versions of the models. This paper presents new tools for an interactive, adaptive, and user-friendly genre for presenting and using complex simulation models. This way to communicate simulation models has hitherto not been possible nor emphasized. The software should not only include the models, but also document them and guide people through a learning process. These issues apply to all kinds of simulation modeling, but illustrations in this paper come from global social and environmental models.

Key words. computer interfaces; education; environmental models; environmental policy; global models; simulation/gaming

Traditionally, global modelers have spent most of their effort building and analyzing global models, and relatively little effort documenting those models and making them accessible to others. This approach has been appropriate for two reasons. First, the process of modeling has been (and is) time consuming and, given their limited resources, researchers have rarely been able to spend the time to make polished documents and computer applications that communicate the models. Second, computer hardware and software have lacked abilities (a) to integrate building and documenting models and (b) to create easy-to-use versions of models without intensive work by professional programmers. These difficulties have caused poor communication of global models. As part of a project to develop educational software based on classic

[1] Department of Social and Decision Sciences, Carnegie Mellon University, Pittsburgh, PA 15213-3890, US; phones 412-268-6851 (w), 412-361-1447 (h); facsimile 412-268-6938; e-mail ks3y+ @ andrew.cmu.edu
[2] Department of Political Science, Massachusetts Institute of Technology, 77 Massachusetts Avenue, Cambridge, MA 02139, US

global models, we are addressing the second difficulty, the need for an easy way to make models accessible to others (researchers, students, the public, policy makers). If global models are to play a role in global policy, they should ideally take a form which policy makers can understand and explore to the depth that they feel is necessary, not the form of enigmatic entities on inaccessible mainframe computers. If global models are to advance research, it will help to distribute those models in forms which other researchers can easily access and which allow them to question and modify the models as they see fit. Greater openness of model software enhances a model's replicability, transforming the software from computerized narratives into scientific, empirical tools. This paper addresses the second problem, the need for methods to develop easy-to-use software that conveys understandings obtainable from simulation models. Methods to integrate model building and model documentation are not addressed in this paper.

The Models in Our Project

For our project, we chose four classic models of global environment and society. The models are:

1. Watson and Lovelock's (1983) DAISYWORLD model that illustrates the Gaia hypothesis
2. A simple radiative-convective model of the greenhouse effect, paired with an energy balance model
3. Forrester's (1973) WORLD DYNAMICS model of population, economic consumption, and natural resource interactions
4. Edmonds and Reilly's IEA/ORAU LONG-TERM GLOBAL ENERGY-CO_2 MODEL (hereafter ENERGY-CO_2 MODEL) of energy supply and demand over the next century (1983).

These models span three intellectual perspectives about Earth:

1. DAISYWORLD represents the Earth as a biosphere that modulates physical conditions for life.
2. The one-dimensional climate model of the greenhouse effect represents the world as a geosphere composed of physical-chemical cycles.
3. WORLD DYNAMICS and the ENERGY-CO_2 MODEL represent the world as a social system in which human behavior determines global phenomena.

The ranges of subject matter and of modeling methods spanned by the project help ensure that the techniques developed in the project are applicable to all sorts of global models. In addition, this selection of famous models allows multidisciplinary teaching about earth systems. In the literature (e.g., Roberts 1978, Alker 1985) they are portrayed as classic models, but most students and researchers have had the chance to read only critical commentary or reviews of them. The fact that knowledge of these models comes from secondary sources (caused by the lack of accessible original models) can be likened to the availability of Shakespearean plays only in summary form. These classic computer models are worthy of being kept in libraries, just as other media (music recordings and films) are regularly archived.

At this time, we have completed the GAIA THEORY AND DAISY-WORLD (hereafter DAISYWORLD), and the energy-balance model and WORLD DYNAMICS are in progress. A version of DAISYWORLD has been translated into Portuguese, for demonstration in Brazil. The Appendix compares our software with other earth-systems software.

Components of the Software

Depending upon its purpose, the software may need any or all of the components shown in Table 1. For details on how to create these components, see Simons K.L. (1991) Software construction for simulation models: A handbook for simulation model software in HyperCard (unpublished material, available from the author).

Program Information

Ideally, how to use the program should be obvious to the user from looking at the computer screen. Screens should be carefully designed to guide a user's attention as desired. Introductory information, help buttons or menus and a paper booklet can provide additional information as necessary.

Model Information: Documentation

In order to be straightforward about the assumptions of a simulation model, we believe that the model should be documented in any program that uses it (Fig. 1). Most current simulation programs do not provide such documentation. Putting these diagrams and equations on the computer ensures that information about the model will always be available when someone uses the model, and it encourages rather than discourages exploration of the model's

Table 1. Components of the software

Program information	Model information	Model use	Education
Cover illustration (optional)	Introduction	Simulation start/ stop	Structured exercises
Summary report (optional)	Documentation of model	Views of the results	Captivating activities
How to use the software	Contextual information	Ways to modify the simulation	Reports for the user on what was learned
Credits	Demonstrations		Rewards for progress
Accompanying booklet (optional)	Source code		

assumptions. Similarly, the software can provide reasons why the model was built, a history of the modeling project, criticisms and answers to them, a bibliography, data that support the model or are used in it, and any other information relevant to the model. Inclusion of source code or files from modeling programs allows other researchers to modify models as they wish and encourages academic debate.

Model Information: User Introduction

Most users will also need a basic introduction to the model: what it is about, why it was built, what are its parts, and what conclusions can be drawn from it. This information is sometimes conveyed in separate books and articles, but it is preferable to make it available again on the computer, where people who use the model can always find it. The introduction could be a reader with textual information and illustrations, it could be animated demonstrations, or it could be a guided series of exercises using the simulation.

Model Use

During the simulation, the user usually takes on some role. For example, with a global environmental model, the user might act as a Special Assistant for the Global Environment to the UN Secretary General. In this case, the goal might be to determine environmental policies that are effective and politically

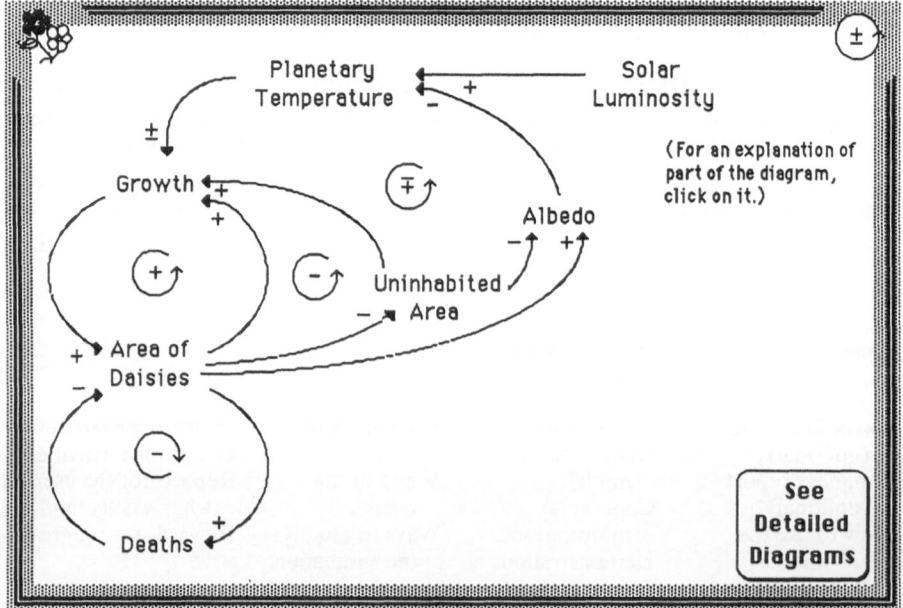

Fig. 1. Model documentation

feasible, and the model would be used to analyze the effects of different policies.

Model use can be either active or passive. In active use, the user makes decisions during the simulation. In passive use, the user makes decisions before the simulation starts, then watches their consequences (Fig. 2). Active use of a model is often exciting, which helps keep users' attention. Passive use can be more understandable: people make one set of decisions and they get back one set of results.

A combination of traditional graph displays, customized animation, and multimedia can benefit educational software. For example, EARTHQUEST, SIMCITY, and SIMEARTH make good use of the color, sound, and animation capabilities of personal computers. Animation is especially well suited for active model use, but traditional graphs and charts should be available too. For example, a world simulation might display a picture of the globe, with pictures of people, factories, and food indicating levels of population, capital investment, and food harvests in different regions of the world. Traditional graphs of, say, population over time, and pie charts of population in different continents, provide information in a form that is valuable but not easily accessible from the animated display.

Education

To communicate the models for education, we use these approaches:

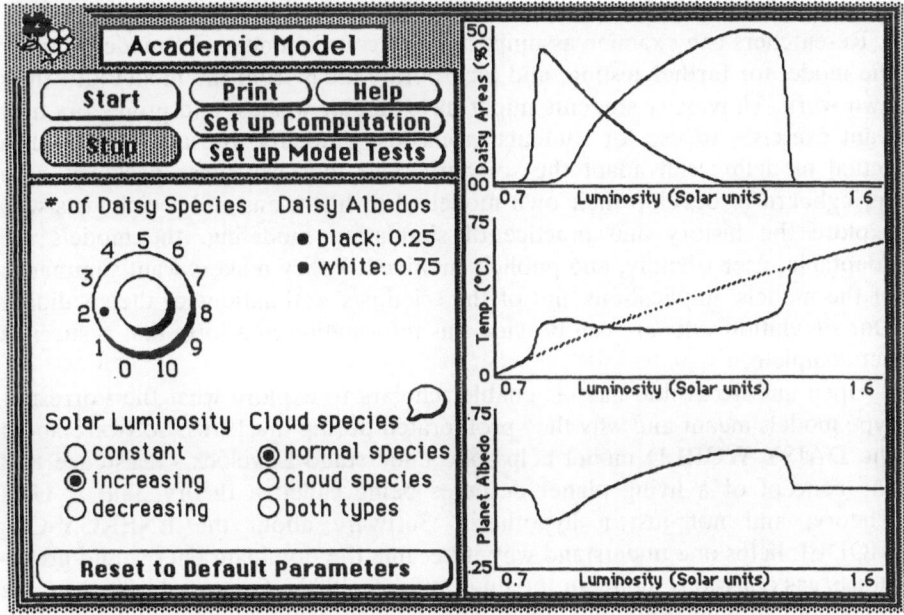

Fig. 2. Model use

1. General descriptions using text; these are little different from a paper, sometimes more difficult to use than a text book, but ensure that the information is available with the model.
2. Series of diagrams that act like overlays, beginning with simple diagrams and progressing to complex diagrams; this helps explain the complex structure of models and important feedback loops. The computer wins over books or blackboard presentations in displaying diagrams this way.
3. Occasional prompts and queries with answers, such as one might find in good educational text books.
4. Exercises with questions and answers that require using the model, such as one would find in a text book.
5. Ability to translate the software into other languages; translation will also become easier with the development of an industry-standard international character set.

The packages are designed so that they can be adapted both by the user and by the instructor. User-friendly programming environments, such as HYPERCARD and STELLA II, ensure this flexibility.

Audience and Purpose

We have developed a flexible and adaptable interface that allows the software to be tailored for particular audiences. We have also created software to fit the needs of multiple audiences. Within our project, using global models of earth systems, we identify four possible audiences.

Researchers can examine assumptions and consequences of a model, modify the model for further testing, and incorporate ideas from the model into their own work. University students might have similar wishes, and professors may want exercises to use for students in classes. Students who are interested in actual modeling can adapt the software, and thus use these packages as a springboard to develop their own models. For historians and sociologists who explore the history and practice of simulation modeling, the models are adaptable, user-friendly, and publicly archived. Policy makers want summaries of the models' implications and of the scientists' estimations of their validity. Our simulation software can provide this information in a form that is succinct yet complete.

Open and accessible models enable scholars to explore what the Forrester-type models meant and why they proliferated during the 1970s. Software about the DAISY WORLD model helps one understand Lovelock's insistence that the concept of a living planet deserves being called a theory, that is Gaia Theory, and not just a hypothesis. Software about the ENERGY-CO_2 MODEL helps one understand why it became the dominant energy and greenhouse gas emissions model in the mid- and late-1980s. For sociologists who are interested in conceptual models of social structures, an open computer software environment like this allows them to probe theories of complexity and to develop empirically testable computer representations of social systems (Luhmann 1988, Maturana 1980).

Testing Effectiveness

Testing the software informally during and after development is important in order to find out what users learn. Formal testing after development can rigorously investigate what people learn, but it is expensive in time and money. Tests of whether a user learns particular concepts, given before and after use of the software, are simplest. Alternatively, the cognitive mapping method used by Bostrum et al. (in press) investigates what concepts people hold and how they think about those concepts, before and after use of training materials. Verbal protocol analysis (Ericcson and Simon 1980, Newell and Simon 1972) can be used to understand how users think about particular parts of the software and to see whether they understand and learn as intended.

Summary

This paper presents a genre of open and flexible software that describes and allows use of global models. Specific software has been developed for teaching global environmental systems from a multidisciplinary perspective. Details of the tools needed for software development are available from the authors (Simons 1991, Software construction for simulation models: A handbook for simulation model software in HyperCard, unpublished material).

Acknowledgments

This project began at the Massachusetts Institute of Technology in 1989. Thanks to the following programs and departments for their financial support: Departments of Civil Engineering, Political Science, and Urban Studies and Planning; School of Engineering; Undergraduate Research Opportunities Program; Center for Global Change and the Houghton Fund. Thanks for advice and assistance especially to Prof. Joseph Ferreira and Dr. Marvin Miller. At Carnegie Mellon University, thanks for advice and assistance especially to Prof. Baruch Fischhoff and Dr. Ann Bostrum. We are also grateful for the hospitality and support of the National Institute for Environmental Studies, Tsukuba, Japan. For work done by Kenneth Simons beginning fall 1990: This material is based upon work supported under a National Science Foundation Graduate Fellowship. Any opinions, findings, conclusions or recommendations expressed in this publication are those of the authors and do not necessarily reflect the views of the National Science Foundation.

References

Alker HR (1985) Global modeling alternatives: The first twenty years. In: Ward MD (ed) Theories, models, and simulations in international relations: Essays in honor of Harold Guetzkow. Westview, Boulder

Bostrum A, Fischhoff B, Morgan G (in press) Characterizing mental models of hazardous processes: A methodology and application to radon. Journal of Social Issues

EARTHQUEST. Earthquest Inc. (1990) Earthquest Inc., Palo Alto (125 University Avenue, Palo Alto, CA 94301, USA)

Edmonds J, Reilly J (1983) A long-term global energy-economic model of carbon dioxide release from fossil fuel use. Energy Economics 5(2):74–88

Ericcson KA, Simon HA (1980) Verbal reports as data. Psychological Review 87(3):215–251

Forrester JW (1973) World dynamics, 2nd edn. Productivity Press, Cambridge, Massachusetts (Wright-Allen Press, 1971)

GAIA THEORY AND DAISYWORLD: A PLANETARY LIFE-SUSTAINING SYSTEM. Simons KL, Poole PJ, Bell M, Rathbun K, Trimble E, Young A (1991) Simons KL, Pittsburgh (Department of Social and Decision Sciences, Carnegie Mellon University, Pittsburgh, PA 15213–3890, USA)

HYPERCARD version 2.1. Claris (1991) Claris, Santa Clara (Box 526, Santa Clara, CA 95052, USA)

IEA/ORAU LONG-TERM GLOBAL ENERGY-CO_2 MODEL: PERSONAL COMPUTER VERSION A84PC. Edmonds J, Reilly J (1986) Oak Ridge National Laboratory, Oak Ridge (Carbon Dioxide Analysis Information Center, Oak Ridge National Laboratory, Oak Ridge, TN 37831, USA)

Luhmann N (1988) Tautology and paradox in the self-descriptions of modern society. Sociological Theory 6 (Spring):21–37

Maturana HR (1980) Autopoiesis: Reproduction, heredity, and evolution. In: Zeleny M (ed) Autopoiesis, dissipative structures, and spontaneous social orders, AAAS selected symposium 55. Westview, Boulder

Newell A, Simon HA (1972) Human problem solving. Prentice Hall, Englewood Cliffs

Roberts PC (1978) Modeling large systems: Limits to growth revisited. Taylor and Francis, London

SIMCITY. Maxis (1989) Maxis, Orinda (2 Theatre Square, Suite 230, Orinda, CA 94563-3041, USA)

SIMEARTH. Maxis (1990) Maxis, Orinda (2 Theatre Square, Suite 230, Orinda, CA 94563–3041, USA)

STELLA II. High Performance Systems (1990) High Performance Systems, Hanover (Suite 300, 45 Lyme Road, Hanover, NH 03755, USA)

Watson AJ, Lovelock JE (1983) Biological homeostasis of the global environment: The parable of Daisyworld. Tellus 35B:284–289

Kenneth Simons researches and teaches about dynamic models of social systems and the use of models for corporate and government policy. He has developed, written about, and created tools for software which uses both managerial and global simulations. When an undergraduate at MIT, he worked for 4 years with its System Dynamics Group. He is a doctoral student and a National Science Fellow at Carnegie Mellon University.

Peter Poole is completing a doctorate on advanced industrial societies and the management of international environmental problems. His work in political science draws on his previous studies in industrial engineering and civil engineering, and field research at the Center for Global Environmental Research of the Japanese Environment Agency.

Appendix

Other Earth-Systems Software

Several other software packages can be used to teach about earth systems. These include the following:

1. Global Data. Data on world issues, collected by the UN, the World Bank and groups like the World Resources Institute have been made available for microcomputers from the World Game Institute in Philadelphia.
2. Games. SIMEARTH, created in collaboration with James Lovelock, has been translated into Japanese by Imagineer Co., in Tokyo. EARTHQUEST integrates historical and geographical information with instruction about local and global environmental problems.
3. Educational software. This category stresses learning and teaching, not gaming just for fun. At an introductory level, the free GLOBAL WARMING STACK prepared by Apple is quite good, though it is oriented for junior high school or high school American children. GLOBAL RECALL, produced by the World Game Institute, presents global data in captivating manner.
4. Adaptable educational software. Our project falls in this category.

Modeling Organizations with Visual Agents

Kenneth A. Griggs[1]

Abstract. This paper describes a Smalltalk-80-based prototype system that uses icons to model an organization. Users communicate complex goals to the system, such as meeting scheduling, by manipulating the icons. Underlying the icons is an elaborate set of data and knowledge structures. The icons are objects that appear to be user 'agents' in the sense that they are seen to contain data and behavior needed to complete the user's goal. The system employs object-oriented programming techniques, expert systems, and other AI techniques.

Key words. agents; object-oriented programming; organizational modeling; organizational simulation; Smalltalk

The value of visual models in interface design is well established scientifically and commercially. The commercial successes of iconic interfaces found in games and other interactive simulations show that, at least for certain classes of problems, iconic representations have much merit.

Visual interface techniques are designed to offload much of the cognitive effort from the user to the system. The ideal interface is transparent to the user—it is an extension of the user's reality and needs no added learning. Sophistication is hidden from the user and seemingly trivial user acts are translated into complex system behavior. These developments lead to the question "can a simulation using graphical objects model a complex phenomenon such as an organization?" This paper describes a Smalltalk-80-based working prototype system that uses iconic objects and agencies to achieve this end.

[1] Decision Science Department, 2404 Maile Way, Honolulu, HI 96822; phone (808)956-7494; facsimile (808)956-3261; e-mail griggs@uhccvx.uhcc.hawaii.edu

Objects

The notion of *objects* and *object-oriented programming* originated with the Simula language (Birtwistel 1973) and was further developed via the creation of the Smalltalk language at the Xerox Palo Alto Research Center (Goldberg and Robson 1983, Goldberg 1984). An object is a packet of information and descriptions of its manipulation; a message is a description of an object's manipulations (Goldberg and Robson 1983). Objects are instances of classes which are themselves part of a class hierarchy. The class hierarchy is an inheritance path for data and behavior. An icon is a graphical object.

Agents and Agencies

Agents provide behavior and data useful in completing a user's goal. An agent has been defined as a "'soft robot' living and doing business within the computer's world" (Kay 1984). Agents act for users and may encapsulate knowledge beyond the user's own. An agent is an object, sometimes represented by an icon which is programmed to use, communicate, and manipulate other objects to complete a complex task. The agent icon may be programmed by a system developer or by the user. When object environments contain multiple, intercommunicating agents, they are termed agencies.

The Visual Agent Approach

The concept of a visual agency is borrowed from computer games such as the legendary PINBALL CONSTRUCTION SET, where users can construct a pinball game from a variety of components represented as icons—plungers, flippers, and so on. The visual agent approach is a kind of "organization construction set" that includes the user as part of the construction. The purpose of the approach is to create an iconic environment in which a user can perform complex, organization-wide tasks with minimal training and effort.

Figure 1 is an example of a hierarchical object classification scheme used to create an object environment for an organization. The classification scheme is also a Smalltalk-80 class hierarchy. Each icon is an independent agent in the sense that specific data and behavior related to each function is encapsulated within the data and knowledge structure linked to the icon. An agent may be left out of a process and yet a goal may be completed through the efforts of other agents. Variations on this theme have been proposed [Minsky 1985, Griggs (1989) GDI (Goal Directed Interface): An intelligent, iconic, object-oriented interface for office systems. University of Arizona, Tucson. Unpublished work].

Specifically, the visual agent approach contains several distinct aspects. The approach (1) postulates a simple model of a user environment having persons, things, and processes to form an agency; (2) represents knowledge in the

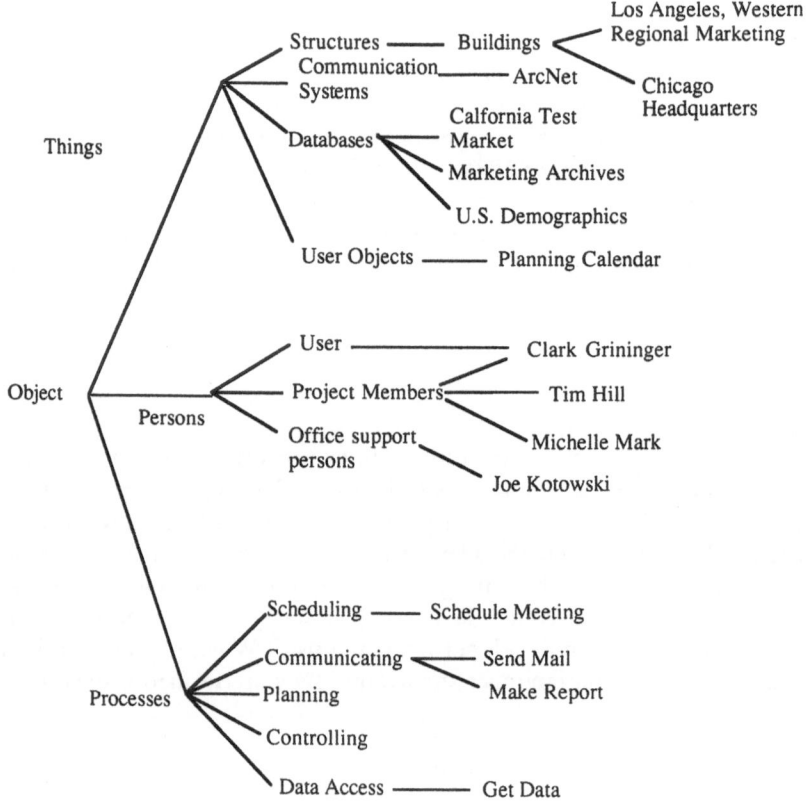

Fig. 1. An organizational class hierarchy

environment through rules, frames and scripts, and object-oriented programming techniques; (3) uses an iconic representation; (4) requires that the user's
own person icon be present for all interactions so that actions appear to happen
in a user-controllable context (the user's icon is, literally, *in the interface*); (5)
provides a selection window through which the user communicates the goal by
grouping relevant icons; (6) uses a rule-based expert system to examine an
icon configuration and, through its knowledge, derives a user goal (despite
ambiguous or faulty icon placements); and (7) tries to complete the user goal
through the use of scripts and multiple expert systems.

The user's personal icon is needed since (1) the user is a necessary part of
the goal completion process and is the principal actor within it, (2) engagement
is enhanced when the user is visually represented, (3) system activities are
viewed from the user's perspective and are represented as such, and (4) data
and knowledge about the user and problem domain are linked to their visual
representation.

An Example of a Visual Agency—A Scenario

Clark Grininger is a marketing executive for a consumer products company headquartered in Chicago. The team consists of Clark Grininger (Assistant Director, Marketing), Tim Hill (Director, Market Analysis), and Michelle Mark (Analyst, Systems Support). Joe Kotowski, an accounting student intern, is not a team member, but will act as a support person for the project. Much of the communication for the project will be via an internal, corporate-wide network (Arcnet) which handles electronic mail and database queries. A variety of geographically dispersed databases will be consulted and updated during the project. Much of the project work will consist of extracting data

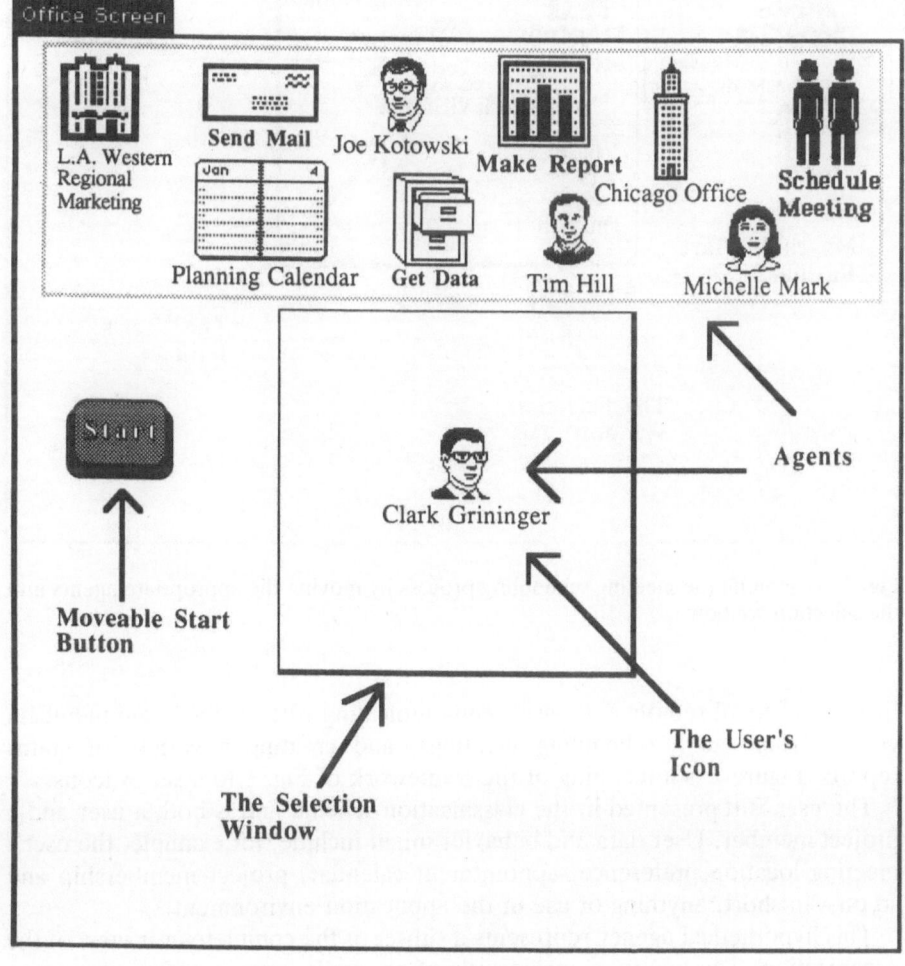

Fig. 2. Organizational visual agency

70   *Griggs*

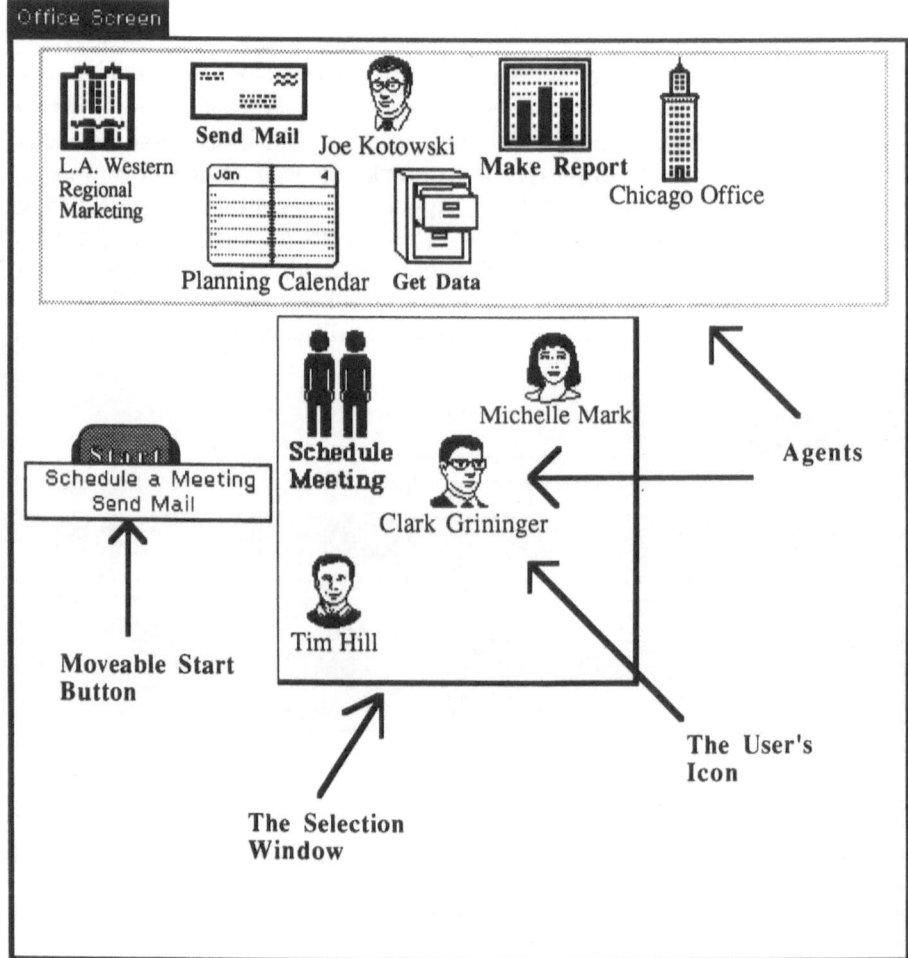

Fig. 3. Beginning the meeting scheduling process by moving the appropriate agents into the selection window

from a variety of remote databases, communicating with project team members via electronic mail, scheduling meetings, and creating a variety of status reports. Figure 2 is a mapping of the framework of Fig. 1 to a set of icons.

The user is represented in the classification scheme and is both a user and a project member. User data and behavior might include, for example, the user's meeting location preference, appointment calendar, project membership and so on—in short, anything of use in the application environment.

This hypothetical agency represents a subset of the complete user view of the organization. The environment consists of an iconic representation of persons (Michelle Mark, Joe Kotowski, Clark Grininger, Tim Hill), things (Chicago Headquarters Office, Los Angeles Western Region Marketing Office, the user's Planning Calendar), and processes (Send Mail, Get Data, Schedule Meeting) representing a particular agency environment. The user (Clark Grininger) is both a user and a member of a project team and thus inherits properties of both.

Figure 3 depicts a screen in which the user has moved two person icons (Michelle Mark, Tim Hill) and a process icon (Schedule Meeting) into the selection window. An expert system containing knowledge about icon configurations attempts to determine a user goal.

Conclusions reached by the expert system are fed back to the user in the form of verification prompts having a list containing one or more possible goals ranked by certainty. In the example, the user wants to schedule a meeting with Michelle Mark and Tim Hill. Based on the type of icons in the selection screen, the Schedule a Meeting goal has the highest likelihood of being the goal and is placed at the top of the list. Accordingly, Send Mail is placed in second position. The user then picks the proper goal from the list and the process is carried forward. This technique provides a degree of user forgiveness. Figure 4 shows the sequence of events in the process.

An initial evaluation of the icon configuration is needed since an actual organization model world might have hundreds of icons. In such an environment, the number of combinations of icons quickly becomes explosive and unmanageable. For example, assuming an average configuration to be four icons out of a possible 30, the number of potential icon configurations is roughly 27,405.

The Current Prototype System

Currently, the prototype system consists of a small model world with three persons, two processes (send mail, schedule meeting), and two things (a building, a planning calendar). Using several sets of rules, the system successfully discriminates between user goals, intelligently schedules meetings (suggests meeting times and places, resolves conflicts), and automatically sets up e-mail protocols between users. The primary motivation for creating the prototype was to explore the feasibility of larger simulations involving complex goals and objects. Additional functionality will be added to the system and an empirical test of the validity of the concept is planned.

Summary

This paper reports on a prototype iconic environment, created in Smalltalk-80, that is a simulation of an organization. The simulation employs a set of icons (including the user) that represent organizational entities and uses rule, script, and frame-based knowledge structures to achieve complex user goals.

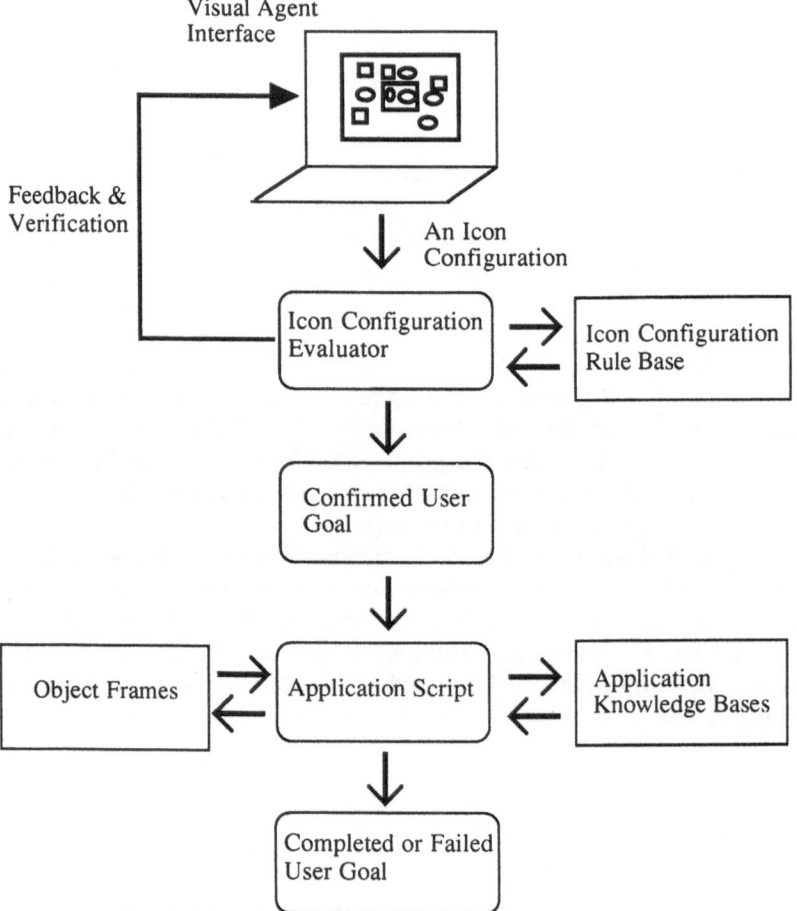

Fig. 4. Event sequence in the visual agent approach

References

Birtwistel GO (1973) Simula begin. Auerbach, Philadelphia
Goldberg A (1984) Smalltalk-80: The interactive programming environment. Addison-Wesley, Reading, MA
Goldberg A, Robson D (1983) Smalltalk-80: The language and its implementation. Addison-Wesley, Reading, MA
Kay A (1984) Computer software. Sci Am 251(3):52–59
Minsky M (1985) The society of mind. Simon and Schuster, New York
PINBALL CONSTRUCTION SET. Budge B (1983) Electronic Arts, San Mateo (1820 Gateway Drive, San Mateo, CA 94404 USA)

Kenneth A. Griggs is an assistant professor in the Decision Science Department at the University of Hawaii at Manoa. His interests include all aspects of object-oriented programming as well as the coupling of artificial intelligence techniques with object environments.

Future Perspectives for Global Modeling

Hiroharu Seki[1]

Abstract. This paper explains the origin of simulation studies for global modeling, and looks at three different types of simulation. It also examines the shift towards cultural aspects of simulation studies. Finally, a new proposal is made for a research program to develop an infrastructure of person-computer simulation.

Key words. all-computer simulation; cultural difference; gaming; global modeling; intellectual social developmental genom; person-computer simulation; subjective-objective nexus; world order scenarios

I will begin by summarizing a short history of simulation research related to the recent increased capabilities of personal computers, multidimensional real network formation in the social system even in international fields, and the salient aspects of transnational global problems which are interconnected with each other. These problems are difficult to solve without alternative efforts at transforming the previous antagonistic relations among nations into radically different forms of intimate cooperation between these nations. This step would surely mean developing future research programs for simulation and gaming in the twenty-first century.

Origins of Simulation Studies for Global Modeling

The person-computer simulation study, originally devised by Harold Guetzkow, was the result of efforts to integrate actor orientation to surrogate decision makers of sovereign states with systems orientation to the traditional Westphalian International Relations.

Guetzkow himself tried to transform person-computer simulation to all computer simulations. His efforts were partly influenced by Forrester's System

---

[1] Faculty of International Relations, Ritsumeikan University, 56 Tojiin-Kita, Kita-ku, Kyoto, 601 Japan; phone 075-465-1111; facsimile 075-465-1214

Dynamics which were successfully applied to the Club of Rome type research for the early studies on *The Limits of Growth*.

GLOBUS was developed as an extension of this direction and was a research program for computerizing global modeling through time-consuming data base construction. However, GLOBUS was not able to forecast the recent drastic change in European politics including German unification and the formation of the EC. GLOBUS' failure to forecast international relations has resulted in a decline of global modeling studies as illustrated at the ISA Convention held in Vancouver, March 1991.

The post-cold war period and the post-gulf war period have raised significant new issues for world order scenarios and their theories covered by the ordinary description of international politics. It is urgent for us to develop global modeling simulations and gaming studies for the exploration of world order scenarios.

Three Types of Simulation in Comparison and Perspective

The comparison of all-computer simulation, gaming, and person-computer simulation was attempted by Hayward Alker in the 1960s. However, this was only a kind of pragmatic comparison; it was neither a syntactic nor semantic comparison. Pragmatic comparison is mostly based upon the criterion of computer methodology. Syntactic comparison is based primarily upon the analytical contents of interrelationships among the parts of systems. Semantic comparison should be based on epistemological and ontological issues. Alker tried to compare the accuracy, repeatability, speed of calculation, degree of freedom, and flexibility in three types of simulations. All-computer simulation is extremely strong in the areas of accuracy, repeatability, and speed of calculation. However, it is clearly weak in the sense that it might very frequently fall into the trap of analyzing fictitious systems. In this case, the computer will only play games which have no relationship with reality. On the other hand, gaming has strength in freedom and flexibility. It is weak in accuracy, repeatability, and speed of calculation, but this weakness is an inevitable characteristic of reality, and gaming could simulate this reality of weakness by its fundamental similarity to the real world. Thus, the potentiality of person-computer simulations could be achieved only if the strength of both types of simulations could be successfully integrated.

Present developments in the uses of personal computers and computer networks creates new possibilities. The dramatic development of computer capabilities led to the availability of natural language processing in different cultures of the world. Global modeling research programs using the method of person-computer simulations could be started as a sincere effort of human self-understanding, that is, such simulations could successfully become a mirror of reality. It could also become a kind of intellectual social developmental genom if we could compare it with a biological genom on global modeling exercises.

The conditions for conducting this kind of research program by constructing infrastructures of person-computer simulations based on global modeling are being matured, given the present development of computers with network formation. Within the complex network system of human brains, new types of computer networking are being introduced in the form of transforming the existing global social system. Using this computer network system to mirror the real global system, our understanding of reality can be further developed, as illustrated by the new creative definition of the intellectual social developmental genom. The global modeling research program using person-computer simulations is surely revolutionary at the global level of identity formation and new stages in the history of human intellectual development. The development of the fifth generation computer was tried in order to approach the human brain in a manner different from the ordinal type of high-speed, real-time fourth generation computer. If such computers could be developed, the present incompatibility with the human brain in person-computer interfacing would become easier to resolve than in the previous stage of computer development. Person-computer simulations which try to advance the self-awareness of the global system to the next level of self-understanding should be conducted.

Culture-Based Differences of Simulation Philosophy

GLOBUS was not successful in achieving high validity in comparison with European reality. Identifying the outcome through simple observation is easy. Doing so using behavioral science methods is not. Thus, the reasons for reviving gaming types of simulation have come to the forefront within this context. Cultural comparison of the gaming environment can be considered in light of its comparative relationship to three types of simulation. The gaming type of simulation would provide opportunities for comparison of different cultures via the human brain through learning or multi-cultural gaming runs. Of course, the learning of different cultures by the human brain is generally very difficult and it takes a lot of time even in the case of the human brain to adapt to a new social context. The gaming method could easily simulate cultural differences by the human brain through comparisons or multi-cultural simulation runs even in the absence of learning by each brain. Harold Guetzkow visited five continents after the 1980s. He was surprised by his observation of how different ideas for designing various simulations co-existed with each other in different cultures and thus he wrote a report based on his travel experience. However, he did not explore the fundamental reasons why such differences exist. What sort of influence on science and technology would be created in the different cultural environments? Such questions will become salient in the case of interface between gaming and computers. In the exercise of gaming practice, cultural or even individual differences could be exposed. Such differences are generally not recognized in models or in indicators, both of which are easily computerized. Exchanged documents during gaming practice or autobiographies written by surrogate decision makers in the post-gaming practice could be a source for exploring individual or cultural differences. Person-computer simulations which tried to simulate tension reduction mostly in the Asia Pacific region in the 1990s were conducted in the latter part of 1969. It was more or less successful in forecasting the present situation of the Asia Pacific in the sense of relative drop of American hegemony, turbulence of politics in China and the Soviet Union, improvement of Sino-Soviet relations, and Korean unification. Forecasting the 1990s through simple person-computer simulations using the computer center of The University of Tokyo in 1969 was far more successful than the extremely complicated GLOBUS which was unable to forecast political changes in Europe within even short period. A comparison of GLOBUS with our Asian Pacific Gaming raises the question of issues of cultural differences as well as the three different forms of simulation.

Other types of gaming in the 1970s and 1980s in Japan were also successful in forecasting Korean unification in the 90s or German unification before it really developed. Have cultural differences influenced the outcome? Or has the gaming type of exercise influenced the outcome? These are significant questions that should be raised and answered.

Research Program for Person-Computer Simulation: Three Perspectives

Person-computer simulations could become the most appropriate method of scientific research of the three types of simulation. The background to this situation is the radical transformation of social systems into the network-based computerization of the present global society. Person-computer network is internalized in the social system. How to see the social system as a whole through the window of computer networks is related to the fundamental reconsideration of the methodology of the scientific research. All computer simulations tend to show a strong intellectual bias toward rejecting the human factor even in the highest stage of development of contemporary social systems. The so-called fifth generation computer might revise this strong intellectual bias. But in what degree is such revision going to be realized? Gaming seems to be a primitive type of simulation in the highest stage of development of social systems. It is different from all computer simulations in the sense that it reflects human factors, but the analysis of the gaming process and of the outcomes of gaming is extremely difficult. Gaming is not appropriate for the purpose of scientific inquiry compared with all-computer simulations on the epistemological dimension. Exploring alternatives in the policy sciences demands clear understanding of alternative causal relations. However, in gaming methodology it is nearly impossible, except for the narrow case of mathematical game theory.

In the case of person-computer simulation, the most significant question is how it can revive the strengths of both types of simulations in order to rectify their weaknesses. Once innovative research programs construct the kind of infrastructure through which the above conditions could be satisfied, it will surely become a pioneer type of "human frontier" research project. Here is the first perspective for person-computer simulations. When the models for internal understanding of total social systems are established as prototypes in a multi-dimension human-computer-interface created by the exercise of person-computer simulation, such models could be defined as the set of intellectual social developmental genom analogically compared with the biological genom. Of course, such prototypes have varieties of variations. And the variations would be well illustrated by the combinations of different modules. We have already experienced such possibilities in designing our person-computer simulation for global modeling.

Hiroharv Seki is a professor and Chairman of the Institute of International Relations and Area Studies at Ritsumeikan University. He is also the Executive Chairman of the Japan Council on International Affairs and an Executive Member of the Organizing Committee of IPRA '92.

Behavioral/Social System Simulation

Charles M. Plummer, and David W. Hollar[1]

Abstract. The Behavioral/Social System Simulation (B/SSS) is a computer-based simulation providing capability to simulate relationships between and among individuals and groups. The simulation is programmed in the object-oriented language Smalltalk/V and operates on a Macintosh llcx personal computer. The B/SSS provides the user the opportunity to apportion varied levels of resources among individuals and groups and establish varied probabilities of individuals and group success, with flexibility for the operator to change the rules and objectives governing the simulated behavioral/social system to explore alternative possibilities.

Key words. Behavioral/Social System Simulation (B/SSS); computer simulation; human-computer interaction; object-oriented programming system (OOPS); Smalltalk/V Mac; social psychology; time perspectives

Social system simulations, as surrogate environments, can potentially provide powerful tools. They may model cross-cultural experiences, transcend time by providing future scenarios or recreating historical events, model interactions in an ideal system which would otherwise be impossible to study, and demonstrate interactions in infrequent, distant, or dangerous situations under controlled conditions.

Social systems consist of individuals with unique personalities pursuing multiple individual goals, and are difficult to simulate. An object-oriented computer language has advantages over previous approaches. It is based on classes of objects, each having the potential of interacting with one another, and it can thus be used to form a more natural representation of real social systems.

The Behavioral/Social System Simulation (B/SSS) provides a user-friendly simulation which could be used for research, evaluation, and instruction.

---

[1] Simulation Systems Laboratory, Rochester Institute of Technology, City Center 05588, 50 West Main Street, Rochester, New York 14614-1274, US; phone: 716-475-5292; facsimile 716-475-5595

Social Systems and Simulation

A system could be seen as a group of individual entities or organizations in interaction that serves a common purpose or function. Plummer (1989b) defines a social system as "... an organized and structured individual or social entity, comprised of individuals engaged in specific patterns of interaction, in a structure that has boundaries and ongoing duration, which has developed to achieve one or more specific goals or purposes."

A simulation of a social system would be a controlled, operating model of some real-world social system imitating relationships between individuals and groups. A desirable simulator would have the following characteristics: (1) varied levels of resources and effort could be apportioned among individuals and groups, (2) probabilities could be altered to produce different levels of individual and group success, (3) the user would have flexibility to change both the "rules" governing the system, and (4) the user could also change the duration of the simulation.

Structure of the Social System Model

The underlying model for this simulation is based on a social interaction (non-computer) simulation designed to model a social system (Plummer 1978). In this study, simulation was used as an experimental treatment to study the impact of the stressors "acceleration of change" and "information overload" on the foreshortening of human temporal perspectives ("future shock"). Subjects ($n = 179$) were randomly assigned to high, medium, or low resource conditions in a simulation, and various measures of time perspective were administered following the experiment. Resources were represented by tokens, and bonus points were awarded for acquiring tokens of the same color.

The simulation experiment directed individuals to trade with one another, with a trade consisting of the exchange of tokens of unequal value between two individuals. The prototype computer-based version of the B/SSS described here is based upon these primary underlying structural characteristics.

Model Design Specifications and the User Interface

The simulation provides for modeling the following structural characteristics of the interaction of individuals within a social system:

1. Primary purpose or goals of individuals and groups
2. Rules upon which the system functions
3. Alternative roles available to individuals
4. Alternative physical locations and rules governing movement and interaction

5. Information which pertains to the social system
6. Resources available to individuals and groups to influence desirable outcomes within the social system

An example of the model design specifications for the interface of the B/SSS with the user appears in Fig. 1 (see also Plummer 1989a, 1990). The *Operational Control* component at the top center displays alternatives available to the operator, to execute the functions of communication, command, and control within the simulation. The *Design System* component at the upper left-hand center displays design options to the user. At the lower left of center appears the *Display Status* component, which provides indicators of the status of the simulation system at a selected time or phase. The center stage is analogous to the stage of a theater. It provides the environmental context in which all action occurs. At center right, lower right, and bottom locations appear the *individual*, *groups*, and *organizations* that the user has created.

The simulation user creates individuals (Fig. 2) who have varied levels of characteristics in each of six areas: intellectual, ethical/reflective, social/

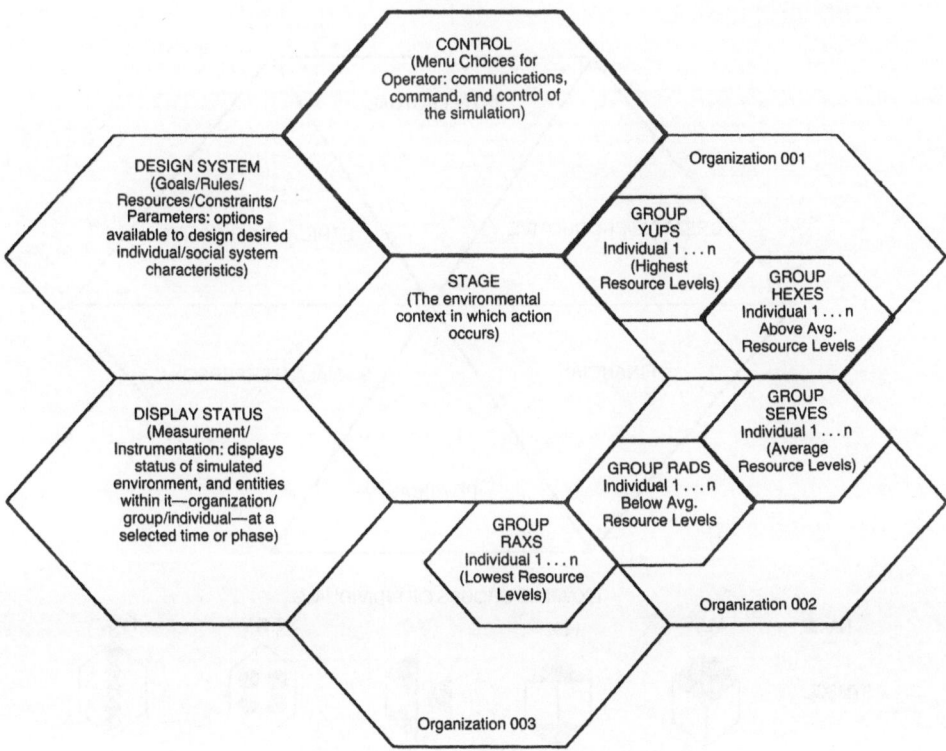

Fig. 1. Behavioral/social system user interface

interpersonal, physical, financial, and creative/productive. These individuals pursue strategies to achieve certain levels of goals in each of these six sectors. Groups are defined as consisting of individuals of shared similarities in one or more sectors, at the discretion of the simulation user. Individuals, groups, and organizations interact in response to tasks specified by the user. Status display can be used to indicate the status of both the task and the individuals, groups, or organizations operating on the task at a given time or phase. The B/SSS has been developed to run on a Macintosh llcx in Smalltalk/V. The original model of the simulation was a non-computer, social-interaction-based version, named "Future Changing Society Simulation" (Plummer 1978).

Behavioral/Social System Simulation Version in Smalltalk/V Mac

The B/SSS contains features that provide the user with more flexibility for establishing alternative social systems. First is the capability to have five groups or social classes with varied resource levels. From highest to lowest resource levels, these groups are named the Yup, Hex, Rad, Serv, and Rax groups.

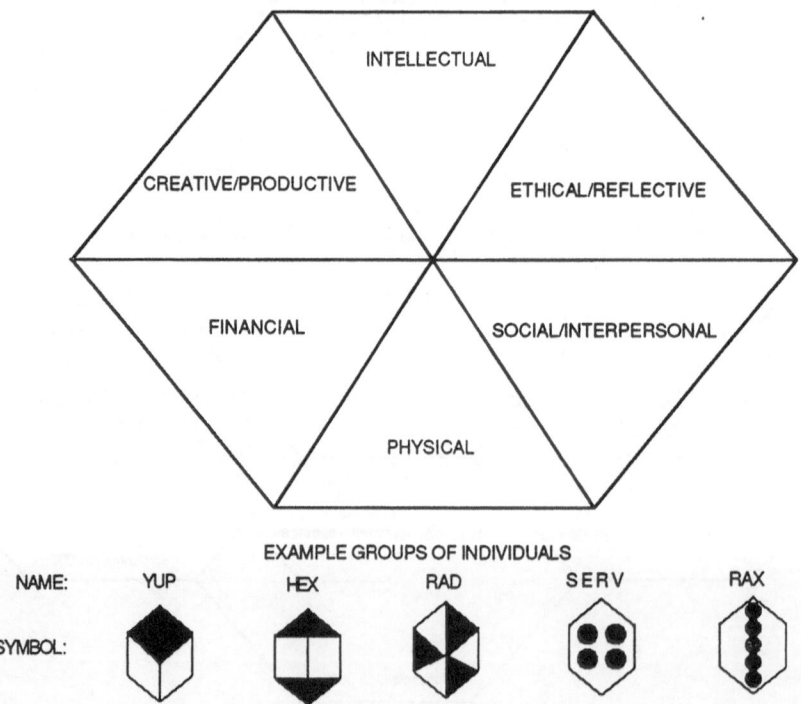

Fig. 2. Operationalization of sectors within one individual in behavioral/social system simulator

Second is the capacity to modify the extent of interaction between groups, to provide for modeling "open" or "closed" social systems, or for studying bias and prejudice. A user can decide whether a group will try to avoid trading with members of other groups and only seek to trade with other members of the same class or whether they will seek to trade with members of any other group. Third is the ability to change the strategy for token trading. Trading strategies available to users include random or artificially intelligent trading decisions. With the intelligent strategy, each entity seeks to make a trade which will yield at least a minimal profit. The random trade strategy is not concerned with the value of the token given or received. Together, these features provide for representing some critical attributes of social systems in the real world.

Object-Oriented Approach to Modeling the Real World

Object-oriented programming offers a powerful, more natural approach to real-world simulation than originated from an earlier simulation language (Simula). In the traditional programming, paradigm procedures function on and manipulate data structures. In contrast, object-oriented programming simulates the way the real world works and therefore represents a better choice for such a simulation. Smalltalk is an object-oriented language that offers an excellent environment for simulation. Figure 3 shows the hierarchy of the classes that were used for the Social System Simulator.

Real and Simulated Experiments

Comparisons were made between the Smalltalk/V B/SSS simulation experiments and the experiments Plummer (1978) performed using the Future Changing Society Simulation. Here there were 7–9 members from each of the

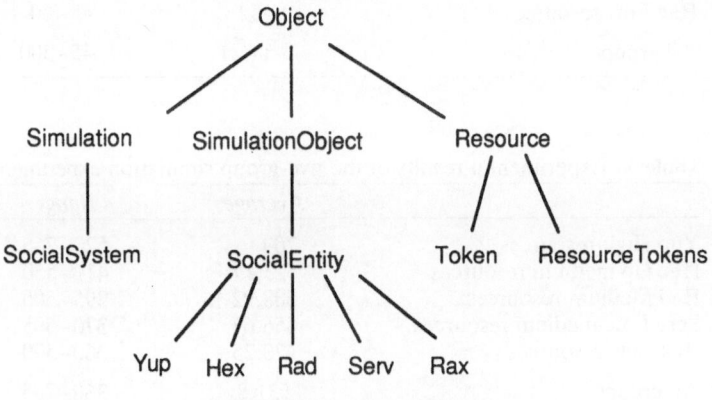

Fig. 3. Hierarchy of classes

three resource levels within four groups under predictable conditions, and four similar groups under an unpredictable condition ($n = 179$). These simulations lasted for 65 min. The Smalltalk/V simulation prototype was operated using three groups with varying parameter settings with seven members in each group for a total of 84 entities during the experiment, with entities moving at a step speed of five and experiments lasting 10 min of simulation time. Both human and computer entities were started with comparable resource levels (high = 150, medium = 100, low = 50). Table 1 presents a comparison of the final resource levels of human subjects with computer-generated results, disclosing very similar results.

A second simulation experiment was also run with differing parameter settings using five resource levels. The groups started with the resource levels illustrated in Table 2. In this experiment there were four entities in each group. The step speed was again at the default of five and each simulation lasted 10 min. The resulting averages are presented in Table 3.

Table 1. Experimental results of three group simulation experiments

	FCSS Human results	B/SSS Computer results
High resources	546.43	543.1
Medium resources	421.84	438
Low resources	393.98	411.4
All groups	421.84	464.17

Table 2. Initial conditions of the five-group simulation experiment

	Average	Range
Yup High resources	240	205–300
Hex Up medium resources	130	115–155
Rad Medium resources	89	80–105
Serv Low-medium resources	69.5	60–75
Rax Low resources	52	45–60
All groups	116.1	45–300

Table 3. Experimental results of the five-group simulation experiment

	Average	Range
Yup High resources	700	520–765
Hex Up medium resources	523.17	410–550
Rad Medium resources	508.92	395–505
Serv Low-medium resources	456.67	370–505
Rax Low resources	470.25	360–500
All groups	531.8	360–765

Desirable Future Enhancements

There are several useful features that could be added to the B/SSS to assist the user and also provide more realism, such as adding more menu choices, adding graphic and statistical reporting options, providing additional alternative intelligent strategies for interaction, and more alternatives for resource allocation. With entities operationalized in all six dimensions, one could simulate varied levels of financial resources and also have varied levels of attributes and goals in intellectual, ethical, physical, and social/interpersonal dimensions. Entities could then trade resources in one area for resources in another area.

Summary

The experimental comparison of simulated and real data from human subjects showed that the B/SSS gave results similar to those produced by real people in these preliminary studies. Future applications of the simulation are planned for studying individual and group behavior, including principles underlying cooperative and competitive behavior, processes of attitude formation and change, human decision-making under uncertainty, and environmental change as a source of social stress.

Acknowledgments

Grateful appreciation is expressed to Robert Gayvert, RIT Research Corporation for his advice and technical assistance to this project.

References

BEHAVIORAL/SOCIAL SYSTEM SIMULATION (B/SSS). Plummer CM, Hollar DW (1991) Simulation Systems Laboratory, Rochester (Rochester Institute of Technology, City Center 05588, 50 West Main Street, Rochester, New York 14614 US)

Knapp V (1986, 1987) The Smalltalk simulation environment, Parts 1 & 2. Proceedings of the Winter Simulation Conference. Society for Computer Simulation, San Diego

Hedry B (1989) Distributed object-oriented discrete event simulation. MS computer science thesis proposal, Rochester Institute of Technology

Plummer CM (1978) Impact on time perspectives of differing levels of environmental change and individual resource control in social systems simulating components of 'future shock.' PhD dissertation, Indiana University

Plummer CM (1989a) Social system simulator design specification 0.0. Paper presented at the North American Simulation and Gaming Association Annual Conference, Indianapolis, Indiana

Plummer CM (1989b) Design and evaluation of computer-based behavioral/social system simulations. In: Klabbers JHG, Scheper WJ, Takkenberg CA Th, Crookall D (eds) Simulation-gaming: On the improvement of competence in dealing with complexity, uncertainty and value conflicts. Oxford, Pergamon Press

Plummer CM (1990) Design and application of a behavioral/social system simulation version 1.0. Simulation Systems Laboratory, Rochester

Smalltalk/V Mac: Object-Oriented Programming System. Digitalk, Inc. (1989) Digitalk, Inc., Los Angeles (9841 Airport Boulevard, Los Angeles, California 90045)

Unger BW (1986) Object-oriented simulation—Ada, C^{2+}, Simula. Proceedings of the Winter Simulation Conference. Society for Computer Simulation, San Diego

Charles M. Plummer is Director of the Simulation Systems Laboratory and a principal scientist at Rochester Institute of Technology. He has designed several simulations, authored publications, and conducted several externally-funded projects involving the use of simulation technology. He chaired the Simulation Systems Special Interest Group of the American Educational Research Association, served as an Associate Editor of *Simulation & Games: An International Journal of Theory, Design, and Research*, and is currently an elected member of the Board of Directors, North American Simulation and Gaming Association.

David W. Hollar, laboratory assistant in the Simulation Systems Laboratory has just completed a Master of Science degree in Computer Science at Rochester Institute of Technology, with emphasis on artificial intelligence, object-oriented programming systems, and data communications. He is now with International Business Machines in Washington, DC.

Section 2
Communication
and Culture

Communication and Understanding Around the World

Linda Costigan Lederman[1]

Abstract. Communication and understanding is a global issue. This presentation takes a systems theoretic perspective to the discussion of communication. It begins with the description of complex information processing systems which are at the heart of communication and their various features. Human beings are identified as these complex information processing systems. The presentation goes on to discuss the importance of recognizing that communication is often taken for granted. The advantages of viewing human communication from a systems perspective, the differences between a linear information transfer model and a transactional model of communication are pointed out. The implications for intercultural communication and understanding, of modeling communication in this way, and more about the applications of these considersations for people concerned with simulations and games are also included.

Key words. communication; communication models; culture; global modeling; information processing; intercultural; language; linear model; meaning

The title of my presentation is global as to befit this conference, which is concerned with global modeling for solving global problems. Certainly effective communication between people and between nations has become a problem of global proportions. The subject, however, is far greater than I could possibly address in the time allotted. I shall limit myself, therefore, to one critical aspect of communication: information processing. My purpose today will be to discuss with you how a systems theoretic perspective on communication and understanding can provide insight into the most complex, challenging, and formidable of information systems to be found anywhere on the globe. Let me begin my remarks by describing to you briefly these information processing systems. I'll take these first few paragraphs to describe them, and then tell you what they are called.

First, I can tell you that everyone knows *something* about these information systems, although some know more than others. Those of you who know more

---

[1] Department of Communication, Rutgers University, New Brunswick, NJ, US

about these information processing systems, know how little you know about them. You know that the more you know about them, the more you know how little you know because these information processing systems are far more complex than they appear to be on the surface. I hope I have not lost you in the paradox I am trying to paint with words. What I am getting at is that the information processing systems I have in mind are complex enough to be worthy of a lifetime of study. Thus even though these systems are familiar to us, they need to be described in detail to remind us of the complexity which we may take for granted due to our familiarity with them.

The information processing systems about which I am concerned today are multidimensional and multifaceted. They can be and in fact usually are programmed to handle a variety of languages—English, Japanese, Russian, French, German, Spanish, to name but a few. Some of them are capable of handling as many as several of these languages. Along with the languages I just mentioned, all of them are programmed to handle other complex symbol coding systems, such as mathematics, logic, decision making, analytic processes, or logistics. These systems, once created and programmed, participate interactively in vast networks of other similarly programmed information processing systems, all of which have the capacity to generate information as well as to classify it, sort it, store it for future use, modify it, update it, transmit, and retrieve it, usually upon demand. These systems are designed to be user friendly, in the sense that they take into account the needs of the users who interface with them. They are capable of communicating with other systems even with seemingly incompatible features.

The systems to which I am calling your attention have a life span of decades and can usually, although not always, can be counted upon to improve with age, at least before they begin to become out of date and less efficient. But even when they can no longer function in some ways as they did when they were newer, they have capacities in their later years that go beyond that for which they were originally programmed. Thus they have the capacity to do more than operate in the transfer of information or data from one place to another. They can assimilate information from various sources and systems within the environment in which they are located. These sources of information can vary from the most complex to the most elemental and still be manageable by the information processing systems I am describing. But they can, too, take raw data from the environment, gather it up, make sense of it, interpret and decide what to keep and what to discard, what to store for immediate access, and what to keep for future retrieval. These systems have the capacity to process information, and are endowed with rich and continuously upgrading processes for interpreting data.

In a word, these are remarkable systems—systems which in one way go beyond anything any of us here might design, create, or invent—yet in another sense, these information processing systems *are* the products of you and me, or others like us. They are dynamic, interactive, mutually affected by, and affecting other systems with which they interface and interact. Let me say one more thing about these systems before I label them. They are able to react to other systems. To respond and create new and different information based on the information provided by other systems and their processing and interpreting of that information. Beyond processing of information, therefore, these systems themselves create information and information about information, and react to this information based on the complex, reiterative, multiphased processes of information input-output.

You may have begun to surmise that the information processing systems to which I am referring are human beings. People can be viewed as information-processing systems. Viewed as such, people appear to have all of the richness of systems such as those engaging in communication I have been describing and more. In sum, people as information-processing systems are complex, multifaceted, and multidimensional. They are programmed to handle a variety of languages, as well as to handle other complex, symbol-coding systems. They participate interactively in vast networks—other people all of whom, too, have the capacity to generate information as well as to classify it, sort it, store it for future use, modify it, and update, transmit and retrieve it, usually upon demand. As systems, people do more than operate in the transfer of information or data from one place to another. They assimilate information, and/or take raw data from the environment, gather it up, make sense of it, interpret it, and decide what to keep and what to discard, and so on. Beyond the processing of information therefore, people, as systems, create information and information about information, and react to this information and to others based upon the complex, reiterative, multiphased processes of information input-output.

Why, you may wonder, especially when the title of my presentation is "Communication and Understanding," have I chosen to talk about people as information processing systems? I have not made this choice because I wish to present people as machines. People are not machines. Quite the contrary. Because people are not machines, it is often possible to take for granted the complexity of design and maturation that goes into the evolution of each human being's capability to participate in communication, to be, as I have been describing it, a system capable of sensing, sorting, interpreting, and responding to information.

I have made this choice to describe people as information-processing systems for three important reasons, each of which I want to discuss. The first reason for thinking of human beings as information-processing systems, and viewing the process in which they engage from a systems perspective, is that communication is such an on-going and ubiquitous experience that it is easy to take it for granted. While most of us may remember when we first learned to read or write, few, if any, among us can remember our first words or sounds. Few can remember beginning to engage in the process of communication. Engagement in communication begins at least at the moment of birth. As Vickers (1967) put it "Insofar as I am human, it is at birth I was claimed by a system which systematically programmed me for operation in itself." Communication as a process is so inherently part of the human experience that it may appear as if it were "natural"and therefore be taken for granted. It is not natural. Our instincts are natural, and communication certainly relies on the use of our instincts. But communication relies upon the process of socialization in which we learn about those instincts and about ways that are socially acceptable expressions of them. So the first reason for viewing communication from a systems perspective is to provide a reminder of the complexity of the process and all that has been and is continuously programmed into us to allow us to participate in it.

There is a second reason for thinking about communication from a systems perspective. The second reason is to emphasize communication as a vital life process, what Thayer (1968), among others, has called the metabolism of information. Viewed this way, there are two vital life processes. The first is the metabolism of matter or energy. The second, is the metabolism of information. As humans, we can no more function in the society if we fail to metabolize information than we can function in the physical world if we cannot metabolize matter. If we cannot breathe, we cannot live. If we cannot ingest, digest, and egest matter, we cannot support life physically. So, too, if we cannot ingest information, process it and eliminate that which we do not need, we cannot support life socially. If we cannot process the information we see here in Kyoto and tell the difference in directionality between the way cars move on the roads of Kyoto from the roads with which many of us are familiar, we cannot even cross a street here.

By thinking of humans as information processors, I am shifting levels of analysis. Rather than viewing people as part of vast social systems, which we are, I am reminding us that as people we are also in and of ourselves information-processing systems. This individual level of analysis is required if one is really to discuss communication and understanding. For understanding does not take place at the level of social systems. It takes place, if at all, at the level of individuals. Groups do not understand one another. Individuals understand one another. If understanding between groups, nations, and cultures is to exist at all, then understanding takes place on the individual, one-to-one basis. Thus a second reason for discussing communication from a systems perspective is to reinforce the complex nature of individual human beings, let alone the social systems which individuals form, and therefore simple answers will not be found in human global issues.

Finally, I have selected to approach the topic of communication from a systems perspective because there are a number of faulty assumptions often made about human communication that become obvious when we think of human beings as systems—systems which take in information and process it. The three most problematic and faulty assumptions are that (1) meanings are in words, (2) the messages sent equal the messages received, and (3) communication equals the transfer of information. Let us examine each of these faulty assumptions and see wherein each is erroneous and is avoided by a systems perspective.

The first faulty assumption is that meanings are in words. It is easy to believe this. Today as you listen to my words—or to the simultaneous translation of them—it is easy to think that words carry meanings and that therefore we need only learn words to communicate. But only the literal meaning of a word exists in the symbol itself. The connotations and the various associations of language in use are derived in people not in words. Words are symbols; it is people who bring the meanings to those symbols. If we view people as information processors, one way of taking into account that part of the human information-processing activity is the interpretative function. That is to say, that part of processing information is bringing meaning to it. Meanings exist more in people than in the words they select to convey their meanings.

The second faulty assumption about the way communication works is that the message sent equals the message received. When we view communication as an event in which one person sends a message to another, we think that good communication occurs when the message sent equals the message received. Viewing communication from a systems perspective reminds us that the messages are input and output of systems and that these are interpretative systems. Therefore, what something means to one human being is not inherently and necessarily what it will mean to the next.

The third faulty assumption is that communication equals the transfer of information. If we assume that meanings are in messages, that messages sent are equal to those received, then we see communication as a linear, information-transfer process. But while getting the information from one place or person to another is a necessary condition for communication, it is not a sufficient condition. Along with the transfer of the information are the interpretative processes which each human being brings to the communication transaction.

A systems theoretic perspective allows us to conceptualize and model communication in such a way as to avoid the misconceptions I have been discussing. Most of these misconceptions grow out of what is called in the communication discipline the linear model of communication. This model, originally put forth in the work of Shannon and Weaver (1949), works well when the object of concern is information and the transfer of information from one place to another. Those familiar with the work of Shannon and Weaver know that they were electrical engineers and that their concern was the fidelity of phone lines for transporting electric impulses with accuracy and without distortion. To this day, this is a concern with technologists and others concerned with telecommunications. The model is a good one for machines. It is an important one when we want to assure that the information generated at point one is the same as that transmitted to point two. Unfortunately, this kind of linearity was applied to the message exchange between people. It is represented in a linear fashion. Figure 1 represents this model.

In the model, S stands for the sender of the message, M stands for the message, C for the channel or pathway for that message, and R stands for the receiver of the message. The model was put forward by Berlo (1960). It is based on the linear notion of the work of Shannon and Weaver and assumes that the sender and the receiver of the messages are people and that people can function like transmitters and receivers.

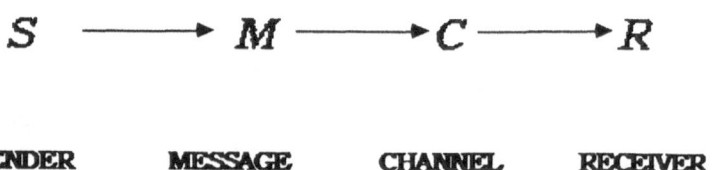

Fig. 1. Linear model of communication

In the information-transfer paradigm, information and its transfer equal communication. Viewed this way, communication is simply the amassment of information. But beyond information, communication exists as a complex process involving interpretations of information. It is this interpretative aspect of communication which accounts for the creation of meaning. Communication is not simply a process in which a message is sent and received. It entails the interpretation of that message and the attribution of meaning to the message by its creator (the encoder of the message) and its recipient (the decoder). Philosophically, it is possible to mount the same arguments about messages to which no meanings are attached as to trees falling in forests where no people are present.

But an apparent anomaly presented itself often enough as to discourage communication scholars in the application of this notion. What became evident and impossible to explain with this linear model was that the message sent did not always equal the message received. Speakers did not always convey to listeners what they wanted to, even when the speakers were good. This led to a variety of changes in the conceptualization of communication and how it works when it involves people and human factors. The most compelling of these is the systems approach, in which people are viewed as systems and messages as the input and output of each of these systems; messages which are exchanged between them. Figure 2 presents a model of communication based on this systems approach. It is referred to as the transactional model of communication, and is a synthesis of the work of many communication scholars, beginning with Newcomb (1961).

In this model the sender and receiver of the message are represented by A and B. The change to the symbols A and B is to indicate that all people in the process of communication are both senders and receivers of messages, and that it is artificial to the point of misrepresentation to divide the functions as if some people only send and some only receive messages. In the model, X represents the message. You can see that the model is not linear. The arrows in the model are two way, to represent the exchange between participants and the impact of people on messages and messages on people.

What are the advantages of seeing communication from this perspective? In the contribution by Dennis Meadows New Earth 21 and education is discussed. Meadows discusses the six elements in that project and the need to educate the public. It is that need that highlights this way of thinking about communication. If communication were merely the transfer of information, then the

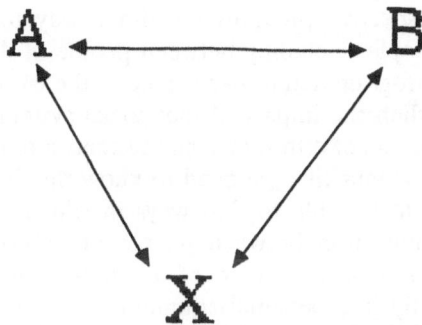

Fig. 2. Transactional model of communication

project would not require reeducation. It is that people are not merely the passive recipients of information that in this case presents the challenge. It presents the challenge of attempting to get those messages through. Of trying to understand sufficiently the people to whom the messages are intended in order to know how to send the messages in ways that are likely to be meaningful to them.

This is the challenge to those of us using simulations and games. It is our understanding of the experiential that allows us to find avenues to create transactions between people that are rich in their potential for educating and for the transfer of paradigms of how to get information to people. We know from our own experiences, especially here at ISAGA, that communication is complex and that it is intercultural. That communication is a product of culture and that it is therefore culture-bound. As early as the 1960s, Marshall McLuhan, the mass media theorist, wrote about the global village. As early as that time, he envisioned the change in our world in which we could know one another like the residents of one village because of the capacity of electronic media, especially television, to let us "travel" to places from our armchairs. Thirty-odd years later, the technological societies in which many of us live and the potential for immediate and further technological change abound more than even the mind of a brilliant thinker like McLuhan could anticipate. We see this in our everyday lives when we can communicate with each other before, during, and after ISAGA through electronic, technological means. We have telecommunications, computers, televisions, fax machines to name but the most obvious used today.

But at the same time, people have not changed. Each of us is born and raised in a particular culture. And that culture teaches us many things, including the socially-defined right and wrong ways of doing most things, including communicating. From a systems perspective, we can think of communication and culture as a dialectic relationship. That is, we can look at the ways in which cultures create communication and communication creates culture. We can look at ourselves here today and see that we are at once the products of our culture and the producers of that culture.

Thus, a systems perspective provides us with a way to think about communication between people, including between people of different cultures. If each of us is a system programmed to participate in the culture in which we are raised and socialized, then the impact of that social system needs to be taken into account when members of different social systems attempt to communicate with one another. This means that we need to know the differences that make a difference; we need to be able to find ways in which to reduce the stress associated with communication between people of different cultures. Intercultural communication requires a kind of adaptation that goes beyond the accommodations of daily interpersonal communication. It requires a kind of self-reflexivity and self-altering, an ability to be creative and find new and meaningful ways of sharing meanings with others. Those of us who work with simulations and games are familiar with these kinds of needs. We design and use simulations and games for many of these same purposes. If we are to find a way to create understanding between people that is as advanced as the technologies we are creating to carry messages between people, we need to find ways in which to educate people about themselves and others, as well as about the differences between them that make a difference.

The challenge, then, in finding applications for the solutions to global problems is to find ways to get human beings to want to make the changes necessary to deal with global problems. Meadows addressed us about the changes necessary in global systems. I am referring here to the changes necessary in human information-processing systems. I am referring here to the need to get people to process information differently and thereby to be willing as well as able to make the changes they need to ensure that the earth and its inhabitants survive.

Thinking about communication from a systems perspective is a major step in the direction towards change in people. It has advantages over the more linear notions about human communication. First, this way of thinking allows us to identify the elements of communication, to describe and explain them in relation to one another, and to create a model and work it to see what it will take to create change for humans. Second, thinking about communication from a systems perspective allows us to identify and list the assumptions underlying that model, to examine them, and to be prepared to challenge them. Finally, this approach allows us to examine the ways in which humans need to be approached and worked with if we are to maximize their ability to communicate well with one another—to find the basis for understanding between them.

One of the most obvious means through which this understanding can take place is language. Language and culture have an inseparable relationship. Both language and culture are dynamic and complex, but patterned. Spoken communication and its cues, are constrained by culture, as is the interpretation of the linguistic cues. Misunderstanding related to intercultural differences in language use can only be prevented by an awareness of how a culture is operating on the production and interpretation of communication. There are a myriad of differences in intercultural communication which can influence how meaning is interpreted. These differences can be explored systematically by examining the dynamics associated with the elements of the communication process: people (acting as encoders and decoders), messages (words and actions), media (pathways), and context (framework). Let's look amongst ourselves gathered here today, for example. The aggressive Westerner and the self-effacing Easterner have learned different styles of self-presentation. These differences are open to misinterpretation since they are differentially valued in the cultures in which they have been learned.

Related to presentation of self is rapport/style or the degree of friendliness or politeness which is deemed desirable. These, too, are culturally influenced. Most speech activities—discussing a topic (collaborative or combative); negotiating (how people agree or disagree); conversation management (turn taking, interruptions, length of pause, use of silence, feedback); ways of structuring information (getting to the point, digression, sender- or listener-centered); ways of speaking (voice tone, intensity, volume, pitch); definitions and interpretations of words and their use, and so on—are culturally derived.

The same holds true for other fundamental elements of communication: media and context. The selection of the appropriate medium for a message—is it a handshake or a wave, a kiss or a smile, words or non-words, written or spoken?—is culturally derived. So, too, is the understanding of the context. The definition of the situation and what is expected and appropriate to it are part of the element of the context of communication. These are culturally derived and therefore likely sources of differences amongst people from various cultural backgrounds.

You and I here today come from different cultures and different disciplines; we have different goals and different skills. What we have in common is our interest in simulations and games and our concern with their application towards making the world a more liveable place. As we heard in the presentations this morning, the solutions cannot be merely technical. The solutions need to take humans into account. And this is what I have been talking about this afternoon. Communication and understanding *between* people requires understanding *of* people. It requires understanding of what people are and how they function in communication. It requires facing our assumptions and not falling victim to them—not falling prey to the assumption that because we have created new technologies for the transfer of information that we have automatically created new capacities in human beings for the processing of that information or for more effective communication. We as simulation and game people are concerned with the ways of knowing. We are skilled at designing and using systems which represent those issues, concerns, and ideas we want to model. We can model global issues. It is human response to those issues to which we also need to attend. There is a story of an elephant at the circus which would help me make my point of the pitfalls to avoid. It was chained to a small wire from the time it was an infant elephant. When it was huge and full grown it remained tied to that tiny chain. A visitor to the circus remarked to the elephant trainer that he was afraid the elephant might break loose and hurt people. The trainer assured him that it was impossible. "That elephant learned in infancy that he couldn't escape from that chain no matter how hard he tried.

He doesn't know he's changed." said the trainer, "He doesn't know how much he has grown." Let us not in our design of new and complex systems forget about people. Let us not assume that they have changed, for even if they have and they know it not, they, like the elephant at the circus may not be able to take the new steps of which they are capable.

In our work with simulations and games, it is our job to make sure that we help the humans for whom they are designed. We have to encourage them to realize the capacity to change and to do what they perhaps could not once do. That is the challenge that faces us.

References

Berlo D (1960) The process of communication. Holt, New York

Newcomb T (1961) The acquaintance process. Holt, New York

Shannon C, Weaver W (1949) The mathematical theory of communication. University of Illinois Press, Urbana, IL

Thayer L (1968) Communication and communication systems. Irwin, Homewood, IL

Vickers G (1967) The multivalued choice In: Thayer L (ed) Concepts and directives. Spartan, Washington, DC

Linda Costigan Lederman, Chair, Department of Communication, Rutgers University, US, specializes in instructional and interpersonal communication with an emphasis on experiential learning. Dr. Lederman received her education at Brown University (BA), Columbia University (MA), and Rutgers University (PhD). She has written 4 books, 20 book chapters and/or journal articles; presented more than 100 conference papers and professional seminars; and designed/published a variety of communication simulations and games including L & L ASSOCIATES, and with Lea P. Stewart, SIMCORP, LINDLEE ENTERPRISES, THE MARBLE COMPANY GAME, and PASS IT ON. Dr. Lederman is immediate past Editor of *Communication Quarterly* and Associate Editor of *Simulation & Gaming: An International Journal of Theory, Design, and Research*. Her research focuses on the use of qualitative methods, most especially, simulation/games and focus group interviews, to generate qualitative data about affective and behavioral dynamics associated with verbal communication.

Synchrony and Intercultural Communication

Young Yun Kim[1]

Abstract. The concept of synchrony presented at the conference is elaborated here as a foundation of effective intercultural communication. Synchrony refers to a state of congruence and harmony in verbal and nonverbal communication patterns of two or more interactants. Synchronic communication naturally occurs when the interactants share common cultural norms and whose psychological orientation toward each other is one of harmony and cooperation. Because synchronic communication, by and large, follows cultural scripts shared by the interactants, creating synchrony across cultures requires a special awareness and effort by at least one party. A three-person role-play that is designed to enhance an awareness of synchronic communication is presented at the end of this paper.

Key words. adaptation; accommodation; congruence; harmony; intercultural communication; kinesics; paralinguistics; role-play; synchrony

Culture has been compared to computer software because it provides a "program" for our perceptions, attitudes, and behaviors (Hall 1976). Cultural programming for communication goes far beyond words. It includes many nonverbal acts, upon which judgments are based of what is transpiring and from which conclusions are drawn as to what has occurred and how to react to it. What words we say and how we say them are largely within the perimeters of our cultural program, including the time and timing of speech acts, the tone of voice appropriate to the subject matter, the facial and bodily expressions accompanying speech, and the physical distance separating one speaker from another (Gudykunst and Kim 1992).

Because members of a given culture share a common set of communication norms, they are able to understand, respond to, and coordinate each other's verbal and nonverbal acts and messages with fidelity. When this interpersonal "coordination" (Pearce and Stamm 1973) occurs in a high-level fidelity, a state of synchrony or harmonic convergence is achieved between the participants'

---

[1] Department of Communication, University of Oklahoma, 610 Elm Avenue, Norman, Oklahoma 73019, US; phone 405-325-3111; facsimile 405-325-7625; e-mail aa0101 @ uokmvsa.bitnet

communication styles—just as the sound from a radio station becomes crystal clear as we tune in at the correct frequency level. Although synchronic communication generally occurs without the conscious awareness of the participants, the participants themselves do sense the "togetherness" that results from the smooth and cohesive flow of their interaction.

This paper advances a view that the concept of synchrony is equally relevant to situations of *inter*cultural communication in which interactants do not share common communicative norms and styles. Even casual observers recognize that Americans, for instance, do not move or vocalize the way Japanese do, and that each group's common behavioral patterns help define its cultural distinctiveness in communication. The present discussion of intercultural synchrony emphasizes, however, that, unlike *intra*cultural communication situations, participants in intercultural encounters must make conscious efforts to create synchrony, and that a synchronic state, when created, is likely to serve as a crucial enhancer of the quality of communication between cultural strangers. Learning a language may be necessary but not sufficient to communicate with those in other cultures.

Synchrony

Synchrony takes various forms, both in spoken language and nonverbal activities. Hall (1976), in studying synchronic patterns in various cultures, focused his analysis primarily on kinesic and proxemic behavior patterns such as bodily, posture, and facial expressions. He used the term, "syncing," or "being in sync," to describe the phenomenon in which people when communicating move together, in part or in whole. In doing so, each party *mirrors* or *complements* the other, in whole or in part, in a kind of dance. Such is the case of the Japanese bowing, in which two people's lowering of the head and the upper body follows a symmetric or complementary form and rhythm of movement.

Along with bodily activities, synchrony occurs in the paralinguistic aspects of speech. When two friends confide in each other, for example, they often speak "in one voice" each converging to, and echoing, the other's tempo, loudness, pitch, tone, and pauses—just as singers in a chorus sing the same melody together or sing two different melodies at once in harmony. Research in Speech Accommodation Theory (Giles 1977) and Communication Accommodation Theory (Gallois et al. 1988) provides an explanation and empirical evidence for such paralinguistic convergence. Studies have shown that, in cooperative and trusting social contexts, speakers tend to shift their speech styles to increase perceived similarity between them. It has also been demonstrated that speakers diverge from an interaction partner's speech style when they find their identity threatened by the encounter or the other party.

Synchronic communication can occur in other aspects of verbal messages as well. It has been observed that people adjust their verbal message strategies and contents in ways that are more appropriate to particular encounters. Bernstein (1966, 1981) has suggested that people tend to focus on the lexical and syntactic levels of others in terms of elaborated and restricted codes, and then adapt their linguistic choices accordingly. Higgins et al. (1982) found that speakers were more likely to "stick to facts" relating something about a target person, when they believed the listener did not have the same information about this person. Speakers have also been shown to assess listeners' common sense for additional information inferences, and do not feel the need to verbalize what listeners can figure out from something they hear. (See Cheng 1991 for a review of related studies.)

The various forms of synchrony—kinesic, paralinguistic, and linguistic—are most evident in communication encounters in which the interactants are conjoined with mutual interest, concern, and cooperation, as explained by the Principles of Cooperation for conversation (Grice 1975, Levinson 1983). In this state of psychological convergence, as in the case of two good friends sharing an intimate moment, synchronic communication naturally unfolds and manifests itself in harmoniously-orchestrated bodily rhythms and movements and cohesive verbal exchanges. As such, synchrony provides an *optimal context* in which interactants experience minimum interpersonal strain and maximum communicative fidelity.

Creating Intercultural Synchrony

Given that synchronic communication is based on a common, culturally-programmed, behavioral repertoire and that this phenomenon is most salient when the participants are in a cooperative mode, difficulties of creating synchrony across cultures are unavoidable. Intercultural communication typically begins in an *asynchronic* state with limited psychological and communication coorientation between participants. Much of one's taken-for-granted cultural assumptions can be a hindrance to this situation, and kinesic, paralinguistic, and linguistic practices are only partially readable to each other. More than likely, the interactants themselves recognize this "out-of-sync" situation and feel the confusion and discomfort it brings. Even an American and an English person may experience at least some level of difficulty in this regard, and the chances of one's being unable to synchronize with the other increases as cultural distances increase.

To the extent that intercultural synchrony does not occur naturally, the communicators have to "work" to *create* it. They need to make a conscious effort to establish a congruent and harmonious interaction rhythm if they are to enhance their chances of succeeding in their communication activities. At least one party has to make adjustments in his/her own "normal" communication style to accommodate and adapt to those of the other person. Without such deliberate attempts, the participants' communication experiences are likely to remain awkward, uncomfortable, and unsuccessful.

Preparing for Intercultural Synchrony

Achieving intercultural synchrony is a process that requires experience. To facilitate this learning process, a newcomer to intercultural communication activities should work to (1) to recognize that each interactant is an individual and (2) to remain calm. First, synchrony is facilitated when the communicator approaches an intercultural encounter with a *personalized orientation*, or a view that takes into account the fact that the interaction partner is as much a unique person as s/he is a member of a cultural group. No two persons in any group are alike; yet we often let our stereotypes dictate the way we deal with a specific person with whom we are communicating. Social identity theory (Tajfel 1974, Turner and Giles 1981) articulates this general tendency that people generally have of seeing outgroup members in light of their group membership, and not by their individual identity. This stereotype-based group orientation contributes to the phenomenon of "intergroup posturing" (Kim 1991), which is manifested in exaggerated "we-they" distinctions, in hasty and premature judgments of outgroups, and in impersonal treatments of the individual outgroup members.

Second, when dealing with strangers from a different and unfamiliar culture, our communication experiences are characteristically stressful. Stress is a natural human reaction whenever our internal capabilities are not adequately prepared for the demands of a situation. Much research attention has been given to the stress-related phenomena of intercultural communication such as "intergroup anxiety" (Gudykunst 1988, Stephan and Stephan 1985), "culture shock" (Furnham and Bochner 1986), and "adaptive stress" (Kim 1988, 1991).

Yet, to create synchrony interculturally necessitates that we are able to *relax* and manage stress. It is this ability to relax in the potentially stressful situations of communicating with cultural strangers that helps us remain flexible and not rigid or closed-minded. As we relax, we are better able to focus our attention fully on the stranger without being distracted by our internal anxiety and without making premature judgments about him/her. In doing so, we can more accurately recognize the stranger's new and unfamiliar communication style, which then "frees" us from the control of our cultural norms of communication and allows us to make appropriate adjustments in our behaviors.

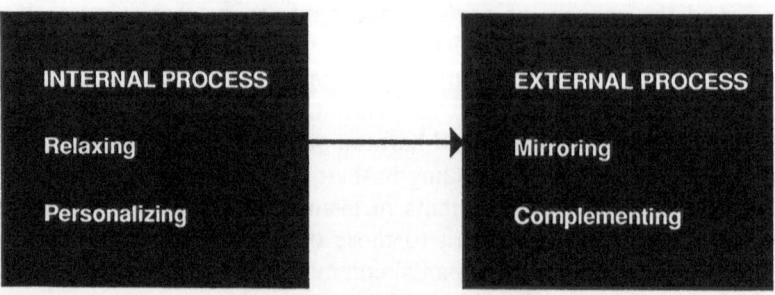

Fig. 1. The synchronic process

It is clear, then, that we must overcome the tendency of intergroup posturing and learn to give each cultural stranger a chance to fully reveal himself/herself by paying special attention to his/her uniquenesses. The personalized orientation, along with our ability to relax in the face of cultural unfamiliarity and differences, should facilitate our behavioral flexibility to mirror and complement the cultural stranger's verbal and nonverbal communication patterns (Fig. 1.) In light of the significance and utility of synchrony in intercultural communication, a three-person role-play is presented in the Appendix as an initial attempt to develop useful learning activities.

Conclusion

In sum, this essay has forwarded the concept of synchrony as vital to enhancing the quality of intercultural communication in subtle but profound ways. Synchrony, when achieved, should provide a harmonious interpersonal milieu, in which psychological and behavioral differences can be bridged and cooperative interactions can be facilitated. Our ability to create synchrony, in turn, serves as a basic communication competence crucial to developing mutually satisfying relationships with cultural strangers—one at a time. To this end, we need to improve our ability to stay relaxed and flexible and to attend to the other person's uniquenesses while being mindful of his/her cultural characteristics. The question, then, remains: How *willing* are we to give each intercultural encounter a full chance it deserves and the best effort it requires?

References

Bernstein B (1966) Elaborated and restricted codes: Their social origins and some consequences. In: Smith AG (ed) Communication and culture. Holt, Rinehart and Winston, New York, pp 427–441

Bernstein B (1981) Codes, modalities and the process of cultural reproduction: A model. Language in Society 10:327–363

Cheng L (1991) Topic management as communication accommodation strategies in intra- vs intercultural interactions. Doctoral thesis, Ohio State University

Furnham A, Bochner S (1986) Culture shock: Psychological reactions to unfamiliar environments. Methuen, London

Gallois C et al. (1988) Communication accommodation in intercultural communication. In: Kim YY, Gudykunst WB (eds) Theories in intercultural communication. Sage, Newbury Park, pp 157–185

Giles H (1977) Language, ethnicity and intergroup relations. Academic Press, London

Grice H (1975) Logic and conversation. In: Cole P, Morgan J (eds) Syntax and semantics 3: Speech acts. Academic Press, New York, pp 41–58

Gudykunst WB (1988) Uncertainty and anxiety. In: Kim YY, Gudykunst WB (eds) Theories in intercultural communication. Sage, Newbury Park, pp 123–156

Gudykunst WB, Kim YY (1992) Communicating with strangers: An approach to intercultural communication, 2nd edn. McGraw-Hill, New York

Hall ET (1976) Beyond culture. Anchor Books, Garden City, NY

Higgins et al. (1982) The "communication game": Goal-directed encoding and cognitive consequences. Social Psychology 1:21–37

Kim YY (1988) Communication and cross-cultural adaptation: An integrative theory. Multilingual Matters, Clevedon, UK

Kim YY (1991) Intercultural communication competence: A systems-theoretic view. In: Ting-Toomey S, Korzenny F (eds) Cross-cultural interpersonal communication. Sage, Newbury Park, pp 259–275

Levinson SC (1983) Pragmatics. Cambridge University Press, Cambridge

Pearce B, Stamm K (1973) Communication behavior and coorientation relation. In: Clarke P (ed) New models for mass communication research. Sage, Beverly Hills, pp 177–203

Stephan W, Stephan C (1985) Intergroup anxiety. Journal of Social Issues 41(3): 157–175

Tajfel H (1974) Social identity and intergroup behavior. Social Science Information 13:65–93

Turner R, Giles H (1981) Intergroup behavior. University of Chicago Press, Chicago

Young Yun Kim is a professor of Communication at the University of Oklahoma. She has written *Communicating with strangers* (with W. Gudykunst) and *Communication and cross-cultural adaptation: An integrative theory*. Among her edited books are: *Interethnic communication* and *Theories of intercultural communication* (with W. Gudykunst). Her current research foci include the role of communication in interethnic and interracial relations and the dimensions of communication competence that facilitate intercultural adaptation and relationship building. She serves on several journal editorial boards including *Communication Theory, International Journal of Intercultural Relations* and *Simulation & Gaming: An International Journal of Theory, Design, and Research.*

Appendix: A Role-Play on Synchrony

The specific synchronic communication activities involved in this role-play is limited to kinesic behaviors. This role-play serves, however, as a protocol, based on which similar versions can be developed for paralinguistic, linguistic, and other aspects of communication. Debriefing activities are incorporated into the instruction for each player. Once the role-play is completed, members of each three-person team share individual experiences, followed by a discussion among the entire group.

Instructions for Person A

Standing up, B will be initiating a 10-min conversation with you introducing himself/ herself. Engage in this conversation by asking appropriate questions to learn about B, and observe and identify B's facial expressions, hand gestures, body postures, and body movements. As you recognize B's unique kinesic style, make necessary adjustments in your own style in such a way that is congruent to B's. Throughout the conversation, try to be relaxed. Be prepared to discuss the following after the role-play.

1. What specific patterns of B's facial, hand, and body movement were you able to identify?

2. What adjustments did you try to make in your own facial, hand, and body movements to make them more congruent with B's?
3. Did your efforts to synchronize with B's movements make any difference in the conversation process? If yes, how? If no, why not?

Instructions for Person B

Standing up, initiate a 10-min conversation with A introducing yourself (your name, background, current work, etc.). Speak and act naturally. After the conversation, be prepared to discuss the following.

1. Do you feel good about the conversation with A? Why? Why not?
2. How do you feel about your partner A? Did you notice any specific things A did that helped you to carry on the conversation?

Instructions for Person C

Standing up, observe a 10-min conversation between A and B. Position yourself on the sideline where you cannot interfere with the conversation. B will initiate a conversation introducing himself/herself to A. A will attempt to synchronize with B by adjusting his/her kinesic styles to that of A. Closely observe how A synchronizes his/her hand gestures, body postures, facial expressions, and body movements with those of B. Also observe how A's synchronizing helps B's behavior and the flow of the conversation as a whole. Be prepared to discuss the following after the conversation.

1. What specific styles in B's body language were you able to identity?
2. How successful was A in synchronizing with B's body language?
3. How did A's synchronic effort help the flow of conversation?

Global Network Simulation:
An Environment for Global Awareness

David Crookall[1] *and Patty Landis*[2]

Abstract. Two multi-site, network simulations are described—ICONS and IDEALS. Many overlaps exist between the two projects, both in form and substance, but the differences are important too. Project ICONS focuses on foreign policy and foreign language translation; Project IDEALS concentrates on international understanding and cross-cultural communication. Both projects use specialized simulation management software to allow teams, situated in different parts of the world, to send messages to each other and to hold real-time teleconferences.

Key words. computerized simulation; cross-cultural communication; environmental concerns/issues; foreign language; foreign policy; global concerns/issues; international understanding; network gaming; network simulation; software; teleconferencing; translation

Computer-assisted simulations in international education have a variety of applications which span academic disciplines. The earliest simulations to take advantage of computer networks were pioneered by Professor Robert C. Noel of the University of California at Santa Barbara (Noel 1969, 1979; Noel et al. 1987). The POLIS simulations utilized ARPANET and later Telenet to link several universities in the United States and abroad, via connections through US military bases. These simulations use the computer as a tool to facilitate and organize communications among teams situated around the globe. Students relate to each other and respond to human situations, with the computer as a background aid, as opposed to reacting to a computer program that largely

---

[1] Project IDEALS, English/Morgan, Box 8670244, University of Alabama, Tuscaloosa, AL 35487-0244 US; phones 205-348-9494 (w), 205-752-0690 (h); facsimile 205-348-5298; e-mail crookall@ua1vm (Bitnet), crookall@ua1vm.ua.edu (Internet), crookall@igc.org (PeaceNet/EcoNet)
[2] Project ICONS, Government and Politics, University of Maryland, College Park, MD 20740 US; phone 301-405-4171; e-mail patty@umd2.umd.edu (Internet)

determines the course of the simulation (for a discussion, see Crookall et al. 1986).

The International COmmunication and Negotiation Simulation (ICONS) and the International Dimension in Education via Active Learning and Simulation (IDEALS) are two simulation projects which developed on the foundation laid by POLIS. Project ICONS evolved from POLIS, and Project IDEALS can be considered as a third generation of network gaming. Both ICONS and IDEALS offer large-scale, multi-site, interactive educational experiences in a variety of fields related to international and global concerns. In a word, Project ICONS provides a range of simulation exercises, particularly in the areas of foreign policy and foreign language, while Project IDEALS uses simulation as a tool to foster international understanding and develop cross-cultural communication skills among people of diverse cultural and ethnic backgrounds.

In both projects, the simulation experience is grounded in the principles of experiential learning. The experience itself motivates students to become fully involved by giving then an opportunity to take responsibility for their own learning. As in many simulation/gaming activities, learning by doing becomes a reality, or as Confucius once said:

> I hear and I forget
> I see and I remember
> I do and I understand

Computers and Networks

The central component of both projects a is large-scale simulation assisted by computers and telecommunications. Students take on the roles of high-level officials representing various countries involved in negotiations or attending an international conference. The country teams are situated at different campuses (usually one team per campus) and communicate using computer networks and specialized simulation management software called Polnet II (containing over 40,000 lines of C code). This has been designed specifically to allow participation in simulations by teams of students located at various sites across the globe.

In order to participate in ICONS or IDEALS, a team must have access to a personal computer (PC), a word processor, a printer, a modem, a telecommunications package, and a phone line. Any PC can be used, including IBMs, clones, Apple family PCs, and Acorn BBCs or Archimedes. There are several alternatives for connecting to Polnet II, housed at the host institutions (Maryland or Alabama). The main options are Internet and SprintNet. Internet is a worldwide telecommunications system, the US-based part developed by the US National Science Foundation for educational and governmental use. Many universities have interactive connections to Internet and allow local high schools to tie into the system. The principle advantage of

Internet (NSFnet) is that there is no cost to the end user for communication. If an institution does not have access to an Internet connection, a commercial communication service called Sprintnet (previously Telenet) may be used to make the connection for the ICONS simulation. Local participants can also dial in using a regular phone line.

Messages are exchanged over the computer network in two ways. A daily mail system allows teams to send texts to other teams, which may be situated on the other side of the planet. In addition, real-time teleconferences are scheduled on a regular basis and focus on a particular issue. In a teleconference, students in Japan, France, Argentina, and the US, for example, would actually be communicating simultaneously with each other. As the team in France sends off its message in the dialog, it automatically appears on the screens of the teams in Japan, Argentina, and the US.

Structure

Both projects have a basically similar three-part structure. Teams spend some weeks preparing for the simulation. This may involve conducting research into the issues, studying background documents, preparing position papers, refining their country's foreign policy, looking at cross-cultural communication, learning how to use the computer, developing familiarity with simulation methodology, and organizing their team.

The actual simulation generally lasts 4–6 weeks, during which time participants become immersed in their learning environment. They hold formal and informal meetings, make decisions, negotiate, draft texts, do library research, take part in teleconferences, use computers and telecommunication networks, videotape sessions, and keep a log or journal of their experiences. This phase may contain pauses for feedback (formative debriefing) on such aspects as group work and cross-cultural communication skills.

The third phase, following the simulation, lasts 2–3 weeks and is dedicated to the most important aspect of the learning experience: debriefing the simulation experience. Debriefing is conducted primarily on each individual campus, although pilot networked debriefing sessions have proven successful at both the faculty and student levels. Participants analyze the entire experience, drawing upon the on-line questionnaire data as well as additional critical-appraisal forms, viewing videotapes, taking part in structured discussions, examining their personal logs, and conducting searches on the messages sent during the simulation.

One aspect of the debriefing process is looking at the messages which were sent during the simulation. The message bank is opened up at the end of each simulation to permit students to track messages that were exchanged between other countries. This facilitates research into negotiation styles and a comparison of strategies. In addition, students are encouraged to discuss and write about the simulation experience in relation to real-world issues.

Project ICONS

Project ICONS was developed in 1981 at the University of Maryland. The simulations enable students to obtain a realistic sense of the complex relations among international issues and an understanding of the role that language plays in international affairs. Groups of student decision makers at participating institutions are each assigned to represent the government of a particular country. The country-teams spend several weeks negotiating real-world problems, such as global warming, international debt, international trade, the nuclear non-proliferation treaty, human rights, and drug trafficking. The issues that will be covered are defined in detail in a scenario tailored for each individual simulation.

Messages are generated in the foreign language of the country being represented. For example, a team in Argentina or a team in the US representing a Spanish-speaking country, sends its messages in Spanish. Language students at the receiving end translate these messages and pass them along to the foreign policy students for appropriate action. Thus far, ICONS simulations have been conducted in Spanish, French, English, German, Russian, Japanese, and Hebrew. Language students are given the opportunity to apply their language skills in situations where their peers depend upon the accuracy and promptness of translations.

Participation in an ICONS simulation permits students to create and test negotiation strategies, improve communication skills in several languages, understand the interdependence of international issues, appreciate cultural differences in approaches to world problems, work in teams to solve problems, and use computers for multinational communication. ICONS is a program which brings students and faculty from various disciplines together to deal with issues that bridge traditional departmental boundaries.

In the 1990–91 academic year, ICONS conducted ten simulations. These included two university programs and eight simulations for high school students. Twelve simulations were conducted in the 1991–92 academic year, with a pilot program in Latin America concentrating on illicit drug issues. ICONS currently services more than 100 schools and universities from many countries, including Britain, Japan, Argentina, Chile, France, Germany, Finland, Korea, the US, and other countries.

Project IDEALS

Project IDEALS is based on ICONS and offers a learning environment in which the primary objective is to bring students together from different cultural backgrounds, essentially foreign (or international) and native (or host) students, both in the US and abroad. Like ICONS, IDEALS is a multi-site, semester-long, socially-interactive simulation.

Teams are made up of native (host) and international (foreign) students. Each team develops an identity, usually by taking on the role of a hypothetical country. Participants within each team develop cross-cultural awareness and skills through interaction with others in their team, as well as through tele-communication with other distant teams. They also develop skills in decision making, negotiation, and leadership.

The ultimate goal of the simulation is for teams to negotiate an agreement related to some international situation. For example, the task might be to hammer out the text for a treaty governing some aspect of the global environment, such as CFC emissions, the use of the ocean's resources, and the future of Antarctica. The technological, economic, social, cultural, and ethical dimensions of these issues are hotly debated, from several cultural perspectives, within and among teams.

Participating institutions are to be found in Australia, Finland, France, Hong Kong, Japan, Mexico, Netherlands, Singapore, the UK, and the US. Software design and development currently under way consist of additional modules for Polnet II, including an on-line questionnaire authoring system, a text analysis facility, and a simple mathematical modeling capability.

Workshops

Both IDEALS and ICONS provide 2-day training workshops for team facilitators. The workshops provide an overview of the simulations, computer training, and sessions on the content and curriculum of the specific simulation. Workshops are usually held in the semester preceding participation in a simulation. Information on how to participate in ICONS or IDEALS can be obtained by contacting one of the authors. For further information, see Crookall et al. (1988), Crookall and Wilkenfeld (1985), Klobusicky-Mailander (1991), Rawson (1990), Tammelin (1991), and Wilkenfeld (1983).

Conclusion

ICONS and IDEALS offer two related, but very different simulation experiences designed to improve students' personal grasp of international issues and communication. Participation in these programs gives students the opportunity of interacting on a substantive level with students on their own campus and at other institutions around the world. Through computer-assisted, interactive network simulation, students are given the opportunity to learn about global interdependence and to take responsibility for that learning.

Acknowledgments

Support for ICONS has been provided by the US Institute of Peace (USIP), the Organization of American States (OAS), the US Department of Education, IBM, the Maryland State Department of Education, the state of Maryland, the state of Pennsylvania, and the Toronto Board of Education. Support for IDEALS has been provided by the Fund for the Improvement of Post-secondary Education (FIPSE), US Department of Education, by the University of Maryland, and by the University of Alabama.

References

Crookall D, Wilkenfeld J (1985) ICONS: Communication technologies and international relations. System: An International Journal of Educational Technology and Applied Linguistics 13(3):253–258

Crookall D, Martin A, Saunders D, Coote A (1986) Human involvement in computerized simulation. Simulation & Games: An International Journal of Theory, Design, and Research 17(3):345–375

Crookall D, Oxford R, Saunders D, Lavine R (1988) Our multicultural global village: Foreign languages, simulations, and network gaming. In: Crookall D, Saunders D (eds) Communication and simulation: From two fields to one theme. Multilingual Matters, Cevedon, Avon

Klobusicky-Mailander E (1991) Iconomics 101: A telecommunications simulation on Europe 1992. Simulation/Games for Learning 21(1):43–53

Noel RC (1969) The POLIS Laboratory. American Behavioral Scientist 12:30–35

Noel RC (1979) The POLIS methodology for distributed political gaming via computer networks. In: Bruin K et al. (eds) How to build a simulation/gaming. Leeuwarden, Netherlands

Noel RC, Crookall D, Wilkenfeld J, Schapira L (1987) Network gaming: A vehicle for intercultural communication. In: Crookall D, Greenblat CS, Coote A, Klabbers JHG, Watson DR (eds) Simulation-gaming in the late 1980s. Pergamon, Oxford

Rawson JH (1990) Simulation at a distance using computer conferencing. Educational & Training Technology International 27(3):285–292

Tammelin M (1991) Project ICONS: Using a multinational computer-assisted simulation in a language class. In: Savolainen H, Telenius J (eds) EUROCALL 1991: International conference on computer-assisted language learning. Helsingin kauppakorkeakoulun julkaisuja

Wilkenfeld J (1983) Computer-assisted international studies. Teaching Political Science 10:4

David Crookall is Director of Project IDEALS at the University of Alabama and a tenured faculty member in the French university system. He is past president of ISAGA and Editor of *Simulation & Gaming: An International Journal of Theory, Design, and Research*.

Patty Landis is the ICONS Simulation Director at the University of Maryland. In this capacity, she writes simulation scenarios, conducts simulations, and plans and organizes the ICONS simulations. In previous lives she has been a US Foreign Service Officer and a government instructor.

Global University for Global Peace Gaming

Takeshi Utsumi[1] and Arturo Garzon[2]

Abstract. This paper describes the construction of an infrastructure for global education and peace gaming, in particular on the issue of environment and sustainable development in Third World countries. The games are intended to train would-be decision makers in crisis management, conflict resolution, and negotiation techniques. Experience shows that the expertise necessary to participate in peace games does not exist in many parts of the world. To help educate future participants and to promote the cause of peace by enhancing exchanges of education and joint research, the Multinational Project on Secondary and Higher Education (PROMESUP) of the Organization of American States (OAS) is joining efforts with the GLObal Systems Analysis and Simulation (GLOSAS) Project to create a Global/Latin American (electronic) University (GLAU) as an initial step in that region towards full implementation of a Global (electronic) University (GU) consortium. This paper provides a brief account of the steps taken over the past 12 years which have led to the development of the GU so that Latin American institutions can meet the challenge of global issues.

Key words. conflict resolution; crisis management; distance education; global environmental peace gaming; global lecture hall; global neural computer network; global university; globally distributed decision-support system; GLOSAS project; OAS; REDLAED

The need to understand the economic, social, and environmental issues in different regions of the planet and the need for the peoples of the world to learn to communicate and to cooperate has never been more pressing. Economic, ecological, and political issues today are global, and they must be faced in all of their complexity. It is imperative to develop an authentic sense of planetary citizenship to confront planetary issues that endanger the life of our species. This task is too large for government regulation, aid agencies, or

[1] GLOSAS/USA, 43-23 Colden Street, Flushing, NY 11355-3998, US; phone 718-939-0928 (w/h); telex 386131 (GIS USA); EASYLINK 62756570; SprintMail TUTSUMI/GU.USA/ASSOCIATES.TNET; Internet utsumi@cunixf.cc.columbia.edu
[2] Department of Educational Affairs, Organization of American States, 1889 F Street N.W. Washington, DC 20006, US; phones 202-458-3309 (w) 301-983-8762 (h); facsimile 202-458-3149; SpintMail AGARZON/GU.USA/ASSOCIATES.TNET

development banks alone. Restoration of the environment must engage all citizens of the globe, and yet sustainable development is ultimately a local activity.

To support the struggle for the preservation of our ecological heritage, we propose a worldwide telecommunications network for education and non-profit purposes: a Global (electronic) University (GU) consortium. One initial step of this would be the establishment of a Global/Latin American (electronic) University (GLAU), cooperating with the recently created Latin American Network for the Development of Distance Education (REDLAED).

GU can facilitate the operation of existing distance educational enterprises by developing a cooperative and worldwide infrastructure and by bringing the powers and resources of telecommunications to ordinary citizens around the world. The quality of education for those unable to attend conventional universities in disadvantaged countries could be greatly enhanced.

Connections between departments of economics, sociology, and political science in various countries are being established to explore conflict resolution and for new-world-order alternatives to war, with the use of global teleconferencing. Faculties, researchers, would-be decision makers, and students of those institutions and universities can be the players of a global peace game.

Background

In 1972, the GLObal Systems Analysis and Simulation (GLOSAS) project on energy, resources, and environment (ERE) systems for global peace gaming was started (Utsumi 1977, Rossman and Utsumi 1986, Utsumi et al. 1986). With computerized simulations and a combination of advanced telecommunication channels, such gaming will enable experts in many countries to collaborate in finding new solutions to the problems that have previously been the causes of war.

Over the past 12 years, GLOSAS played a major role in making possible the extension of USA data communication networks to other countries, particularly to Japan. GLOSAS helped the deregulation of Japanese telecommunication policies for the use of electronic mail and computer conferencing through USA-Japan public packet-switching lines. This enabled cost reduction of telecommunications and the European Economic Community (EEC) and Latin American countries have followed suit.

Multipoint-to-multipoint multimedia interactive teleconferencing technology which GLOSAS/USA has developed and demonstrated for the past several years uses audio, data, text, computer, and slow-scan TV teleconferencings, audio-graphic, facsimile, packet-radio and packet-satellite, and full-color, full-motion video teleconferencing. GLOSAS/USA has conducted many demonstrations of a global lecture hall, in which participants in several countries can hear, talk to, and see the other participants, using inexpensive methods for Third World countries. Demonstrations included uplinking to satellites combined with audio and slow-scan teleconferencing, global computer conferencing, and facsimile for question-and-answer exchanges. The most ambitious demonstration had 14 sites linked together, from the east coast of the USA to Korea, Alaska, and Australia.

In the particular case of Latin America, their educational systems have not been able to provide the quality and quantity of education needed for self-sustained development. To confront this situation, Latin American and Caribbean governments, with the support of the Regional Program for Educational Development (PREDE) of the Organization of American States (OAS), commenced the implementation of distance education projects with the purpose of improving and expanding educational opportunities for a growing population of students who could not attend the traditional education system.

In order to support the efforts of Latin American distance educators, GLOSAS/USA organized a demonstration of the large-scale interactive satellite teleconference, global lecture hall with the use of various inexpensive global telecommunication media to show the possibilities of global education. This was done on the occasion of the XVth World Conference of the International Council of Distance Education (ICDE) in Venezuela, 1990.

Global University

The GU will distribute education from all the world's best sources to all the students who crave knowledge, wherever they are, so as to enlarge and expand the present exchange of courses into a worldwide educational system. The system will provide a specially tailored educational program for each individual, bringing to his/her home an array of resources that can empower individuals and bring new wealth to the Third World.

GLOSAS/USA was established as a publicly supported, nonprofit, educational service organization in 1988 for quality and availability of international educational exchange through the use of computer, telecommunication, and information technologies. It seeks to create a Global (electronic) University (GU) Consortium which will become a more permanent organization of the international education exchange via various telecommunication media.

To help educate future participants of peace games and to promote the cause of peace by enhancing exchanges of education and joint research, GLOSAS is attempting to create a Global/Pacific University (GPU), a Global/Latin American University (GLAU), and a Global/European University (GEU) consortia. These, along with a Global/Indian University (GIU) (Charp 1988), can become part of a true Global University.

Global Peace Gaming of GLOSAS

The global peace gaming of GLOSAS is a computerized gaming/simulation to help decision makers construct a globally distributed decision-support system for positive sum/win-win alternatives to conflict and war. The idea involves interconnecting experts in many countries via global value-added networks (VANs) to collaborate in discovering new solutions for world crises, such as the deteriorating ecology of our globe, and to explore new alternatives for a world order capable of addressing the problems and opportunities of an interdependent globe (Mische 1988).

The globally distributed peace gaming/simulations will be for policy analysis, conflict resolution, cooperation, and training in negotiation techniques. Gaming/simulation is the best tool we have for understanding the world's problems and the solutions we propose for them. The distributed mode with autonomously maintained and updated databases and simulation models will not only give credibility and integrity to the databases and models but will also motivate local people.

One of the largest and perhaps most successful demonstrations was held at the conference on Crisis Management and Conflict Resolution given by the World Future Society in New York, 1986. A global gaming/simulation session with a multimedia teleconference on the USA-Japan trade and economy issues was demonstrated. Nearly 1,500 people took part, in New York, Tokyo, Honolulu, and Vancouver.

Some countries of Latin America and the Caribbean have been experiencing environmental deterioration due to the mass impact of a rapidly growing and poorly educated population. A GLAU can take full advantage of the potential that telecommunication networking offers for education, information, simulation, exchange of ideas, cooperation, and problem solving. Education and socioeconomic system simulations are the warp and woof in the fabric of projects that GLOSAS/USA is weaving in collaboration with voluntary associates in several states and overseas countries. In Latin America, education and system simulations are two of the many activities needed to save the environment while pursuing industrial development.

GLOSAS/USA submitted a project plan to the United Nations Development Program (UNDP) that will expand and exploit telecommunications systems within Latin America and outside to the United States and Japan. The specific objectives are:

— To organize the operation of telecommunication networks for sharing experiences and reporting regional issues
— To build databases on environmental issues in distributed mode
— To implement training and educational courses
— To construct a globally distributed decision-support system with distributed interactive computerized gaming/simulation systems for problem analysis, policy formulation, and assessment, which will be used for training of decision makers in conflict resolution and negotiation

These are to be done with distributed computer conferencing, databases, and simulation systems among several Latin American countries to provide globally distributed peace gaming/simulations focusing on environmental issues. Several systems will be interconnected to form a global neural computer network in such a way that the total system will act as if it were a single system with parallel processing of those subsystems in individual countries—here, each game player with his submodel and database corresponds to a neuron of a global brain.

A comprehensive model of global resources, ecology, and economy is needed for the rational management of ecology and for economic cooperation among nations and economic blocs. We propose a public open modeling network (OMN) which will consist of models developed by local experts interconnected by global VANs (Utsumi et al. 1986). Interconnection of dissimilar computers and models for peace gaming on ERE systems, architectures for linking heterogenous computers were outlined by Utsumi and DeVita (1982).

An outline of the hierarchical structure and distributed components of an integrated, interactive peace gaming/simulation system for energy, economics, foreign trade, and so on is depicted in Fig. 1 (Utsumi 1974a). Each block in the figure represents dissimilar computers of the public VANs in those countries. These computers include simulation models designated in each block. All models will be executed simultaneously and concertedly via satellite and terrestrial telecommunication links. For example, suppose pollution in Japan exceeded a certain allowable level, say, around 1977 on Fig. 2 (Utsumi 1974b), the Japanese expert watching it on the display unit will stop the entire simulation. All participants, wherever they are located, will then try to find, with the use of the conferencing system, a consensus on a new set of pseudo-alternative policy parameters which will be executed until a new crisis appears, say, around 1984 on the figure. The process will be repeated for rational policy analysis, based on facts and figures, and with international cooperation of experts in both countries.

The purpose of an interactive gaming mechanism is to help find appropriate alternative policies by establishing consensus among participating parties. It is suggested here that globally distributed computer simulation should be tested interactively with the game player inserting pseudo-policy parameters into the models whenever necessary. This is called peace gaming/simulation (Utsumi 1977) similar to war games practiced by military strategists (Schram et al. 1971). With the advent of global VANs and standard interface protocols for interconnecting various dispersed, dissimilar host computers, the potential exists for ensuring the coordination of international efforts by providing more frequent communications and an environment for shared development, enabling more credible simulation study than was previously possible.

It is now possible to combine existing technologies to make sophisticated and more holistic explorations of various scenarios for solving global social problems. Many small computers in different countries can be interconnected, through globally distributed network and information processing, into modeling and simulation instruments for playing peace games on the scale of Pentagon war games (McLeod 1987).

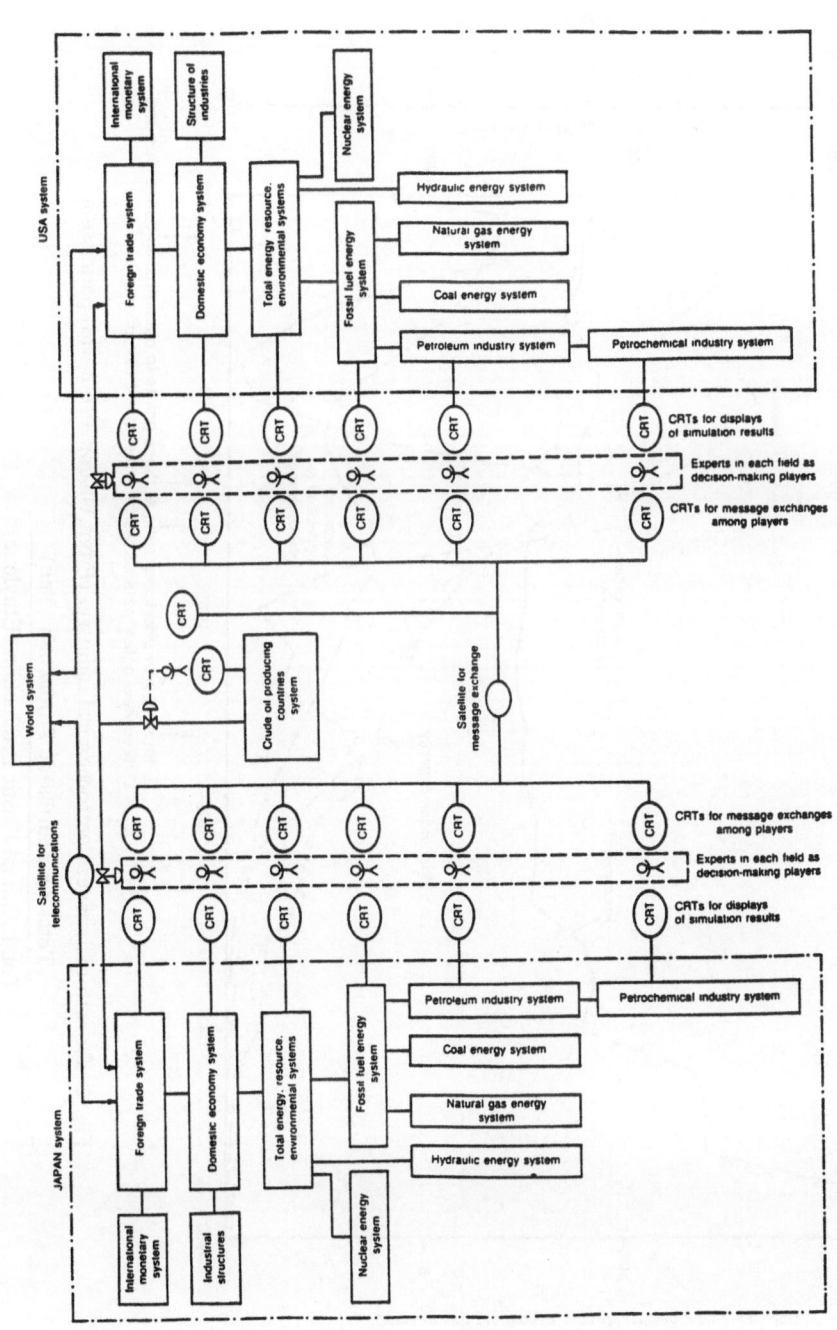

Fig. 1. Structure of integrated models and communication network. *Boxes, dispersed, dissimilar computers around global VANs.* (From Utsumi 1974a with permission)

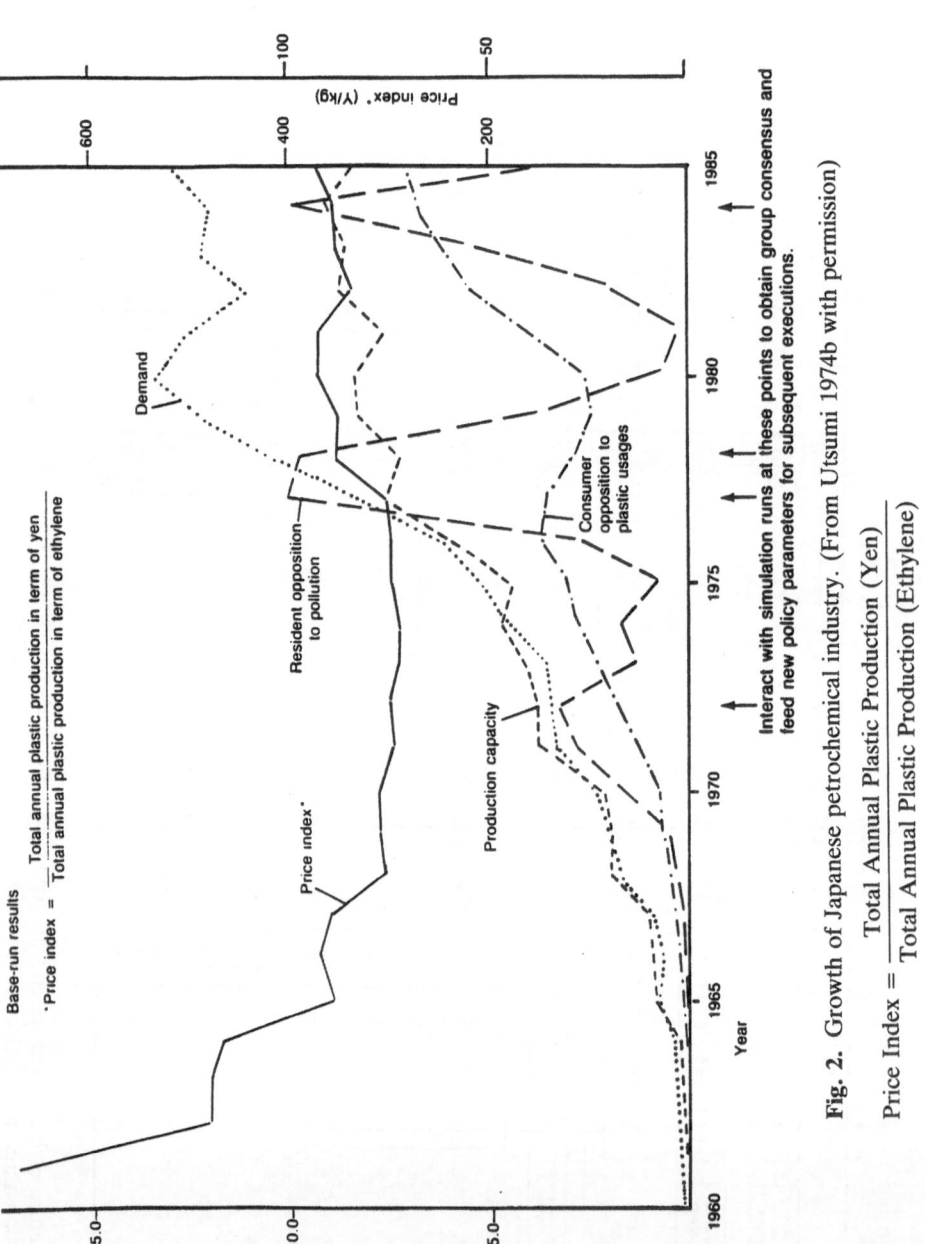

Fig. 2. Growth of Japanese petrochemical industry. (From Utsumi 1974b with permission)

$$\text{Price Index} = \frac{\text{Total Annual Plastic Production (Yen)}}{\text{Total Annual Plastic Production (Ethylene)}}$$

Conclusion

Global education via satellite and other telecommunication media is the way towards the twenty-first century "Age of Knowledge," laying a social infrastructure for global citizenship of the global village. Extending communications through a global network and sharing ideas and educational opportunities with other locations is of paramount interest. The exchange of knowledge among countries can make major contributions to world peace. Developments in global electronic education can transform education at all levels around the world, and can enrich and transform human society.

GU is an evolutionary concept with no global precedent. It can now take shape gradually through parallel steps and many kinds of initiatives in many regions, encouraging a sense of universally shared responsibility and a spirit of participation, in an enterprise truly global in scope.

The world is shrinking in the electronic sense and all people and all educational programs are becoming increasingly interconnected. With this interconnection, however, there comes the potential for escalating regional conflicts, so the need for global education with global peace gaming has never been greater. Senator Fulbright once said that "learning together and working together are the first steps toward world peace." The time is ripe for global education. The technology is now available. What we need now are people who are eager to face the challenges of our time and to forge ahead toward the education of the twenty-first century.

References

Charp S (1988) Editorial. THE Journal 8 August

McLeod J (1987) TAK is TICKING. Simulation December: 273–274

Mische G (1988) Partners for world-order alternatives. Breakthrough 9:1–3, 18

Rossman P, Utsumi T (1986) Waging peace with globally-interconnected computers. In: Didsbury HF Jr. (ed) Challenges and opportunities: From now to 2001. World Future Society, Bethesda, MD

Schram S, Marks H, Behrens W, Levin G, McLeod J, et al. (1971) Macro-system simulation. Panel discussion session at the 1971 Summer Computer Simulation Conference (SCSC). In: 1972 SCSC Proceedings. Society for Computer Simulation, San Diego, pp 1491–1502

Utsumi T (1974a) Joint US/Japan project on global systems analysis and simulation (GLOSAS) of energy, resources and environment (ERE) systems. In: Proceedings of the conference on energy modelling and forecasting, Berkeley California, 28–29 June 1974. National Technical Information Service, Springfield, VA, pp 121–144

Utsumi T (1974b) Japan petrochemical industry model for the GLOSAS project. In: Proceedings of 1974 SCSC. Society for Computer Simulation, San Diego, pp 318–325

Utsumi T (1977) Peace game. Simulation November: 135

Utsumi T, DeVita J (1982) GLOSAS project. In: Schoemaker S (ed) Computer networks and simulation II. North Holland, Amsterdam, pp 279–326

Utsumi T, Mikes PO, Rossman P (1986) Peace games with open modeling network. In: Schoemaker S (ed) Computer networks and simulation III. North Holland, Amsterdam, pp 267–298

Takeshi Utsumi is Chairman of GLOSAS/USA which he created in 1988; a member of the Board of the University of the World in California; Director of the World Association for the Use of Satellites in Education (WAUSE) in Bari, Italy; a board member of the Institute for Educational Studies in Atlanta, Georgia; an advisor to Electronic Information Exchange System (EIES) of the New Jersey Institute of Technology; and Technical Director of the GLOSAS/Japan.

Arturo Garzon is Principal Specialist of the Department of Educational Affairs in the Organization of American States, and the Editor of the *Interamerican Review of Education*. He majored in science teaching and earned a masters degree in educational technology at Florida State University. He joined the OAS in 1972 to provide technical assistance to member States in introducing educational innovations and promoting technical cooperation.

Establishing Cross-Cultural Connections in the Classroom

Arlen Gargagliano[1]

Abstract. In this paper I will describe my presentation at the 1991 ISAGA conference. This session demonstrated how role-plays based on contemporary videos or critical incidents can be used for increasing cultural awareness and generating communication in a variety of learning situations.

Key words. critical incidents; cross-cultural awareness; cross-cultural communication; debriefing; role-play

My presentation at the 1991 ISAGA conference contained possible activities for (1) generating communication among multicultural, non-native English speaking participants/students and (2) increasing cultural awareness. Both activities use situations of misunderstanding due to cultural differences as a basis for role-play. The first activity was a role-play based on video segments from a contemporary US television show. The second activity was a role-play based on a critical incident. Both activities explore the use of different options in particular situations, and inspire discussion on the varied interpretations of the encounters by the students/participants.

There are a couple of reasons why I began to explore the area of role-play. Role-play in a limited form has been a part of my English-as-a-Second-Language (ESL) classroom for some time. I say limited because initially I was creating situations for role-play that were often too contrived, and students were not participating in the way I had hoped they would.

Another reason for working on role-play activities was that for the past several years my classroom population has consisted primarily of Japanese students. In general, these students have studied English for many years, but have virtually no experience in speaking the language. Though these students were continually encouraged to participate in class discussions, they appeared reluctant to join in and were often overpowered by their non-Japanese class-

[1] English Language Center, Concordia College, 171 White Plains Road, Bronxville, New York, 10708 US; phones 914-337-9300 ext. 2134 (w) 914-235-6922 (h); facsimile 914-395-4500

mates. Then I began experimenting with assigning specific roles for role-play activities. I found that if my students were assigned a new identity, student participation improved dramatically. In addition, these activities gave us a chance to explore stereotypes and prejudices, while generating a great deal of communication in English.

The first role-play was based on a video segment from the popular US television show *Thirtysomething*. The segment was divided into four separate scenes. All the scenes were about the wedding of two people who were from two very different cultural backgrounds.

Initially I described how a theme-related discussion should be used before viewing a video segment. For example, because this segment deals with the marriage of two people from very different cultural backgrounds, we could address such questions as: Are intercultural marriages common in your culture? What are some of the problems encountered by people in intercultural marriages? How does one propose marriage in your culture? Participants agreed that this type of discussion serves as an introduction to the topic of the segment, as well as to the vocabulary.

In the first scene of the video segment we see several family members eating dinner. They are discussing something in Polish and are speaking loudly, simultaneously, and using many gestures. One person, a female guest, has no idea what is going on. She, Ellen, appears frightened by what she views as a major family argument. Ellen is seated next to her boyfriend, who is introducing her to his family. When the conversation changes to English, Ellen learns that they were discussing Sylvester Stallones' not winning the academy award for the movie Rocky. Ellen is quite surprised to learn the topic of discussion.

We then viewed a second segment in which Billy, Ellen's boyfriend, stands up to make an announcement in Polish. By the parents' reaction, we can see that he has announced he has asked Ellen to be his wife. Ellen appears confused as Billy's mother repeatedly kisses her hands (a congratulatory gesture) and his family smiles at the exciting news.

Immediately following this segment I stopped the tape. At that point I assigned roles. The students played the roles of Billy, Ellen, Billy's father, his mother, and his grandmother. They were told to sit at desks arranged in the position of a large dining room table, and to continue the discussion immediately where the tape left off. The rest of the participants were observers.

The role-play went on for about 5–10 min. Their conversation was based on themes introduced by the video. They elaborated by bringing up additional questions such as where they would live, and how often they would be able to come back to visit. Some participants spoke using their own accents, while others used an accent they thought their character would use.

When we stopped the role-play, one of the observing participants suggested that we change it so that the situation be more realistic according to the participants' cultural backgrounds. We changed the role-play around so that Japanese participants played Japanese parents in a situation as described in the video. This role-play did not flow as well; in fact, it just stopped. The participants said that they weren't sure how Japanese parents would react. They said they had felt more comfortable in the more unfamiliar roles.

A few different points came up during the debriefing. One participant said that he was really into the character, but suddenly he started thinking about the role-play itself and lost his train of thought. We briefly discussed how people can fall out of character, which makes it sometimes difficult to get back into the role-play. When I commented on one participant's use of an accent different from his own, he said he had adopted it as part of making the role more vivid.

We also discussed how much people draw on their personal experiences when acting within a role-play. One participant suggested that it is the reality of certain situations for participants that make role-plays such effective learning experiences. We talked about the applications of this type of experiential learning to business training and the use of video as a tool for generating communication about a variety of situations.

The second role-play we worked on was given in written form to all of the participants. It was based on the following critical incident:

> A group of students from country X attend a basketball game in the United States at the college where they are currently studying. While watching the game, the referee makes what they consider a bad call, so they start whistling loudly. A young professor turns to them and says "Hey, what's the matter with you guys?" Two of the students reply "Nothing, what's the matter with you?" The professor then replies "I'll see you two at the Dean's office at 8:00 a.m. Monday morning." Surprised, the students he addressed begin to laugh. The young professor adds "I was only kidding before, but now I mean it. You better be there Monday morning."

After reading the incident, the participants were divided into groups of four. Within each group they were assigned roles: the Dean, the young professor and the two students. They were asked to imagine that they were in the Dean's office on the following Monday morning.

After several minutes of role-play, the participants switched roles and continued. Then the role-play was discussed. During the debriefing, several points where raised. One point that came up was how people interpreted the role of the Dean differently. For example, in one group the Dean was completely authoritative, while in another, the Dean was much more diplomatic in terms of asking those involved what they thought should happen next. We also discussed whether in that case the role of the Dean was that of student advocate. Participants pointed out that because the critical incident is somewhat vaguely described, the role-play has a lot of flexibility and it becomes a richer vehicle for exploration by people involved in it.

Perhaps the most fascinating part of the debriefing was the participants' discussion about applications of these role-plays to their own varied situations. For example, I never would have imagined using these type of activities in a business situation. However, several of the participants were involved in cross-cultural awareness training for businesses and said that they were very excited about using these types of role-plays in their work.

Acknowledgements

The author thanks David Crookall for his suggestions and support, and Margaret Pusch for introducing her to the idea of using critical incidents.

Arlen Gargagliano teaches English as a Second Language at Concordia College in Bronxville, New York, where she is also International Student Advisor. She conducts workshops for students, educators, and educational consultants on cross-cultural awareness. She is currently exploring the use of video as a means of both stimulating and recording role-play situations.

Collaborative Creation of Adventure Games in the ESL One-Computer Classroom

Douglas W. Coleman[1]

Abstract. This is a report on a workshop conducted at ISAGA '91 in Kyoto, Japan. The format of this workshop was as follows. First, two short simulations used in an ESL teacher-training course were demonstrated to participants: one of a "traditional" (grammar-focused) language-learning exercise, the other of a "communicative" activity. A debriefing was conducted in which participant reactions were discussed, as well as the apparent effects of the simulations on actual teacher trainee behavior. Finally, the presenter showed how generic computer software can be adapted to a novel use in the ESL classroom, describing an activity in which ESL students collaboratively create a computerized text-adventure game. The activity is intended to promote self-reliance in writing tasks by taking students from an environment of peer support to one of increasing independence. At the same time, it creates a "bridge" for the novice teacher from a highly teacher-centered (traditional) classroom to one in which the teacher's role is quite different.

Key words CALL; computer-assisted language learning; English as a Second Language; ESL; simulation/gaming; teacher training; TESL

In this workshop, two short simulation-activity demonstrations were followed by a brief explanation of a generic system adapted for the collaborative creation of adventure-game-type, interactive fiction in the one-computer English as a Second Language (ESL) classroom. The first short demonstration was of a simulation of a "traditional" foreign-language-learning exercise focused primarily on grammatical explanation rather than on communication in the target language. The second demonstration simulated a brief language lesson in the form of a "Direct Method" variant resembling Asher's (1977) Total Physical Response (TPR) method. Slightly extended forms of these simulations are used in a graduate ESL teacher-training course. Invariably, students in the course recognize that they frequently can perform with 100% accuracy on a traditional exercise even in cases (like the one in the simulation) in which they do not understand any of the target language forms with which they are

[1] Department of English Language and Literature, University of Toledo, Toledo, OH 43606-3390 US; phones 419-537-2318 (w), 419-475-8301 (h); facsimile 419-537-4940; e-mail fac3079@uoft01.bitnet

Model: Tu piuro.
Response: Tu-chi piuro.

Model: Ta swoike.
Response: Ta-chi swoike.

Model: Response:

1. Tu ofizino.

2. Ta skelpe.

3. Ta zione.

4. Tu meanso.

5. Tu tramo.

Fig. 1. "Traditional" language exercise

dealing. They also realize that they have learned from the second type of lesson something about how to communicate in the target language despite its complete lack of grammatical explanation. Even after experiencing such demonstrations themselves, however, the same teacher trainees, when in the role of language teacher, still tend to continue to use classroom techniques based almost exclusively on the traditional type of lesson. A writing activity couched in semi-traditional terms, but involving collaborative creation of adventure games in the one-computer classroom, is proposed as a "bridge" to help novice language teachers learn more about simulation/gaming in language learning. The activity itself is described in the second part of this write-up.

Simulation of a Traditional Language Learning Exercise

I prepared the material shown in Fig. 1 on the blackboard before the workshop actually began. (Note: both this and the second simulated-language lesson are based on hypothetical "languages.") I explained the following points to the participants: (1) *tu* and *ta* are Type X words; (2) *piuro, swoike, ofizino,* etc. are Type Y words; (3) the grammatical order requires Type X + Type Y; (4) adjacent Type X and Type Y words must have grammatical agreement; (5) to form your response from the model, change Type X + Type Y to Type X + *chi* + Type Y.

After this explanation, I led participants in providing a choral response to each item. They were obviously quite familiar with the format of the exercise, as they performed without hesitation despite only a brief exhortation to "provide the appropriate response—everyone together." Participants showed a marked degree of confidence that they were giving the "correct answers," as in fact, they were. Accuracy was essentially 100%. Movement through the simulated lesson was brisk and very smooth.

Simulation of a Direct Method Variant

Immediately after this first demonstration, I presented a brief language lesson in the form of a Direct Method variant resembling Asher's (1977) TPR method in some ways and Gattengo's (1972) so-called Silent Way in others. As props, I used a ballpoint pen, a paperback book, and a 1-yen coin. Offering the pen to one participant, I said *Koi dwugop*. When she hesitated, I requested more emphatically *Prosh, Koi dwugop*. When she took it, I said *Tienko*. Then I said to her *Prosh, Dei dwugop*—at the same time holding out my empty hand to take back the pen. When I had the pen in my hand once more, I responded *Tienko*. With another participant, I repeated much the same procedure, but this time with the book, saying first *Koi liber* and after that *Prosh, Dei liber*. Then I performed these actions (with yet another participant) with the coin, this time accompanying them with *Koi monet* and *Dei monet*. After another trial or two, I gave one (female) participant the coin (saying *Koi monet*) and then, pointing at another (male) participant, said *Prosh, Duo tan dei monet*— stressing the elements duo tan while pointing. When the male participant had received the coin, I said to the woman *Tienko*. Then, turning to the man, I said *Prosh, Duo tani dei monet*—this time stressing the elements *duo tani* while pointing at the woman. He gave her the coin, I said *Tienko*, and the second simulated language lesson ended there.

Overall, there were significantly more hesitations during the second simulated lesson than during the first. However, it should also be noted that the accuracy of the responses required of the participants was still essentially 100%.

Debriefing/Discussion

First we discussed the traditional sort of lesson. Everyone seemed to agree that the vast majority of our collective language-learning experiences most closely resembled that type of presentation. The point that none of the participants knew what any of the items on the blackboard meant was raised (though they might make some guesses), and it became clear that this seemed to have no effect whatsoever on their ability to do the exercise. A few of the participants seemed surprised; undoubtedly, familiarity with the format was so great that some never noticed this not-so-minor detail. One participant asked me directly at this point what the model-and-response sentences meant; I replied (as is in fact the case) that I had no particular meanings in mind for them when I created the simulation. As another participant pointed out in the discussion, much, if not all, of the grammatical explanation was itself superfluous to the exercise, since the sample model/response pairs probably provided the necessary information to participants in order to allow them to produce the required permutations for the response column.

Comments from participants relative to the second activity tended to focus on its more communicative nature. They noticed that they came to understand words of the (albeit hypothetical) language very rapidly. What soon became clear was that they also were learning rules of word order (e.g., *koi* or *dei* followed by *dwugop* or *liber* or *monet*) and different forms for masculine/feminine pronouns (*duo tan* vs *duo tani*), as well as socially important language functions such as how to emphasize a request (with *Prosh*) or express appreciation (with *Tienko*) for an action performed on one's behalf.

At this point, I mentioned my observations of teacher-trainees who had experienced simulated lessons just like these, later in their own ESL teaching. As I said in the introduction to this article, their own teaching styles made extensive use of materials strongly resembling the traditional language exercise in the first simulated lesson. We briefly discussed why this might be so. There seemed to be general agreement among those present that two causes were most likely. First, novice teachers are often lacking in confidence (falling back on a familiar teaching style provides the comfort of at least knowing that one is engaging in an activity with a broad social acceptance, among colleagues as well as students). Second, most teacher trainees themselves have several years' prior experience as language learners, most of it in the traditional mode (it is unreasonable to expect the intellectual "ah-ha!" that comes with their participation in a set of short simulations in a one-quarter course to completely overcome these years of exposure to something else).

A Bridge Activity

What I am proposing is that collaborative creation of adventure games by students in the one-computer classroom is one way to acclimate novice ESL teachers to simulation/gaming by having them lead students in an activity in which the end result is a piece of interactive fiction. This activity takes place over a few class periods, as described below.

For the activity, we have an MS-DOS PC system outfitted with a color LCD display unit that sits on any overhead projector to produce a wall-size image. To build the piece of interactive fiction, a shareware hypertext program called HYPE works quite well for displaying the story; it requires a word processor that can produce "straight ASCII" files (GALAXY is one possibility, PC-WRITE, another; both are also shareware).

Via the computer, the teacher shows students the opening scene (of only a few sentences) of a piece of interactive fiction. This scene ends with two branches; the teacher selects one alternative—advancing the story by one scene. The second scene, like the first, ends with two branches. The teacher returns to the opening scene and asks the students to predict "What if?" the other alternative is chosen. The teacher this time selects the second alternative, but the students see that this version of the second scene is blank. At this point, the teacher leads the students in creating one. The activity up to this point is a fairly traditional teacher-centered one. (Note that the students' new second scene is also set up to end with two alternatives.) When the new version of the second scene is completed, the story file is modified, so that the students can see the effect of having two alternative second scenes.

With our typical maximum class size of 16 students, the students can then be divided into four groups of four working on alternative third scenes. Each of these teams writes and checks the language of its scene, and turns in the result at the end of that class period. (All scenes end with two alternatives, unless it is necessary to accommodate a class size of greater or less than 16.) The teacher enters the new third scene alternatives before the second class. In that session, students are divided into eight pairs, each working on a different fourth scene; as before, each new scene typically ends with two alternatives. The teacher collects the pairs' productions for incorporation into the story file before the third class session. In this third class period, students work individually to write and edit their own endings to the story. They are then able to explore and discuss each others' various endings in the fourth (and final) class period devoted to the activity.

This particular activity resembles a "frame game," in the sense that it allows a new group to create a completely original interactive story with only minimal additional preparation by the teacher, who must write the opening scene and one alternative second scene. My motivation in designing this activity was to foster increased self-reliance on the part of ESL students engaged in writing and editing tasks. Over a period of four class sessions, it takes students from an environment of peer support to one of increasing independence, although they are engaged in a cooperative effort throughout. It should be noted, however, that it also takes the teacher *and* students from a classroom environment that is highly teacher-centered (in a very authoritarian sense) to one in which the teacher's role is quite different. While this is happening, the novice teacher gets to see how an adventure game might be created, and how it fosters student discussion and leads to writing that is both creative and communicative, not mechanical. In sum, as the students and teacher trainee are constructing an interactive adventure game, they are in the process of constructing a bridge that leads to greater student independence in writing and greater teacher confidence in non-traditional approaches to language learning.

References

Asher JJ (1977) Learning another language through actions: A complete teacher's guidebook. Sky Oak Productions, Los Gatos, CA

GALAXY ver. 2.3. Schauer S, Foster B (1987) Omniverse, Renton (Omniverse, PO Box 2974, Renton, WA 98056 USA)

Gattengo C (1972) Teaching foreign languages in schools: The silent way. Educational Solutions, New York

HYPE. Thompson B, Thompson B (1985) Public domain hypertext program. Available from the authors c/o AI Expert Magazine, 500 Howard St., San Francisco, CA 94105 USA

PC-WRITE (ver. 3.02). Wallace B (1988) Quicksoft, Seattle (Quicksoft, Inc., 219 First
Ave. N, #224, Seattle, WA 98109 USA)

Douglas W. Coleman is Director of English as a Second Language in the
Department of English Language and Literature of the University of Toledo
(Toledo, OH USA). During 1990 and 1991, he has guest edited three special
Reports and Communications sections of the journal *Simulation & Gaming:
An International Journal of Theory, Design, and Research* on "Computerized
Simulations and Language Learning".

Chinese Word Games for School Children

Gee Kin Yeo[1]

Abstract. Mastering the Chinese language is becoming a serious problem for school children of ethnic Chinese in Singapore, where English is the working language and very often the home language as well. While it is believed that learning through play can be achieved to some extent with computer games, developing computerized word games in Chinese is not easy because of problems of input/output and the internal representation of Chinese characters. Many Chinese systems depend on additional hardware cards installed on standard PCs to increase the processing speed. Games developed on such systems will not be readily portable. To achieve portability, some features of game interaction are sacrificed. There are two components of the game system we have designed: one for teachers who prepare the word database and one for the children who play the games. Graphics and sound effects are also added to improve the appeal to small children.

Key words CALL; Chinese computing; interaction mode; portability; word games

Singapore is a multiracial society where English is the working language. In many homes of the over 70% Chinese descendants, English is also the mother tongue of small children. Surveys by the Ministry of Education show that it has increased from 9.3% in 1980 to 23.3% in 1989, and has become the main language of communication in the homes of school children of ethnic Chinese. On the other hand, more and more Chinese school children are finding the learning of Chinese language in schools a tedious task. In the majority of primary schools, Chinese is being taught as a second language, utilizing up to only 5 h of formal lessons in a week. Thus, teaching Chinese effectively in the classroom alone is becoming more and more difficult. This provides our motivation in developing computer games to help children learn the Chinese language at their leisure and in a fun way.

Games have been used very often to stimulate learning. Many Singapore children are exposed to a variety of computer games. Earlier computerized

[1] Department of Information Systems and Computer Science, National University of Singapore, Kent Ridge, Singapore 0511, Republic of Singapore; phones 065-772-2908 (w), 065-469-3238 (h); facsimile 065-779-4580; e-mail iscyeogk@nusvm.bitnet

language games were aimed mostly at vocabulary building and grammatical rigor. More recently, computerized adventure games are also being looked upon as effective communication development aids (Baltra, 1990). Adventure games may exploit Chinese history, such as in ROMANCE OF THE THREE KINGDOMS, the BANDIT KINGS OF ANCIENT CHINA, or GENGHIS KHAN, which are popular among school children in Singapore. However, the presentation and interaction media are still in English. Because of intrinsic technical difficulty, computerized Chinese language games are far less numerous than their counterparts in alphabetic languages. Even if they were already available in abundance elsewhere, it is necessary for Singapore to develop games more attuned to its unique social and educational settings.

Chinese Computing

Chinese is not alphabetical. A Chinese character, as opposed to characters in the Latin alphabet, has not only a definite shape in a two-dimensional ideogram, but also its own sound and meaning. A Chinese character is made of radicals, while a radical may itself be an independent character and take up more than one different form. In pronunciation, a sound can be varied in five different tones. In the Kang Shi dictionary compiled in the Qing Dynasty, there are 49,030 characters, 214 radicals, and 412 sounds. A word with a distinctive meaning is made up of between one, and four characters in general, with the majority having one or two characters.

The problem of representation and processing in computers of some 4,000 Chinese characters in common usage has been the preoccupation of many Chinese linguists and computer scientists for the last 3 decades or so. With the appearance and large scale application of low-cost microcomputers, research activities have intensified, resulting in the development of more than 500 input methods and some desktop publication and typesetting systems (Lua et al. 1990).

Inputting a Chinese character into a microcomputer using the Qwerty keyboard requires two distinct steps. The first involves a mental mapping of the two-dimensional ideogram into an input code. The second is the physical process of converting this input code into the internal code of a Chinese character within the computer system. Just as there are many different input methods, there has also not been a consensus on the codes for internal representation. Essentially, the extended GB codes are adopted in mainland China, while Big-5 is generally used in Taiwan.

System Design

We started simply, requiring the word games to be mainly drill-and-practice. Since they were meant to supplement classroom exercises, we wanted to give the teachers more control over the content of the word database. Thus, there was to be a wordbase management system for the teachers, separated from the game system. Developing such a system with Chinese software alone poses severe memory limitation problems. Thus, we used a Chinese system with a hard card, which also facilitates the input and output for the wordbase management system. The wordbase management system allows the teachers to create different databases for children in different standards. A requirement of the resulting games is that they must be portable. This allows every child to play the games on any microcomputer anywhere, and so they must not rely on the Chinese hard card.

Word recognition is one of the main problems for school children in learning Chinese. For example, character x is often confused as character y as they make up xa and yb, two distinct words. Here x could be similar to y in pronunciation, or share some similar radicals in their ideograms. Two of the three games we designed aim to familiarize children with the correct characters in the formation of words—one game on word formation of two characters, and the other on matching words for idioms. The third game contains riddles.

The technique involved in making the games portable is actually not very difficult. The databases created by the teachers will contain only the internal codes of the Chinese words set up for the games. Thus, besides the databases, the library which contains the bit patterns of the Chinese characters in the internal codes has to be included with the games in order to display the characters on the screen. Randomly generated characters required in the games also come from the library. Graphics and music were also added to enhance the appeal of the games. The software used to develop the whole system includes Clipper Summer '87, dBase III Plus, Turbo C++ Language, and the Graphics Animation System for Professional.

Future Work

There are obvious shortcomings in the word games designed. First, the gaming activities involved are essentially "receptive," in that they merely require players to select from items already given. The major difficulty in allowing other forms of input using only the standard keyboard comes from the fact that there is no standard Chinese input method. The existing input methods are either too intelligent, hence offering far too much assistance for language learning purposes or too difficult for small children. There ought to be a Chinese input method that is natural in the process of learning the characters themselves. This problem is generally acknowledged and is still being actively addressed (Chen et al. 1987).

The only gaming characteristics included in our games are time limitations and scoring boards. Much can be learned from word games in English in the CALL literature, (see, e.g., Jones and Fortescue 1987), where simulation and role-play activities to practice communicative skills can also be incorporated.

As mentioned earlier, children often mix up words containing similar radicals. We actually attempted to design a game challenging children to compose valid characters from some radicals given. The game was not included in the end because a library of radicals would require a much bigger space and it would be hard to maintain a reasonable search-time involved. But the idea to work with radicals should not be abandoned altogether.

Future work is also required in the design of a Chinese dictionary database, incorporating attributes essential for language learning. Different games can then use the same database. The schema and data structures devised can be independent of the representation codes of the Chinese system. Hence, in a special sense, the authoring system of the games can be portable, too.

It is our ambition to provide a wide enough variety of games to satisfy all tastes and abilities. Behind the ambition is our strong desire to help Singapore children learn the Chinese language.

Acknowledgements

Dr. Lua K.T., the Chinese computing expert in the Department of Information Systems and Computer Science, National University of Singapore, has contributed a great deal towards the project as well as this paper.

References

Baltra A (1990) Language learning through computer adventure games. Simulation & Gaming: An International Journal of Theory, Design, and Research 21(4):445–450
Chen A, Chen Z (1987) Methodology of Chinese character coding and symbols for shape-representation of Chinese characters (in Chinese). Computer Research and Development 24(1):20–38
Jones C, Fortescue S (1987) Using computers in the language classroom. Longman, London
Lua KT, Gan KW, Wong YW (1990) Asian language processing. Information Technology 3(2):65–77

Gee Kin Yeo has been responsible for developing MAGNUS, a management game, for many years in the National University of Singaore. Graduated with mathematics honors from Singapore, she did postgraduate work at the University of Waterloo, Ontario, Canada. She is a senior lecturer in the Department of Information Systems and Computer Science and has been teaching courses in operations research and decision-support systems. A member of the ACM, the Computer Society of IEEE, and the International Association of Statistical Computing, her current research interests include model management in decision-support systems and computers in education. She is a member of the Editorial Board of *Simulation & Gaming: An International Journal of Theory, Design, and Research*. She is married with three children, with whom she enjoys solving mathematical puzzles.

Simulation/Gaming for Language Learning in China

Zhipu Qiu[1] and Joanne Velan Dunn[2]

Abstract. This paper reviews research on simulation/games development at Nanjing University, China. The research responded to the need for culturally-based, interactive learning situations which would motivate foreign students to use the Chinese language in different situations with different media. The presentation at ISAGA included a videotaped demonstration as well as descriptions of games and simulations for foreign language communication (Chinese). Included in this paper are design considerations and descriptions of computer, video, and manipulative simulation/games for elementary and advanced students, and international business majors studying Chinese as a foreign language.

Key words business; Chinese; communication; cross-cultural; foreign language; gaming; multi-disciplinary; simulation;

Since 1985, simulation and gaming approaches for teaching Chinese as a second language gained research attention at Nanjing University, China, as an important method for teaching Chinese. Research centered on designing more interactive, efficient, and true-to-life methods for encouraging students of Chinese to use the language in situations which are realistic and meaningful. In addition, the use of multimedia technology extended the possibilities of offering the learner an even broader arena for communication and cultural experience.

Research integrating simulation and games with multimedia to teach Chinese as a foreign or second language is innovative for China. Generally speaking, simulation and games are considered of little importance for instruction. The Chinese usually associate games with children or sports. Games for teaching language have not been highly regarded and little has been reported and researched in the use of simulation for teaching Chinese.

---

[1] Department of Chinese Language and Literature, Nanjing University, Jiangsu Province, Nanjing, People's Republic of China 210008; phone 86-25-634651 ex. 3393
[2] Department of Instructional Communications, Community College of Allegheny County, Allegheny Campus, 808 Ridge Ave. M-523 Pittsburgh, PA 15212-6097 US; phones 412-237-2628 (w) 412-922-5159 (h); facsimile 412-922-5159

The interest in simulation/gaming techniques was stimulated by the influx of foreign students wishing to study Chinese and by the realization that the traditional, structured approach alone (including memorization and pattern drills) may not be enjoyable and effective with foreign students. This presentation outlined the research at Nanjing University and described the involvement with simulation and games as an effective method for allowing students on a number of levels and disciplines to become familiar with and communicate in Chinese.

Design Considerations

Simulations for language offer the following possibilities for elementary and advanced learners:

At the beginning levels, simulation/games:
— Promote familiarity with the language—pronunciation, structure, and meaning of the written character
— Offer the opportunity to explore the language in a non-threatening situation, so students develop concepts about, and skills in using, the language
— Allow play within a cultrual setting, making the language relevant and dynamic

On more advanced levels, simulation/games:
— Allow students to draw on interpersonal and problem-solving skills from their own experiences
— Help students fine-tune communication strategies
— Offer practice in functioning appropriately within authentic cultural situations.

For example, in 1984, a simple computer program was developed at the Applied Linguistics Center at Nanjing University to familiarize the elementary student with simple Chinese characters. More recently, SINOPHONE, developed in 1989, includes video role-plays and simulations. The role-play situations encourage the modeling of appropriate language behavior. Simulations, however, allow more flexibility in communication where the learners draw on their own *language strategies* and *problem solving abilities* within a specific cultural context.

In designing the simulation/games for second language learning, the following points were considered:

— The use of the simulation/game for facilitating skill- getting and skill- using. Skill-getting activities must also be pseudo-communication (Rivers and Temperly 1978)
— The role that simulation/games play as an important part of the process involved in the practice of pseudo-communication
— The use of appropriate instructional media to enhance, make more realistic, and individualize knowledge practice

— The identification of the learner's goals, skills, and strategies which affect their success
— The identification of the problems to be presented, as well as appropriate strategies and formats

Descriptions of Simulation/Games Developed at Nanjing University

CHINESE CHARACTER PUZZLES

These puzzles were created in 1984 to introduce some of the basic concepts of Chinese writing. The puzzles were one of the first attempts in China to use the computer as a tutor for language training. Because many of the Chinese characters are made up of pictograms, and ideograms not unlike computer icons, the computer seemed a natural choice for developing a pilot project on the recognition of Chinese characters.

Unlike English words, which are based on phonetic symbols and the formation of sound patterns, many Chinese characters graphically represent the words or concepts they are communicating. For example, the Chinese character for "man" is written 人. It resembles the figure of a man 人. The ancient character for sun was written ⊙ and then gradually simplified and squared off, changing its shape to 日. It is easy to guess the meaning of many characters and have some fun doing it. So after a short introduction, learners are asked to select ten basic characters one by one from a menu and then to choose a meaning which most closely resembles the character. If they choose an incorrect response, they are given a clue consisting of the ancient or original pictograph of that character. They are then given another chance to select a meaning.

FINDING THE WAY

This simulation using more complex Chinese characters was designed to help students recognize signs on buildings and in public places in China. The participants are shown eight characters and their English equivalents.

电话	telephone dian hua	书店	book store shudian
邮局	post office youju	百货公司	department store baihuo gongsi
银行	bank yin hang	男厕所	men's room nan cesuo
饭店	hotel fandian	女厕所	women's room nu cesuo

After the students have seen the signs they are asked to look at a city map on the next screen. First the cursor moves by itself from place to place, revealing the Chinese name of each place where it stops.

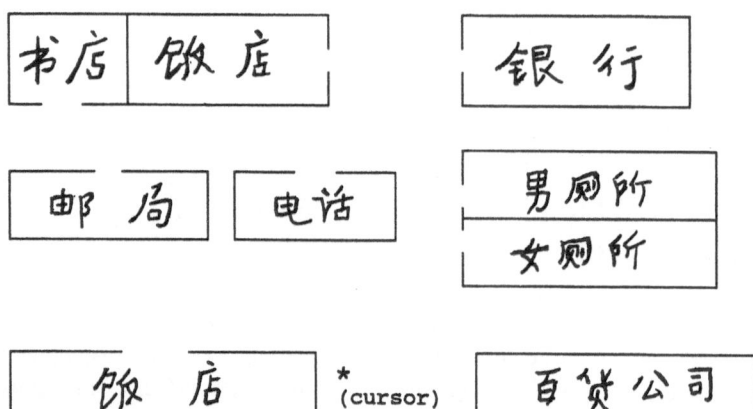

Then students use keys or a mouse to move the cursor by themselves in response to questions like: "Can you find where to go to change money?" On the next screen a new map appears and the places are in different positions.

In 1985, this program was presented to participants in the US and China and was found effective in helping the beginner to recognize the concept of Chinese characters as well as sparking their interest in the written language.

DRAGON BONES

This is a game for elementary students that was designed to help them have fun and feel comfortable using basic Chinese conversational patterns. DRAGON BONES is composed of 36 dice-like cubes coded in seven colors which display more than 200 key words and phrases. Approximately ten of the selected Chinese patterns closely resemble English language structures. This design strategy was selected to facilitate learning for English speakers.

For example:

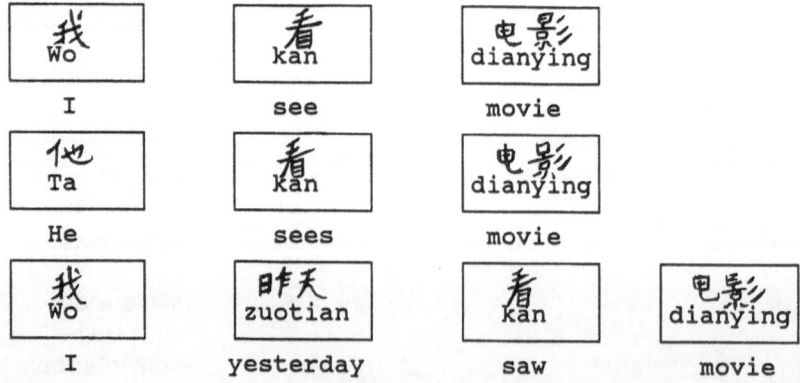

In Chinese *kan* is a verb that means to see. In English there are many forms of the verb to see (see, saw, seen, sees, has seen, etc.) but there is only one form *kan* used in Chinese for every tense, case, person, and gender. Another example is the use of pronouns. In English the pronoun forms change, but in Chinese *wo* (I or me), *ni* (you), and *ta* (he and she) are always the same form in spoken and written Chinese. With this in mind, the primary colors (red, blue, yellow) are used for dice containing nouns, noun phrases, and pronouns.

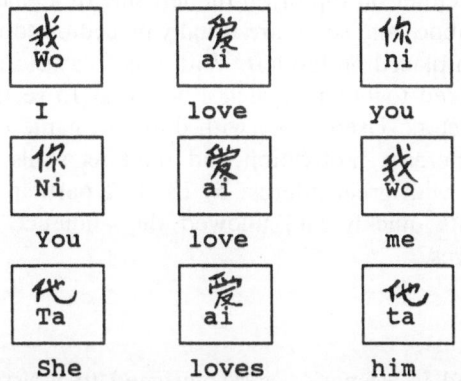

我 Wo	愛 ai	你 ni
I	love	you
你 Ni	愛 ai	我 wo
You	love	me
代 Ta	愛 ai	他 ta
She	loves	him

The secondary colors, purple and green, are used for adjectives, adverbs, and adverbial phrases. Two white dice contain 12 functional words and an orange one contains the most useful daily expressions. There are 36 dice in all and these are divided into ten learning stages. They may be used individually or with others and in teams. Highly motivated students may be able to master all stages in about 25 h while using a participant's manual and an audio cassette.

CHINESE LINKING-LEARNING CUBES

The uniqueness of the Chinese characters, and Japanese kanji as well, offer many possibilities for interesting simulation/game design. CHINESE LINKING-LEARNING CUBES are one design where different character parts can be combined to form new words. The CUBES are a series of rectangular blocks made up of eight small separate dice-like cubes containing printed characters on each of the sides. These cubes are linked in such a way as to allow the students to manipulate the connected cubes to form new words. About 20 different characters may be formed using each block.

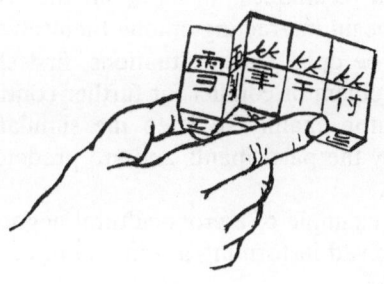

The concept of Chinese character formation has remained the same from ancient times to the present. Most Chinese characters are made up of two parts: the radical which connotes the meaning, and the sound part which connotes the pronunciation. The five elements (gold (metal), wood, water, fire, and earth) form the basic radicals used in Chinese writing. Normally the radicals are placed either on top or on the left side of a character. The radicals for rain, grass, bamboo, and so on are usually placed on top, water, wood, fire, and so on are often placed on the left. With this concept in mind, radicals and sound parts are linked together in such a way that 15 rectangular blocks may generate 300 characters. Used along with different game designs, the instructional goal is to increase motivation and thinking skills. This manipulative game was received with great interest by ISAGA participants because it presented the concepts quickly and allowed the audience to manipulate the characters themselves.

SINOPHONE

This simulation and its scenarios were designed to assist the participants to explore the organizational culture and climate of a Chinese and an American company and to analyze the management styles of each. The goals of the simulation are:

— To explore different negotiation and management styles
— To present interactive models which allow students to analyze cross-cultural consensus and decision-making
— To "participate" in the communication process in a second language (Chinese)

SINOPHONE is a series of scenarios in videotape format which present culturally based background information for use in negotiation training. The scenarios accompany "What to Say and How to Say It" (Qiu 1990), a text for students of advanced Chinese designed to familiarize the readers with the proper use of expressions, not only in everday situations but in more complex ones, such as correct ways to praise, complain, criticize, discuss art, religion, war, marriage, and business negotiations. The book evaluation gave rise to a need for developing video scenarios combining true-to-life cultural situations and new language material into scenes to be used as models for illustrating the correct communication techniques. In addition, the scenes provide a starting point for students to begin interacting among themselves.

The scenarios may be open ended situations, first showing how to interact and then offering a problem or conflict for further continuation. The instructor functions as a facilitator, mainly to keep the simulation running smoothly. Decisions are made by the participants and are predetermined by the amount of resources available.

The following is an example of a cross-cultural negotiation simulation where two countries are involved in forming a joint venture. The original material is in Chinese (Qiu 1990).

— *Description of the scenario*. A Chinese computer company is negotiating with a large American computer company to establish a joint venture in Shanghai. The problems:

a. Representatives from the American company are dissatisfied with the location for the future site of the company. They feel that is too far from the city center and the investment climate is not what was expected. They hope there is a possibility for a municipality near Shanghai to provide a location closer to the center of the city.

b. How can the investment climate be improved? For example, how to cut through the red tape and formalities, how to improve conditions for transportation, and how to simplify customs procedures. Given a map and a video scenario, both sides announce their position and begin the discussion.

— *New expressions for use in the situation*. The expressions provided include those for making agreements, disagreeing, arguing, making excuses, and making assumptions.

— *Videotape scenario*. The scenario is of an informal meeting held at a Chinese computer company.

— *Activities involving participants*. These include reading relevant articles and textbooks, making notes and summarizing points, preparing questions, answering the questions posed in the videotape, reporting main points to the team at meetings, and preparing new rounds for negotiation.

It was found that this type of video scenario for business/advanced language and situational practice is effective in improving communication and thinking skills.

Conclusion

Participants at ISAGA were enthusiastic about the simple and interesting way in which the Chinese games and simulations clarified basic concepts and presented cultural situations. It was surprising that the materials demonstrated were received with interest both by those who had little knowledge of Chinese and by those who were familiar with the language.

It must be kept in mind that the uniqueness of each language, the level of instruction, and the content objectives pose both advantages and constraints in simulation/game development. But, there is no doubt about the value of simulation/games as a communication and learning tool for foreign language and cross-cultural instruction. (For further discussion, see Crookall and Oxford 1990). These kinds of studies and simulation/game use and development are just beginning in China. We shall continue to develop simulation/games to improve not only language skills but also cross-cultural understanding.

Acknowledgments

Domo, Domo, Domo! We wish to thank the ISAGA organizers, especially Professor Kiyoshi Arai, for helping with the international arrangements and all the supporters of the conference. Without your help we could not have attended.

References

Crookall D, Oxford R (eds) (1990) Simulation, gaming, and language learning. Newburg House/Heinle & Heinle, Boston

Qiu, Zhipu (1990) What to say and how to say it. Nanjing University, China

Rivers WM, Temperley M (1978) A practical guide to the teaching of English as a second or foreign language. Oxford University Press, Oxfrod, UK

Qiu Zhipu is one of the few professors in China researching and developing simulation games. He was a Fulbright scholar at the University of Pittsburgh 1981–82, visiting Research Professor at Middlebury College Vermont, and Exchange Professor at Oberlin College, Ohio. He is Professor of Linguistics, Director of Research of Applied Linguistics at Nanjing University and Professor-in-Charge of the Chinese programs at Nanjing/Johns Hopkins Center (1986–88). In 1984 he was elected one of the top 25 teachers of Chinese by the Chinese Ministry of Education. He received a Chinese National Science Foundation Award for writing the first dictionary covering the lexical differences between Mainland China and Taiwan. In the last 10 years he has published nine books. He is a violin maker, writer, photographer, inventor and cultural expert.

Joanne Velan Dunn, Instructional Designer, has been involved with game design since graduate school when she was a conference organizer for ISAGA in 1974 in Pittsburgh. She has worked in China as a consultant over the past 9 years. She received the Norman Linck Award for outstanding contributions to the field of Educational Communications and Technology and the AECT award for building American/Chinese relations. She is an artist and interested in visual and creative thinking.

Training International Commercial Negotiators Through Simulation

José Pavis[1]

Abstract. International trade is spreading fast in every corner of the world. Face-to-face, personal communication between individuals from divergent cultural backgrounds, using different communicative styles, is now a very common phenomenon. How much potential exchange of goods and services is lost due to miscommunication with the other trading partner? A great deal. We need to develop the human resources required to conduct successful negotiations with people from foreign cultures. A negotiation simulation (the CROSS-CULTURAL NEGOTIATING GAME) with export managers from Australia, Saudi Arabia, France, China, and Brazil was designed with the dual purpose of (1) yielding data as close as possible to "real" cross-cultural business negotiation discourse lending itself to systematic analysis and (2) providing material for training future export managers in "the art" of negotiating cross-culturally.

Key words　communicative styles; cross-cultural communication; intercultural effectiveness; negotiations; simulation

What happens when a Japanese interacts verbally and non-verbally with an American, a Korean, or a Zimbabwean? How likely are they to understand (or misunderstand) each other if they do not take their different cultural backgrounds into account? What goes on when on the streets of Windoek, Namibia, a *herero* woman tries to sell a carved soapstone to a passing Japanese tourist? What goes on when a team of high-powered US executives negotiate the sale of aircraft equipment to members of a Chinese trade commission in Shanghai? It would be very valuable to observe and record "live" authentic interactions of this kind, but for obvious reasons this is not possible. Simulating such situations with authentic cross-cultural participants is the best alternative. In this article, I will report on some findings about cross-cultural commercial negotiations which export managers from Australia, China, France, Saudi Arabia, and Brazil carried out in front of three video cameras. The simulation which I designed (the CROSS-CULTURAL NEGOTIATION SIMULATION) was aimed primarily at providing me with data relevant to the study of com-

---

[1] Department of French Studies, Sydney University, Sydney, NSW 2006 Australia; phones 2 692 2381 (w) 2 665 0712 (h); facsimile 2 692 4757 (w) 2 664 2018 (h)

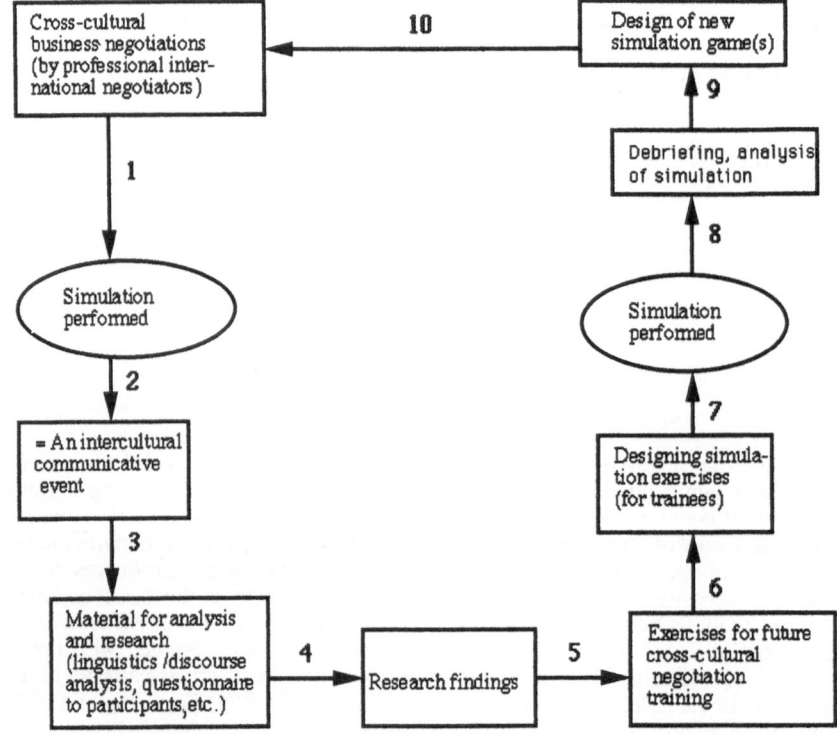

Fig. 1. Overview of the simulation project

mercial negotiations, and subsequently analyzed as an intercultural communicative event (Steps 1–4 in Fig. 1).

Some Definitions

It is important to have a conceptual grasp of the situation which brings commercial negotiators together. Let us examine how some key words are defined.

Negotiating/Bargaining

The following are the minimal features defining a negotiating/bargaining relationship: (1) Two or more parties are involved, (2) there must be a conflict of
interest between the two parties, (3) parties are joined together in a voluntary
relationship (they can push for several possible agreements; they always have
the option to leave the relationship), (4) the activity is aimed at finding an
agreement to the issue(s), and (5) it involves the exchange of offers and
counter-offers in a sequential (not simultaneous) manner.

Negotiating can be seen as a game—a conversational game. The above
statements constitute the rules of the game. These rules are universal and

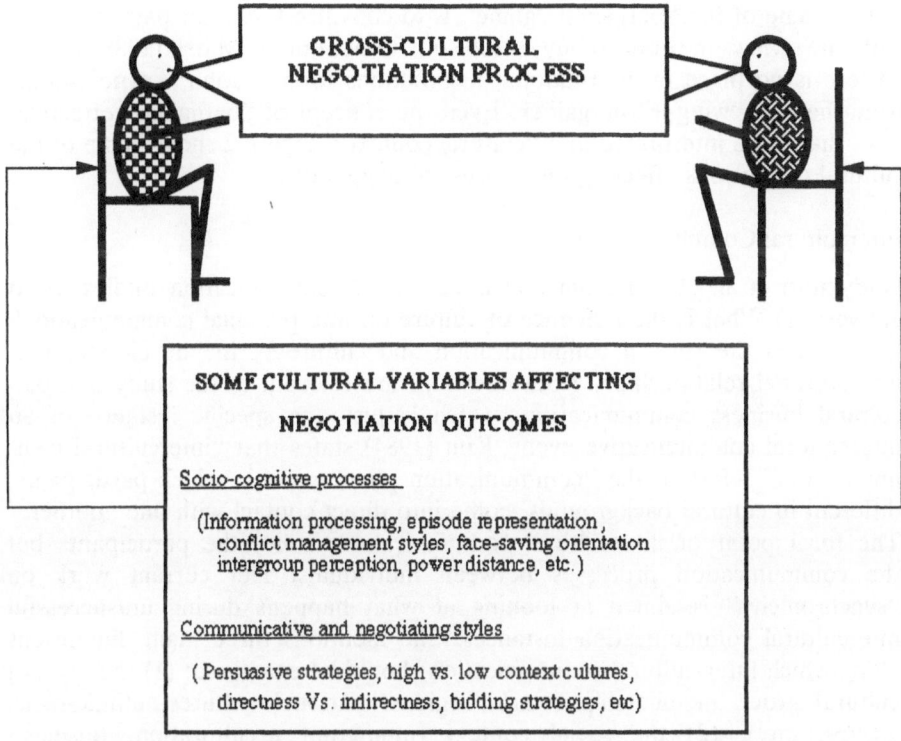

Fig. 2. Some cultural dimensions affecting the negotiating process

cannot be broken. But, as we shall see, in the actual conduct of cross-cultural negotiations, variability rules outnumber these few fixed, rigid rules. Four sets of variables can be said to influence the outcome of a negotiation: (1) the structural context within which bargaining occurs, (2) the behavioral predispositions of the parties involved, (3) the nature and underlying characteristics of the bargainers' interdependence, and (4) the use of social influence and influence strategies in bargaining. For each of these dimensions, "cultural variabilities" will have to be taken into account; negotiators from Australia and Saudi Arabia for instance, have a different mental representation of what is appropriate bargaining behavior. Each bargaining relationship can be characterized by a specific power relationship. This is true for any negotiation context, but "power distance," individual relations to power, "face-saving" and "face-threatening" strategies—in other words "conflict styles"—vary according to cultures. As Ting-Toomey (1988) writes:

> While face is a transcultural concept that governs the active negotiation processes in all cultures, the nuances and subtleties that attach to different facets of "face work management" vary from one culture to the next.

The name of the "bargaining game" is to convince the other party to agree with one's views in order to buy or sell something, but, as Marsh (1988) warns: "There is no place in international negotiations for the John Wayne, shoot-from-the-hip, "wing-it" bargainer. Even the concept of "bargaining effectiveness" has to be interpreted in a "cultural context." Figure 2 shows some of the cultural dimensions affecting the process of negotiating.

Intercultural Communication

Gudykunst et al. (1988) claim that intercultural communication studies try to answer: (1) What is the influence of culture on interpersonal communication? (2) What is the role of communication and culture in the development of interpersonal relationships? Both questions are relevant to the study of cross-cultural business communication, which is just one specific instance of an intercultural communicative event. Kim (1984) states that "intercultural communication" refers to the "communication phenomena in which participants, different in cultural backgrounds, come into direct contact with one another." The focal point of intercultural communication is *not* the participants but the communication processes between individuals. Her current work on "synchronicity" is aimed at looking at what happens during un/successful intercultural communication instances. She mentions three main dimensions along which inter-cultural communication should be examined: (1) the level of cultural group membership (world regions, national cultures, ethnic-racial groups, etc.), (2) the social context (immigrant acculturation, business-organizational contacts, etc.) and (3) the communication channel (interpersonal, media, etc.) (Kim 1984). Nishida (1985) offers the following definition of intercultural communication competence: "The ability to speak a foreign language in an appropriate manner and to demostrate a *knowledge of appropriate communicative behavior in a given situation* in order to *interact effectively* with people from different cultures" (my italics). This tacit knowledge of appropriateness in the use of language is something that we tend to take for granted in the use of our native languages; it is something which comes close to defining the concept of communication style.

Culturally Determined Communicative Styles

Cultures differ to the extent that they create unique forms of human interaction. In addition to providing the rules, schemas, scripts, and values used in communication, cultures—at a very basic level—define the logic of communication itself. Studying the communicative styles favored by various cultures is a prerequisite to the study of intercultural communication. It implies asking: (1) What goals of action are most valued by the culture? (2) What type of communicative strategies are deemed most appropriate for the accomplishment of particular goals? (3) What is the impact of culture on communication? (4) How can we define communication as a set of constraints imposed by cultural rules? According to Gudykunst et al. (1988): "Children do not learn language per se; rather, they learn the various patterns and styles of language

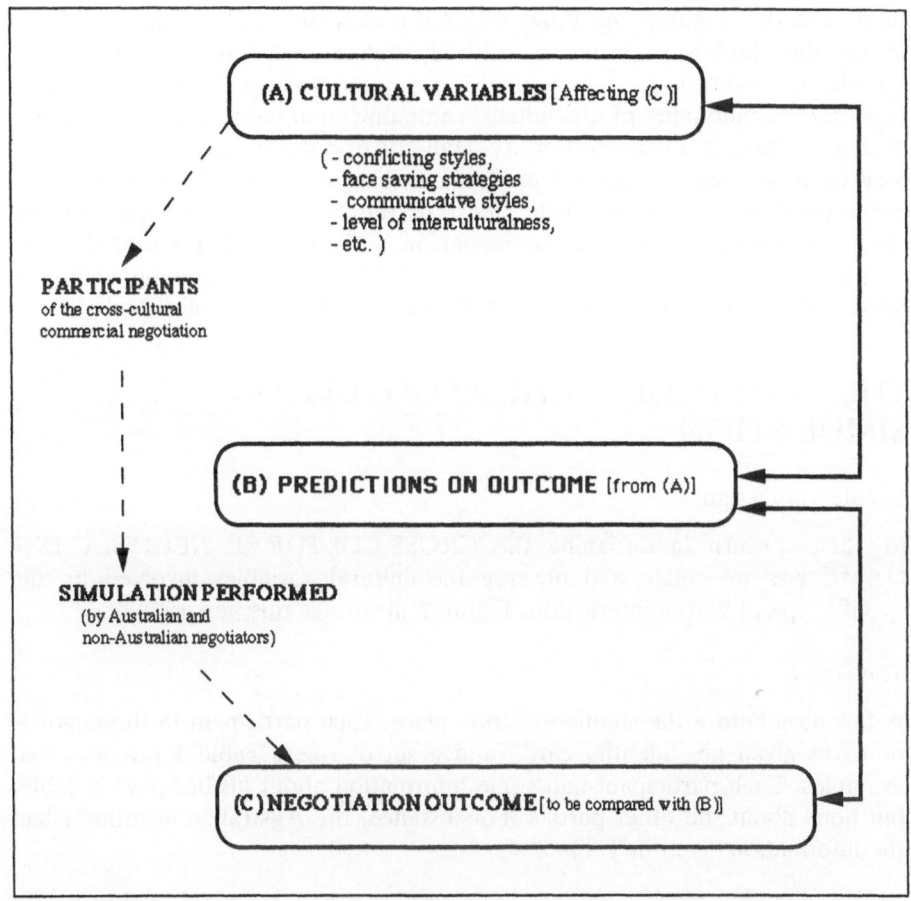

Fig. 3. Goals of the simulation. 1. The simulation was used as a heuristic device to find culture-specific bargaining strategies. 2. It was expected that the following questions would be raised: (a) What kind of communication processes went on during the simulation? (b) What were the participants' mutual perceptions of attitudes, motives, purposes, and feelings? (c) How can we contrast the cultural characteristics of the participants? (d) Which cultural variables affected the outcome of negotiations? (e) How can we set these features against established and normative characteristics of action seen from the perspective of both participants, evaluated against both communities? (f) What was perceived to count as good, effective behavior by the participants? Questions (b) and (f) were investigated through the feedback obtained by questionnaires filled out by the participants, questions (a), (c), (d), and (e) were investigated through my own external analysis. I can only give a brief overview of some of my findings in this paper

interaction that enable them to function as competent communicators in different situational contexts." Nomura and Barnlundt (1983) write: "In learning to interact with others, the child acquires more than vocabulary and grammar; s/he acquires values assumptions that are implicit in these patterns of interaction and that regulate meanings within a particular social system." In other words, the child is acquiring a culturally determined communicative style. Culture is manifested in individuals' communicative patterns. This intersubjective commonality of individuals' communication patterns is what characterizes a culture as a collective entity. Hall (1976) differentiates cultures on the basis of the communication that predominates in the culture. A "high-context" communication is one in which "most of the information is either in the physical context or internalized in the person, while very little is encoded in the explicit, transmitted part of the message." A "low-context" communication is one in which "the mass of the information is vested in the explicit code."

THE CROSS-CULTURAL NEGOTIATION SIMULATION

Simulation Design

My first concern in designing the CROSS-CULTURAL NEGOTIATING GAME was to isolate and observe the cultural variables involved in this specific type of verbal interaction. Figure 3 illustrates this perspective.

The Script

A few days before the simulation took place, each participant in the negotiation was given an "identity card" and a set of rules. Table 1 provides two examples. Each participant had some information about his/her pay-off table, but none about the other party's. For instance, the Australian negotiator had the information given in Table 2.

The Participants

The participants in the CROSS-CULTURAL NEGOTIATION GAME were experienced business people (export managers and general managers of companies involved in international trade in the "real world"). Four dyadic groups were set up: 1st Australian/French, 2nd Australian/Brazilian, 3rd Australian/ Saudi Arabian, 4th Australian/Chinese. Table 3 provides further biographical information about them.

Some Findings

Testing the validity of the script was the first goal. In so far as it allowed the four dyadic groups to come quickly to the point of their encounter, that is, to sell, buy, trade, discuss contracts, and bargain, the script proved successful. Even though they all stemmed from quasi-identical scenarios, all four negotiations yielded quite different results, as Table 4 shows.

Faced with the heterogeneity of these results, the first question to be asked is whether we are dealing with individual variations or "culturally determined" differences. To try to answer this question, I chose to work at two levels: (a)

Table 1. Two examples of identity cards

Saudi Arabian negotiator

- Your name is Maroon Tajer
- Your trading partner's name is John Tradewell
- You work for a company importing and exporting food products all over the world
- You want to import Vegemite[a] from Australia
- You have heard there is a good potential market for it in this part of the world
- You are hoping to negotiate and sign a "packaged deal" contract
- However, three items must be agreed on before such a contract can be signed:
    a. The renewal of an export contract of Saudi dates (due to expire in 6 months)
    b. The Vegemite deal
    c. Setting up a feasibility study for a 10-year joint venture concerning the manufacturing of honey
- A contract can be signed only if *all* of the above are settled
- Each of the issues represents a specific value for your company (see your "pay off" table, below)
- These values may be completely different from those of the other party

Australian negotiator

- Your name is John Tradewell
- Your trading partner's name is Maroon Tajer
- You work for a company exporting and importing food products all over the world
- You want to export Vegemite to the Middle East
- You have heard there is a good potential market for it in this part of the world
- You are hoping to negotiate and sign a "packaged deal" contract
- However, three items must be agreed on before such a contract can be signed:
    a. The renewal of an import contract of Saudi dates (due to expire in 6 months)
    b. The Vegemite deal
    c. Establishing a feasibility study for a 10-year joint venture concerning the manufacturing of honey
- A contract can be signed only if *all* of the above are settled
- Each of the issues represents a specific value for your company (See your "pay off" table below)
- These values may be completely different from those of the other party

[a] Vegemite is a yeast-extract spread used by practically all Australian familes. It is a very "culturally-marked" product (Pavis 1991)

self-analysis by participants (questionnaires and debriefing) and (b) external analysis.

Debriefing and Questionnaires

I had two ways of getting information from the participants themselves: (1) a first post-session questionnaire filled out immediately after the simulation, and (2) a second (more detailed) post-session questionnaire filled out 1 month later, after having watched the video replay of the interaction Some of the salient findings that came out from the first source of enquiry are given in the Appendix. This type of data is very useful for initiating discussion amongst trainees (steps 5 and 6 in Fig. 1).

Table 2. The pay-off table of the Australian negotiator

A = Renewal of the contract (due to expire in 6 months)
B = The Vegemite deal
C = Accepting involvement in research for the feasibility study on a possible joint
      venture.

Australian negotiator			Non-Australian negotiator		
A	B	C	A	B	C
By not renew- ing, you lose	If no agreement to export Vegemite is made, you lose	If no agreement to do the feasibility study is made, you win	?	?	?
300 pts Profit if you renew for	700 pts Profit you make if you export over	100 pts			

1 year =   0 pt   1 year = 750 pts
2 years =   0 pt   2 years = 670 pts
3 years =   30 pts   3 years = 590 pts
4 years = 400 pts   4 years =   40 pts
5 years = 700 pts   5 years =   0 pt

Table 3. Some biographical information about the participants

Biographical information

- Name . . . .
- Nationality . . . 4 Australians, 4 non-Australians (1 French, 1 Brazilian, 1 Saudi Arabian, 1 Chinese)
- Name of the company you work for . . . .
- Position held 4 = export managers
                       4 = managing directors involved in international business
- Number of years of experience in international trade: 3, 4, 5, 7, 10, 15, 20, 22 years
- With which countries do you have most experience?
  Brazil/Portugal (1), South-East Asia (4), Middle East (2), China, US (2), Europe (2)
- In which country do you find it particularly difficult to negotiate?
  Italy, Japan (2), Asia, China, Turkey, Australia, "All different"
- Why?
  (1) Reluctance to compromise, acknowledge shortcomings (in Japan), (2) different in commercial cultures (Asia), (3) high cost, (4) not interested/ignorance of world markets (Australia), (5) culture and language (China), (6) very tough negotiators (in Turkey), (7) you have to be very cautious not to offend people (in Japan), (8) contracts never seem to mean too much (in Italy)

External Analysis

The data consisted of four dyadic negotiation interactions of 45 min each, recorded on videotape. This yielded a very large amount of material. For the purpose of this paper, I will concentrate only on one aspect of the communica-

Table 4. Overview of the negotiation processes

	1st Australian French	2nd Australian Brazilian	3rd Australian Saudi Arabian	4th Australian Chinese
Negotiating one item at a time?	Yes	Yes	No	No
Sequence of issues discussed	1. Renewal 2. Vegemite 3. Feasibility study	1. Renewal 2. Vegemite 3. Feasibility study	1. Renewal 2. Vegemite 3. Feasibility study 4. Renewal 5. Vegemite 6. Renewal 7. Feasibility study	1. Renewal 2. Vegemite 3. Feasibility study 4. Vegemite 5. Feasibility study
Percentage of time spent on each issue	Renewal = 10% Vegemite = 55% Feasibility study = 30%	Renewal = 33% Vegemite = 33% Feasibility study = 22%	Renewal = 11% Vegemite = 39% Feasibility study = 44%	Renewal = 15% Vegemite = 32% Feasibility study = 38%
Bidding early?	No (long discussion before bidding)	Yes (almost straight away by Brazilian)	No	No
Who does initial bidding?	Always French (only once asking the other party to "make an offer")	Always Brazilian	Always Saudi (unless asking the other party to "make an offer")	Always Australian

Table 5. Bidding pattern in two negotiation interactions: General comments

Australian/French	*Australian/Saudi Arabian*
— Long discussion before first bidding	— Fast pace, not getting stuck on one issue
— Bidding mixed with lengthy arguments to convince other party	— Trade-off always on Saudi negotiator's mind
— "Head-on collision," polite but quite confrontational; no one seems to be prepared to give in	— Cyclic pattern of items discussed
— "Yeah but" type of discussion	— Pace set by Saudi Arabian
— Not too many figures used	

Table 6. Concessions made in two negotiation interactions

French/Australian		*Saudi Arabian/Australian*	
Renewal contract		**Renewal contract**	
French	No initial bid	Saudi Arabian	Initial bid = 1 year
	Final agreement = 5 years		Final agreement = 2 years
Australian	Initial = 5 years	Australian	Initial bid = 4 years
	Final agreement = 5 years		Final agreement = 2 years
Vegemite		**Vegemite**	
French	Initial bid = 6/7 years	Saudi Arabian	Initial bid = 5 years
	Final agreement = 5 years		Final agreement = 3 years
Australian	No initial bid	Australian	Initial bid = 1 year
	Final agreement = 5 years		Final agreement = 3 years

tive event: the bidding strategies used by the negotiators (first offer, second offer, size of initial concession, who starts the bidding, when (early or after getting some information from the other party), what sort of counter-bidding follows, being ambitious, "bluff" bidding, strategies used to accept or refuse the other party's biddings). Let us contrast two bidding patterns in two sets of negotiations. Tables 5 and 6 contrast the French/Australian dyad and the Saudi Arabian/Australian dyad.

Overall Progression of the Bidding Interaction

Tables 7 and 8 provide an outline of the content analysis of the bidding interaction. It is clear that the participants of both dyads resorted to very different strategies. Both the French and the Saudi negotiators managed to impose their negotiating styles summed up in the section "general comments" above.

Obviously, viewing the video of the actual bargaining interaction would make this contrastive analysis more meaningful. But the main point to be made here is that different strategies were used by each of the participants. In the post-session questionnaire, the Saudi negotiator confirmed that he was deliberately using a bargaining strategy very much favored by Saudi businessmen. According to him shifting the topic, keeping a fast pace of negotiation,

Table 7. French/Australian bidding pattern

Bidding pattern			Comments
1. Renewal contract			
Initial general discussion ("for" vs "against" long contract)			= Observation round
Initial bid	by	Australian	French asked Australian to make an offer
	for	5 years	She answers the question: "How many years"
Reaction of French		Suggests moving to second item	Australian's bidding = information collecting for French (?)
			No counter-bidding
2. Vegemite			
Initial bid	by	French	Starts high
	for	6/7 years	
Reaction of Australian		Argues to shorten	Long discussion
			No counter-bidding
Second bid	by	French	
	for	5 years at least	
Reaction of Australian		Argues to shorten	Long discussion
Third bid	by	French	We renew the mineral water if you accept 5 years for vegemite
	for	trade-off	
Reaction of Australian		Counterbid	Non-explicit counterbid
		Accepts trade-off principle but for a shorter period of time	No precise time given
Reaction of French		Argues not to shorten Lengthy discussion	
Mutual agreement		Australian gives in Agrees to 5 years for both issues	"I suppose we could talk all afternoon about this"

slightly confusing the negotiating partner are all part of the favorite Saudi ploys when it comes to discussing business transactions. By contrast, the French negotiator was making genuine efforts to convince his trading partner about the universal validity of his arguments. He went as far as drawing the profit curve that the Australian company would make if only his trading partner could agree to his plan. His cartesian style of negotiation was at work.

A lot can be learned from analyzing such strategies in detail. Debriefing this interaction with potential commercial negotiators can be a useful way of developing one's "intercultural effectiveness." Analysis of bidding strategies is only one of the areas of communication strategies that deserves attention. The following issues would be expected to come up: (1) cross-cultural pragmatic contrastive analysis of speech acts typical of a negotiation setting (requesting, asking questions, refusing, accepting, etc.), (2) content analysis of the argument used to encourage or discourage the buying or selling of a specific

Table 8. Saudi Arabian/Australian bidding pattern

Bidding pattern			Comments
1. Renewal contract (1)			
Initial bid	by	Saudi Arabian	"The only thing I can do for
	for	1 year	you . . ."
Counter-bid	by	Australian	
	for	4 years	
Reaction of		Let's put it aside	Not getting stuck in this issue.
Saudi		Let's discuss your other points	Leaves it open for later
Arabian			negotiation ("if we can
			get . . . we might negotiate").
			Bidding to test the ground
2. Vegemite (1)			
Initial bid	by	Saudi Arabian	This follows a long exchange of
	for	5 years	information
Reaction of		No counter-bid	Non-confrontational
Australian		Says what he could do to help	Indirectness
		with promotion	
Reaction of		"Well, let's look at it, maybe	Same strategy: trade-off
Saudi		we can make some money,	Non-confrontational
Arabian		we can compensate the dates	Keeping the negotiation
		for you (. . .) Any other	flowing, changing the issue
		subject you would like to	
		discuss?"	
3. Honey/Feasibility Study (1)			
Initial bid	by	Saudi Arabian	Assertive. Asks for counter-bid
	for	10 years	("How do you feel your
			company can co-operate with
			us in this field?")
Reaction of		We'd like to know a little bit	Non-confrontational
Australian		more (. . .)	Indirectness, argues to
		"I'd like to ask you a few	postpone
		questions"	
Second bid	by	Saudi Arabian = "Be our	
		agent"	
Reaction of		Let's look at global contract	Subtle, indirect refusal
Australian			

Continued

product, (3) strategies used not to lose face in front of the other negotiator, and (4) levels of self-disclosure among participants.

Conclusion

I have shown one example of difference in negotiating styles that can be attributed to different cultural backgrounds of the participants. The study involved in this paper concentrated exclusively on steps 1, 2, 3, and 4 of Fig. 1. Training future export managers and negotiators would constitute the obvious next step. After having observed with the trainees how the simulation was

Table 8. *Continued*

Bidding pattern			Comments
4. Renewal contract (2)			
Initial bid	by	Australian	
	for	4 years (2 years at same price)	
Counter-bid	by	Saudi Arabian	"This is the best I can do"
	for	1st year = same price 2nd year = +5%	(. . .) "You have to understand my position" "You have to give some, I have to give some"
Reaction of Australian		"Are we able to clinch a 4-year deal with a negotiated price?"	Not too sure what is going on
Reaction of Saudi Arabian		"That's fair" . . . Shall we move to Vegemite?"	Not clear about what has been agreed on
5. Vegemite 2			
Initial bid	by	Australian	As in previous "round"
	for	1 year initially (Then long term contract)	
Reaction of Saudi Arabian		"No way. No way, because John. . . . (. . .) John, that doesn't suit us . . ."	
Counter-bid	by	Saudi Arabian	
	for	Minimum of 3 years up to 5 years	
Reaction of Australian		"OK, let's go back and look at the position we are in at the moment"	Tries to have a global agreement Long joint summary and agreement elaborated together

conducted by negotiators from different cultural backgrounds, it would make sense to prepare them for face-to-face negotiations with people using different communicative and bargaining strategies. Making use of games and simulation exercises would undoubtedly constitute the most efficient way of achieving the experiential learning required to prepare oneself for the "real world," where mismanagement of communication can be a very costly matter. The first type of game would involve imposing different sets of communicative rules on participants in a cross-cultural negotiation exercise. The variables which could be manipulated might include: directness vs indirectness, bidding pattens, knowledge (or ignorance) of rules, and norms of interaction governing commercial negotiations in various cultures. For each variable under scrutiny, specific games could be designed and played in front of an "observer" whose comments would be useful during the debriefing phase.

Giving a golden rule about communication in general, Berger (1986) writes: "When we open our mouth to communicate with other persons, we would be wise to keep in mind that what we utter to these persons is based upon an incomplete understanding of who they are and who we are." This advice is even more valid when individuals from different cultural backgrounds are involved in commercial negotiations. Communicators with people from different cultures should also bear in mind Szalay's advice (1981): "The more we consider our views and experiences to be absolute and universal, the less prepared we are to deal with people who have different backgrounds, experiences, cultures, and therefore different views of the universe." I am convinced that the CROSS-CULTURAL NEGOTIATION GAEM can help negotiators become better inter-cultural communicators and can contribute to developing what Smith (1987) calls the "five senses"—a sense of self, a sense of the other, a sense of the relationship between self and other, a sense of setting and situation, and a sense of goal and objective.

Appendix

General Assessment of the Negotiation Simulation

Assessing your bargaining performance
1. Are you satisfied with the result of the negotiation? .... Yes = 88%, No = 22%
2. Did you expect to make more profit than you actually did? .... Yes = 50%, No = 50%
3. Did you expect to reach agreement before the minimal time (45 min)? .... Yes = 50%, No = 50%
4. Would you say that generally speaking, you were controlling the process of the negotiations? .... Yes = 50%, No = 50%
5. Did you have the feeling that your trading partner trusted you? .... Yes = 100%, No = 0%

Your general bargaining philosophy: Do you agree with the following statements?
1. The outcome of commercial negotiations depends more on outside factors than on the negotiators' skills to agree with each other .... Yes = 25%, No = 63%, Both = 12%
2. A good bargainer is someone who is concerned about maximizing the profit s/he has made at the end of the negotiation...................... Yes = 25%, No = 75%
3. A successful negotiator is one ending with distributive justice and equity for the other party .... Yes = 100%, No = 0%
4. A good bargainer is someone who is concerned about making a greater profit than his/her trading partner .... Yes = 25%, No = 75%
5. I was concerned about not hurting the feelings of the other person .... Yes = 62%, No = 37%
6. I was concerned and made some effort so that my trading partner liked me .... Yes = 87%, No = 22%

7. Negotiators in general tend to behave more
   cooperatively with negotiators from the same
   national background ....   Yes = 62%, No = 25%
                                                   Possibly = 12%

8. I tend to be more cooperative when I feel my
   trading partner is being cooperative ....   Yes = 100%, No = 0%

9. I am more a risk taker than a cautious
   bargainer..................................   Risk taker = 12%
                                                   Cautious bargainer = 75%
                                                   In between = 12%

Assessing your trading partner's strategies

1. In your opinion, what was the most important
   issue for your trading partner? ....   Right guess: Australian = 1,
                                                                Non-Australian = 3
                                                   Wrong guess: Australian = 3,
                                                                Non-Australian = 1

2. Do you think your trading partner knew which
   issue was the most important to you?..........   Yes = 62%, No = 37%

3. Did you think your trading partner had a better
   bargaining position than yours? ....   Yes = 0%, No = 100%

4. Were you satisfied by the concessions your
   trading partner was making to your demands? ..   Yes = 87%, No = 12%

5. Did you find that your trading partner was
   making unrealistic demands on you? ....   Yes = 0%, No = 100%

Assessing your trading partner

1. As a whole, did you find it difficult and tedious
   to negotiate with your trading partner? ....   Yes = 0%, No = 100%

2. Would you characterize your trading partner's
   attitude as cooperative or competitive? ....   Cooperative = 100%
                                                   Competitive = 0%

3. Do you think your trading partner was satisfied
   with the result of the negotiation? ....   Yes = 87%, No = 12%

4. Do you think your trading partner was expecting
   to reach agreement before the minimal time
   (45 mins)? ....   Yes = 62%, No = 25%
                                                   Don't know = 12%

5. Would you categorize your trading partner as a
   risk taker or as a cautious bargainer?..........   Risk taker = 87%
                                                   Cautious bargainer = 0%
                                                   In between = 12%

6. Did you ever feel threatened by your trading
   partner's bargaining strategies?..............   Yes = 12%, No = 87%

7. On the whole, did you trust your trading
   partner? ....   Yes = 100%, No = 0%

8. Were you concerned about keeping your trading
   partner's feelings satisfied? ....   Yes = 62%, No = 25%,
                                                   Yes but not totally = 12%

9. Do you assume that how s/he feels is completely
   up to him/her? ....   Yes = 12%, No = 87%

Communication problems
1. Did you find trading partner was clear about communicating his/her position to you? .... Yes = 100%, No = 0%
2. At times, did you have difficulties understanding what your trading partner was saying? .... Yes = 0%, No = 100%
3. At times, did you feel your trading partner had difficulties understanding what you were saying? Yes = 12%, No = 87%
4. Do you think the negotiation would have been easier with somebody from your own culture? .. Yes = 12%, No = 87%

References

Berger C (1986) Social cognition and intergroup communication. In: Gudykunst W (ed) Intergroup communication. Edward Arnold, London

Gudykunst W, Kim, Y (1988) Theories in intercultural communication. Sage, Newbury Park, CA

Gudykunst W, Ting-Toomey S, Chua E (1988) Culture and interpersonal communication. Sage, Beverly Hills

Hall ET (1976) Beyond culture. Double Bay, New York

Kim YY (1984) Communication and cross-cultural adaptation. Multilingual Matters, Clevedon, Avon

Marsh R (1988) The Japanese negotiator. Kodansha International, Tokyo

Nishida H (1985) Japanese intercultural communication competence and cross-cultural adjustment. International Journal of Intercultural Relations 9:247–270

Nomura N, Barnlundt J (1983) Patterns of interpersonal criticism in Japan and the United States. International Journal of Intercultural Relations 7:1–18

Pavis (1991) Exporting Vegemite to Venezuela: Marketing or cultural challenge? Cross-Culture 3(1)

Smith L (1987) Discourse across cultures. Prentice Hall, London

Szalay L (1981) Intercultural communication: A process model. International Journal of Intercultural Relations 5:133–146

Ting-Toomey S (1988) Intercultural conflict styles: A face negotiation theory. In: Kim Y, Gudykunst W (eds) Theories in intercultural communication. Sage, Newbury Park, CA

José Pavis teaches French at Sydney University, Australia. He has previously taught in France, the US, and Southern Africa. He is more and more convinced that language learning goes hand in hand with developing awareness of both one's native culture and the target culture. Most of his publications are related to the field of foreign language pedagogy but his latest research deals with cross-cultural communication and business negotiation.

Global Modelling:
A Game-Generating Game

Elizabeth M. Christopher[1]

Abstract. One method of studying a society is to model some of its processes. A GLOBAL MODEL GAME (GMG) was constructed for ISAGA '91 to illustrate similarities and differences between societies. Three critical aspects of social organization were identified in the GMG for comparison and contrast: the laws and norms of a society, the division of labor in society, and society's assumptions about the nature of family life and individual goals and aspirations. In the ISAGA'91 workshop, the GMG was reconstructed into a game-generating game. This paper describes the principles on which the original model was designed and how workshop members converted it into a gaming device with potential to provide entertaining and thought-provoking insights into players' own and others' cultures.

Key words game-generating game; global modelling

The model was designed as a cube (Fig. 1). Its three planes represent respectively "Individual and Family," "Society," and "Work." These were argued by the designer to be the three essential components of all cultures; and that every society is essentially unique in the ways in which it shapes individual and family lives, imposes social norms and values, and divides its labor. The model was given the tentative title "Building a Global Future: The Model Begins with Me." The intention was to understand the bases and biases of one's own culture so one can begin to transcend it in order to enter a more global mindset.

Each plane of the cube model was then subdivided into four. It can be seen from Fig. 1 that these divisions create eight small cubes from the big one. The intention was to identify specific variables: for example, the cube representing "Marriage and children," "Task," and "Education" might serve to raise questions about the education of women and the position of married women in the workforce of any given society. This then might become a topic for cross-cultural comparison.

[1] Charles Sturt University, School of Business and Public Administration, Bathurst, NSW 2795, Australia

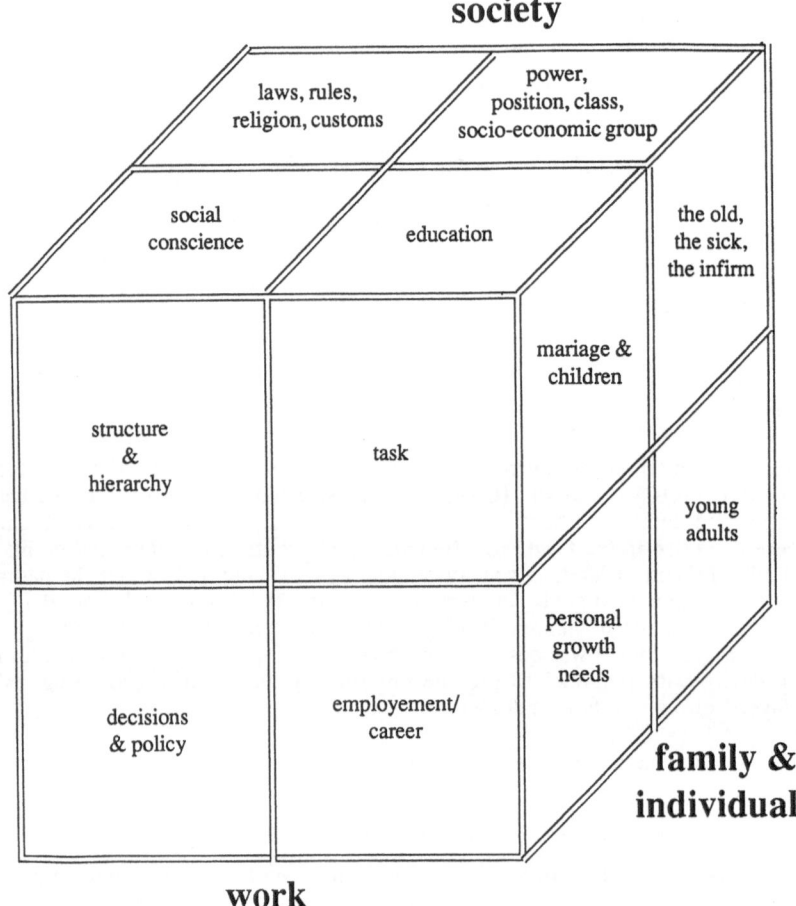

Fig. 1. Building a global future: The model begins with me

This was the stage of development of the GLOBAL MODEL when it was presented to a group of workshop participants at the ISAGA'91 international congress in Japan, July 1991. It was presented as a set of eight cardboard cubes, labelled as shown in Fig. 1.

Participants expressed interest in the model, immediately began to take it apart and started to play with the eight small cubes. One person observed that three sides of each cube were untitled. The designer replied that although labelling the blank sides would emphasize the planar nature of the model (since each title would be the same as that on its opposite side, see Fig. 1.), this would be redundant. At this point, a participant suggested that the cubes could be used as in a game of dice, and someone else remarked that the blank sides could be filled in by the players. From then on, the session became a brainstorm from which the designer was able to construct the game which is described below. Her sincere thanks go to the organizers of ISAGA'91 and all the participants in her workshop. The session was a powerful example of the kind of game synergism that is generated uniquely by ISAGA's international conferences.

Global Modelling: A Game-Generating Game

Objectives

To see the organization and operations of one's own culture with fresh eyes, and to gain new insights into others' cultural behavior in order to improve cross-cultural understanding and communication.

Time Required

About 2 hours.

Number of Players

Eight people for each model set. Others can act as observers, or the game director may want to adapt the game slightly to accommodate fewer or more people.

Materials Required

Copy of GLOBAL MODEL (as Fig. 1); basket or other receptacle to hold the model cubes; four sheets of thin cardboard for each model set, preferably in different colors; felt-tipped pens of various colors; scissors; pencil and ruler; tape.

Advance Preparation

As shown in Fig. 2, cut out eight shapes for each model set from four sheets of cardboard (if there are more than eight players, at least two model sets will be needed). Each of the eight Cubes has labels written as follows:

— Cube 1
  Side a—Individual and family: marriage and children
  Side b—Society: education
  Side c—Work: task (conditions of labor)
— Cube 2
  Side a—Individual and family: the old, sick, and infirm
  Side b—Society: power and position (class, socio-economic grouping)
  Side c—Work: task (conditions of labor)
— Cube 3
  Side a—Individual and family: the old, sick, and infirm
  Side b—Society: rules and religion (laws, customs, tradition)
  Side c—Work: hierarchy (organizational structure)

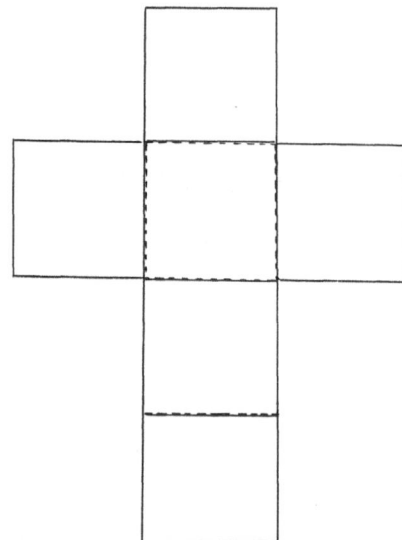

Fig. 2. Each side of a cube should be about 5 cm square

— Cube 4
   Side a—Individual and family: marriage and children
   Side b—Society: health and welfare (social conscience)
   Side c—Work: hierarchy (organizational structure)
— Cube 5
   Side a—Individual and family: personal goals (growth needs)
   Side b—Society: education
   Side c—Work: employment (jobs and careers)
— Cube 6
   Side a—Individual and family: young adults
   Side b—Society: power and position (class, socio-economic groupings)
   Side c—Work: employment (jobs and careers)
— Cube 7
   Side a—Individual and family: young adults
   Side b—Society: rules and religion (laws, customs, traditions)
   Side c—Work: organizational policy
— Cube 8
   Side a—Individual and family: personal goals
   Side b—Society: health and welfare
   Side c—Work: organizational policy

How the Game Is Played

Players sit in a circle and the complete model is placed on display in front of them. The game director talks for a few minutes about the model, explaining its rationale as described above, and initiating a short discussion of about

10 min. The model is then broken into its eight component cubes which are placed in a basket and passed around the circle. Players in turn draw a cube. Each player then writes the following on the three blank sides of the cube:

— Side 1—the word "game", or "Role-play" or some other activity such as "case study"
— Side 2—a few words to describe some kind of scenario, such as "at a bus stop," "in the bank," or "at work"
— Side 3—two roles to match the scenario.

When all players have inscribed the three sides of their cube, all cubes are replaced at random in the basket. The basket is then passed around the circle. Players, without looking, pick a cube each.

Players now form dyads. In turn, a member of each dyad throws the two cubes like dice onto the playing area (table, carpet, etc., depending on how the group is seated). This sequence continues until one dyad has succeeded in throwing an activity, a scenario, two roles, and an aspect of society, work, and individual. This dyad then drops out of the circle and begins to work on the designated activity while the other players continue throwing in turn until all have completed the task. After an appropriate planning time, each dyad presents a role-play, game, case study, and so on to the rest of the group.

Examples

One dyad might eventually throw the following:

— Marriage and children
— Health and welfare
— How people are hired and fired
— Activity: negotiation role-play
— Scenario: company boardroom
— Roles: employer and worker representatives

This combination might result in a negotiation to illustrate the power of workers' representatives (unions) to protect their members from harsh employment conditions such as extremely long working hours away from their homes and families, or dismissal.

Another example might be:

— Young adults
— Power and position
— Organizational structures
— Activity: case study
— Scenario: a personnel office
— Roles: personnel officer, young employee

This combination of variables might result in the creation of a case study to focus attention, for example, on career paths for young people after accepting employment within an organizational structure; or acceptance or rejection, within the organizational structure, of the principle of promotion on merit regardless of age or length of employment within the organization.

Suggestions for Debriefing

Initiate a discussion after each activity by asking players to discuss social, work, and family variables in their own cultures, then to compare their responses with those of others in the group. To play the GLOBAL MODEL GAME, it is not necessary to assemble a multinational group of players: there are many "cultural" differences among groups within a society. For example, in many cultures, women are socialized so differently from men that it can be said they possess different cultures. Employers and employees often operate from different assumptions about the nature of work. The GLOBAL MODEL GAME helps to identify and explore such differences.

Dr. Elizabeth Christopher is a faculty member in the School of Business, Charles Sturt University, New South Wales, Australia. She is known internationally as a designer of games and simulations for management training and has published two books in collaboration with Larry Smith, Director of the Institute of Culture and Communication, East-West Center, Honolulu: *Leadership training through gaming: Power, people and problem solving* and *Negotiation training through gaming: Strategies, tactics, and manoeuvres,* (Kogan Page Nichols). She is a member of the Editorial Board of *Simulation & Gaming: An International Journal of Theory, Design, and Research.*

Simulation Structure and Attitude Change in a High Technology Culture

T.R. Schumacher[1]

Abstract. The organizational culture of a software engineering firm was studied to discover inconsistencies between the existing culture and expressed ideals. The gap between existing and ideal was measured using an attitude questionnaire drawn from statements made during interviews with employees. A simulation game was created to induce attitude change toward the ideals. Subjects sequentially experienced two contrasting simulated cultures as they made decisions as a department manager in a computerized business simulation. The simulation is written in HyperCard and each subject made decisions on a Macintosh computer. The subjects participated in four–person teams, whose computers were linked and shared data. During the ten training classes, 122 employees attended one class and played one of three simulation versions. A control group of 42 people, and 97 of the 122 who played the simulation, completed pre- and post-simulation questionnaires. There was extensive attitude change in the treatment groups.

Key words   attitude change; HyperCard; organizational culture; simulation

Research on the effectiveness of simulations in causing attitude change has focused on the comparison of simulations to traditional education methods such as lectures. The results of these studies provide mixed evidence to support the claim that simulations induce attitude change. Pierfy (1977), Bredemeier and Greenblat (1981), and Butler et al. (1988) have reviewed this research and have raised a number of methodological concerns. It was believed that these concerns could be addressed by conducting a study outside the classroom.

The present study was conducted in a rapidly growing software company. The simulation content addresses how rapid growth influences the organizational culture. It focuses on attitudes about the balance between short-term, task-directed work and long-term, culture-development work. Subjects were employees who were involved daily with the issues in the simulation.

---

[1] Systems Science PhD Program, Portland State University, P.O. Box 751, Portland, OR 97207 US; phones (503) 725-4960 (w) (503) 775-9378 (h)

Methodological Concerns of Previous Studies

Pierfy (1977) describes several methodological weaknesses that are commonly found in studies of simulation effectiveness, including (1) the potential for experimenter bias, (2) teacher effectiveness as an uncontrolled variable, and (3) compressed research cycles where the simulation experience and post-simulation measurements have been conducted within a single class session, typically 60–90 min. The potential for a significant demand effect when teachers are giving questionnaires to their students has also been a concern.

The majority of studies have been conduced using students as subjects and classes to define treatment groups. Butler et al. (1988) consider this a "quasi-experimental design." They surveyed articles in *Simulation & Games: An International Journal of Theory, Design, and Research* and the ABSEL proceedings for an 8 year period. They reported that only 14% of the journal articles (total $n = 170$) and less than 3% of the proceedings articles (total $n = 458$) met their definition of "experimental research designs." These concerns were addressed in the current study as described in the methodology section.

An additional question for any research on attitude change concerns the centrality within subjects' cognitive structure of the attitudes that are expected to change. Those researching attitude change believe that attitudes are resistant to change, in part, because they are embedded in a system of other attitudes and beliefs. The more central an attitude is in a person's cognitive structure, the more resilient that attitude will be because of the support it has from related cognitive elements. For example, it would be far more difficult to demonstrate a change in economists' attitudes about free trade than, say, a change in the attitudes of college freshmen about a hypothetical "Lord South" whom they had never heard of prior to playing a simulation of his life. Clearly, free trade is more tightly embedded in economists' cognitive structures. The attitudes targeted for change in the present study were central to the organizational culture. (Space limitations prohibit a detailed discussion.)

One reason that simulations can be effective tools for inducing attitude change is that they may provide a means of overcoming this "systemic resilience." Play is a special type of cognitive processing in which the adoption of new and artificial realities is expected and done with less evaluation than usually occurs (Bruner et al. 1976). The temporary adoption of a role when playing a simulation can be an opportunity for participants' to learn a new system of attitudes and beliefs which supports a changed attitude. After the simulation experience, the new attitude then has its own basis for systemic resilience. When simulation participants encounter a situation outside the simulation in which the newly learned attitude or role has some utility they may incorporate these into daily life.

Simulation Structure

The simulation structure developed for this study was a blend of two previous simulations. Fletcher (1971) suggested providing a recording mechanism and multiple plays of a simulation. This allows participants to use acquired knowledge to improve their performance. Fletcher assumes that the content to be conveyed by the simulation is contained in a successful strategy for the participant in the simulated environment. The CARIBOU HUNTING GAME which Fletcher used had no roles for participants to adopt as they played the simulation. Fletcher did not suggest a strategy to participants. Instead he allowed each to develop their own, with the result that after seven plays most were still improving their strategies.

In BAFA BAFA, participants are divided into two groups and each group learns a simulated culture (Alpha or Beta) to play when the two groups interact. There is no strategy to play in BAFA BAFA, and there are no winners or losers at the end of the simulation. Discovery of the contrasting culture is an intended outcome. Even with well conducted debriefings some players remain ignorant of the cultural role they did not play.

The WINNING AT DESIGN AUTOMATION (WADA) simulation developed in this study combines the multiple plays and recording mechanism from Fletcher and two cultural roles from BAFA BAFA. Players are instructed to adopt the Alpha role for the first game and Beta for the second. These roles are contrasting strategies for resource expenditure, and different outcomes will be obtained from the computer with the use of these different strategies. Players therefore experience two contrasting role/strategies with the optimal strategy consisting of the attitudes toward which change is intended—the ideals of the organization's culture.

Beyond these elements of gross structure, the WADA simulation includes a number of features to create strong role identification with the two cultural roles. Livingston and Kidder (1973) and Williams (1986, 1987) have suggested role identification may be a factor in causing attitude change during simulation experience. Features included to support the adoption of the cultural roles are presented below.

Role Identification Features of the WADA Simulation

1. Players read the five-page cultural role description at the beginning of each game. Each team discussed the role and then made a 5-min presentation summarizing the major points using a flip chart.
2. Multiple choice questions were asked about the cultural descriptions as a part of the simulation.
3. Colored hats with the name of each cultural role were presented to players along with instructions to adopt that role during the preparations for each play.
4. The cultural roles were carefully constructed to contain attitudes and beliefs that contrast along the dimensions where attitude change is intended. The Beta culture represents the ideals toward which players were expected to move and it was constructed to be consistent with attitudes and beliefs of the existing culture that were not targeted for change by the simulation.

Fig. 1. Study variables (*boxes*) and hypotheses (*arrows*)

This was expected to increase the player's incorporation of this role into their cognitions of normal work life.

5. The design of the model that drives the behavior of the WADA simulation, as well as much of the content of the simulation, was based upon information gathered in the interviews. Terms incorporated in the simulation (e.g., vision, bureaucracy, small company atmosphere) were frequently used by employees during the interviews. One example of the extent of the use of the interview data is that the primary resource in the simulation is time—and not money as in most business simulations—because of the importance of time in the organizational culture. Another example is that employees believe that a strong vision is important to the success of the company. The cultural roles incorporate this belief and players experience this in the contrast between the two games.

Hypotheses

Several simulation versions were created by including or excluding some of the role identification features described above. Simulation structure (version) was an independent variable and groups of questionnaire items represented the dependent variables in this study as shown in Fig. 1.

It was hypothesized (H_1) that different versions of the simulation would create different degrees of identification with the role of the ideal culture (Beta). It was also hypothesized (H_2) that greater role identification would produce greater attitude change toward that ideal.

Methodology

More than 200 ethnographic interviews were conducted with employees to determine issues in the existing culture and ideals that relate to those issues. The principle investigator explained that the purpose of the interviews and questionnaires was to collect data for his doctoral dissertation on corporate culture.

Statements made during the interviews were used to develop a questionnaire that was sent to a random sample of employees 2 months prior to the first simulation class. Questionnaires were numbered so that individual's pre- and post-simulation measurements could be compared. Employees who returned this questionnaire were assigned to one of three treatment conditions or a control group. Those assigned to a treatment group were invited to attend a 1-day class on corporate culture conducted by the vice president of Human Resources. Subjects were not informed that the class was related to the dissertation study.

The classes lasted 6–8 h and the primary activity was playing the simulation. Players were organized into teams of four which competed to earn the most points. Each person took on the role (on a Macintosh computer) of the manager of a department making decisions about how to spend their time—the primary resource in the simulation. The four computers of each team were networked to pass information.

There were during-simulation and post-simulation discussions that totaled about 2 h in each class. A list of questions addressing the relationships in the simulation was provided to the facilitator to guide the debriefings. Subjects were asked to compare and contrast the fictitious Alpha and Beta cultural roles to the organization where they worked. Audio recordings were made of the debriefing discussions.

A total of 122 employees experienced the simulation during ten 1-day classes. Post-simulation questionnaires were sent to each participant 15 days after their simulation class. A sample of participants were interviewed after all questionnaires had been returned and none of these were aware of a connection between the questionnaires and the cultural simulation class.

The Dependent Variables

Individual questionnaire items were grouped according to topic during the design of the questionnaire and cultural roles in order to define the dependent variables. Eight groups with a total of 40 items are discussed here. Cronback's alpha ranged from 0.62 to 0.80 for the eight items indicating an acceptable level of questionnaire reliability.

One issue that emerged early in the interviews was a concern that the organization's culture was changing. People repeatedly stated that the "small company atmosphere" was being lost and that the organization was becoming more "bureaucratic." Most indicated a belief that bureaucracy was the inevitable result of growth though the definition of bureaucracy was vague.

The first dependent variable sought to clarify the definition of bureaucracy. The items in the second variable linked actions that individuals could take to the inhibition of bureaucracy. The third and fourth variables measure opportunities and responsibilities, respectively, for individuals to take the actions described in variable two. Variable five items ask if it is possible that the organization can continue to grow while avoiding bureaucracy, a position opposite to the initial widespread "bureaucratic inevitability" belief. Variable six assesses whether participants believe there is a need to change things in the organization related to the actions in variable two. One of these changes is the formation of "partnerships," a cultural concept addressed by dependent variable number seven.

Table 1. Pre, post, and change scores of the simulation

Treatment group 1 (n = 37)

Dependent variables	No. of items	Pre	Post	Change[c]	t[a]	P
1 Bureaucracy definition	5	5.27	5.73	0.400	2.49	0.0088*
2 Anti-Bureaucracy actions	7	4.86	5.61	0.783	6.38	0.0001*
3 Opportunities to act	4	4.58	4.91	0.362	1.83	0.0379*
4 Responsibility to act	5	5.38	5.87	0.463	4.12	0.0001*
5 Possible to avoid bureaucracy	5	3.62	4.06	0.434	2.91	0.0031*
6 Need for change	6	5.49	6.16	0.673	7.03	0.0001*
7 Partnerships	4	5.82	6.15	0.338	3.16	0.0017*
8[b] Rewards for actions	4	3.94	3.48	−0.410	−2.64	0.0062*

Control group (n = 41)

Dependent variables	No. of items	Pre	Post	Change[c]	t[a]	P
1 Bureaucracy definition	5	5.38	5.32	−0.054	−0.32	0.7493
2 Anti-Bureaucracy actions	7	4.94	5.04	0.101	1.02	0.3162
3 Opportunities to act	4	4.50	4.72	0.211	1.02	0.3135
4 Responsibility to act	5	5.47	5.38	−0.098	−1.12	0.2687
5 Possible to avoid bureaucracy	5	3.57	3.56	−0.005	−0.03	0.9749
6 Need for change	6	5.55	5.78	0.232	2.43	0.0198*
7 Partnerships	4	5.39	5.78	0.369	2.59	0.0133*
8[b] Rewards for actions	4	3.83	3.73	−0.104	−0.619	0.5395

* $P < 0.05$
[a] One tailed paired t-test.
[b] Rewards variable was not targetted for change, see discussion.
[c] Note the change, t, and P values were calculated in the paired t-test routine which eliminates cases when either pre or post score is missing. The average pre and post scores were calculated in separate routines. The change value does not always match the pre and post difference because additional cases may be missing from this calculation.

Variable eight, "Rewards for actions," was not targeted for change in the simulation. It was included to measure attitudes relating actions suggested by the simulation to participants' perceptions of organizational support for those actions. Each asked for an assessment of company rewards, for example, "Mentor Graphics rewards individuals who take risks". Its possible that the organizational culture did not match expressed ideals because the "ideal actions" were not rewarded. If this was true, changing attitudes in some of the items in variables two to five would be more difficult. Variable eight is included in this analysis because there was change in this variable and this is an interesting result for simulation designers.

Results

Different versions of the simulation did produce different levels of role identification, but the highest identification group did not have the greatest attitude change. Hypotheses relating role identification to attitude change were upheld.

Only the treatment group with the greatest attitude change is discussed here. Table 1 presents data for the eight dependent variables for treatment and control groups.

Discussion

There was substantial change in the treatment groups, and participants reported a high level of enjoyment for the simulation exercise. Dependent variable five, measuring whether participants believed it possible for the company to grow and avoid bureaucracy, showed a shift across the midline of the seven point scale. The dependent variable showing the least significant change was "Opportunities to act". In future simulation use the facilitator could modify delivery to emphasize this. In addition, company management may wish to make changes in order to encourage the exercise of greater power by employees.

The question of the duration of effect is often raised. Extensive change was measured three weeks after the simulation classes. This duration of effect allowed substantial opportunity for the organization to reward and reinforce the change that the simulation introduced. In the longer term, it will be organizational support for the changed attitudes, and not "duration of simulation effect" that determines the survival of those changes.

There was change in the control group for two of the eight dependent variables discussed. Possible sources for that change have been identified. Several managers who were contacted after the study indicated that they had discussed their simulation experience in staff meetings because of its perceived value. One manager admitted taking copies of the cultural roles and sharing them with others even though an announcement had been made in each simulation class not to discuss it for fear of spoiling the experience for others who had yet to play. In fact this announcement and the efforts to collect all materials at the end of each class were done to avoid "leakage" to employees in the control group. The change in the control group was much less significant than that in the treatment group.

Items assessing whether certain actions were rewarded by the company measured significant change in the treatment groups. There was no intent to change these. Rather it was a priori considered extremely difficult to change an assessment of the company's behavior as the simulation did not address this. Apparently the experience of the Beta culture (which has high levels of reward giving) was strong enough for participants to reevaluate their assessment of the company where they worked. In the post-simulation questionnaire they judged the level of rewards as lower than they had initially indicated. Unintended change is an issue that simulation designers need to consider.

The evidence reported here indicates that simulations can be effective instruments to introduce attitude change in organizations. Attention to the socially constructed organizational realities—including incorporation of culturally relevant language and concepts—as well as careful design of the simulation experience are considered the keys to successful sessions.

References

BAFA BAFA. Shirts G (1977) Simile II (218 Twelfth Street, P.O. Box 910, Del Mar, CA 92014 US)

Bredemeier M, Greenblat CS (1981) The educational effectiveness of simulation games. Simulation & Games: An International Journal of Theory, Design, and Research 12(3):307–332

Bruner J, Jolly A, Sylva K (1976) Play—Its role in development and evolution. Penguin, Harmondsworth

Butler R, Markulis P, Strang D (1988) Where are we? An analysis of the methods and focus of the research on simulation gaming. Simulation & Games: An Internation Journal of Theory, Design, and Research 19(1):3–26

Fletcher JL (1971) Evaluation of learning in two social studies simulation games. Simulation & Games: An International Journal of Theory, Design, and Research 2(3):259–286

Livingston S, Kidder S (1973) Role identification and game structure: Effects on political activities. Simulation & Games: An International Journal of Theory, Design, and Research 4(2):131–144

Pierfy D (1977) Comparative simulation game research: Stumbling blocks and stepping stones. Simulation & Games: An International Journal of Theory, Design, and Research 8(3):255–268

Williams R (1986) Changing attitudes with "identification theory". Simulation & Games: An International Journal of Theory, Design, and Research 17(1):25–44

Williams R (1987) Levels of identification as a predictor of attitude change. Simulation & Games: An International Journal of Theory, Design, and Research 18(4):471–487

T.R. Schumacher is a doctoral candidate in the Systems Science PhD program at Portland State University. The research reported here is a portion of his dissertation.

Section 3
Environmental and Developmental Issues

Introducing Gaming-Simulations into the Planning Process in a Developing Country

Cathy S. Greenblat[1]

Abstract. Under the sponsorship of UNDP (the United Nations Development Program), a multistage process of introducing gaming-simulation into various stages of the planning process in Ghana has been underway for the past year. The consultation was invited in response to the pressing demand for a larger number of trained development planners in both the public and private sectors in Ghana. Over the course of two 1-month visits, this author worked with staff at the Department of Planning, Faculty of Environmental and Development Studies, University of Science and Technology (UST), Kumasi, with several major goals connected with the overall aim of strengthening the capacity for using and designing gaming-simulations. As a result of the success of this first consultation, a second stage has been proposed. This paper will elaborate on the first phase of the project.

Key words design; gaming-simulation; ISAGA; national and regional planning; planning

In 1988 in Ghana, PNDC (Provisional National Defence Council) Law 207 shifted budgeting and decision-making responsibilities to the 110 districts in the country. This action created a pressing need for a larger number of trained development planners in both the public and private sectors. The United Nations Development Programme (UNDP) created several projects designed to strengthen development planning in Ghana. In the past year, two of these UNDP projects have included components with the goals of increasing and improving the utilization of gaming-simulation in both the training of planners and in the planning process itself. This paper presents an overview of the first of the two gaming projects supported by UNDP, and a discussion of the factors contributing to its success.

Gaming-Simulations for the Training of Planners

The first project, which was completed in February 1991, was a training enterprise. A major beneficiary of the UNDP assistance was the Department

---
[1] Department of Sociology, Rutgers University, New Brunswick, NJ 08903 US; phones 908-932-0498 (w), 212-353-1990 (h); facsimile 212-228-0576

of Planning, Faculty of Environmental and Development Studies, at the University of Science and Technology (UST) in Kumasi. Consultants have been or will be brought in to enhance the department faculty's usage of a variety of teaching techniques (flip charting, video presentation, etc.), to develop an M.Sc. program in social planning, and to strengthen staff capacity in aspects of economic planning.

The department had a history of using gaming-simulations. Several faculty members, including the two prior chairs, had participated in a workshop conducted by Richard Duke in 1978 in Tema and Kumasi. Their enthusiasm for gaming remained high over the subsequent years. Following the Tema/Kumasi workshop, several games were designed and subsequently employed in classroom teaching. Unfortunately, with the serious economic difficulties in Ghana in the early 1980s, many of the supplies for the games that had been left in the university after the workshop and those designed in its aftermath could not be replenished. The continuing enthusiasm for the technique, however, led the department in 1990 to request a consultation for the purpose of upgrading faculty competence in the use and design of gaming-simulations for planning education. In lieu of the 3-month visit initially requested of me, a two-visit program was adopted. In May I made a preliminary 3-week visit to the university; then I returned in January 1991 for a month, accompanied by Diana Shannon, whose visit extended an additional 2 weeks.

There were four main goals for the project:

1. Obtain and demonstrate existing gaming-simulations dealing with development planning to department members.
2. Train department staff to run those gaming-simulations they find relevant to their teaching, for use with undergraduate and graduate students and with workshop or short course participants.
3. Instruct interested department members in game design techniques through a program of design seminars.
4. Produce prototypes of one or two simulations as a result of collaborative efforts of the consultants and staff members who participate in the design seminars and ongoing efforts.

Preliminary Steps: Familiarization and Needs Assessment

In order to advise the department on which existing games might be employed, which might be modified, and what new games might be developed, it was first necessary to understand the present planning situation in Ghana and the UST Planning Department's role in preparing staff for various levels of work in the new situation. I also considered it essential that we become familiar with the salient aspects of social, political, and economic life in Ghana, to better understand the opportunities and constraints that operate for planners in that country, and to understand the factors that must be built into any simulation of the planning process.

The preliminary work of familiarization with the department's activities in teaching and training, an assessment of their needs and resources, and of the specific content areas in which gaming-simulation might be employed, was undertaken during my 3-week visit in May–June 1990. An active program of meetings was established for me, and I was given the opportunity to review numerous written materials. At the UST, individual and collective meetings were held with most members of the teaching faculty and with a number of students. A variety of opportunities arose to meet with others in Kumasi whose experiences and insights were valuable in the formulation of ideas about the use of gaming-simulations for training planners in Ghana, and in understanding some of the dimensions of change in Ghana. In Accra on the first and last days of the consultation, brief meetings were held with several UNDP staff members and with planners in agencies and in the NDPC (National Development Planning Commission) and with faculty members in the Sociology Department of the University of Ghana at Legon.

Many useful documents were provided to me for review during my stay in Kumasi, including descriptions of courses and workshops, social science analyses of contemporary Ghanaian life, and undergraduate and graduate student papers. The latter were useful both for their substance and as indicators of the quality of analysis and writing of the students in the department. Finally, I had several opportunities to increase my understanding of life and issues in Ghana through visits to a village and to several other cities in the center of the country, as well as through reading Ghanaian newspapers and novels.

Goals 1 and 2: Gaming-Simulation Demonstrations and Development of Faculty Skill in Running Them

While the primary aim of my first visit was to do an assessment of needs and resources for gaming-simulations in the planning department, it was anticipated that during this visit some gaming-simulations could be introduced. I prepared a list of several I anticipated would be useful; UNDP ordered them and had them shipped to Accra prior to my arrival. As a result of the discussions I held in the first visit, and of inquiries made subsequently, a few additional items were purchased for the second trip. The materials brought and used in the project included all those listed in the References section of this paper.

The format for the demonstration of games during both visits was the same:

1. A presentation was made to the department members on the nature and focus of each of the available simulations. The staff then made decisions about which ones they wanted to see, and which group of students might participate. Faculty members were also urged to attend these sessions as participants, and they regularly offered to do so.
2. One or more demonstrations of each gaming-simulation were given, run by myself (in May) and later by Diana and me (in January). In subsequent runs

some faculty members took over some operator's functions, under our supervision.

3. Follow-up discussion focused on how the materials might be employed in existing courses and in anticipated workshops. In addition, we reviewed the operators' instructions with those faculty members who indicated an interest in running the game in the future, assuring that there was at least one "expert" per game in the department.

Goals 3 and 4: Design Seminar and Design of New Gaming-Simulations

Recognizing that few simulations existed which adequately treated the development planning topics of concern to the UST faculty, their interest in design was high. The first visit revealed that there were both important resources and important limitations for such an endeavor.

While the faculty was strong in terms of their knowledge and their interest in the design workshop, it was also obvious that they were extremely busy; preparation for courses and other activities push the staff heavily, and many have outside responsibilities (e.g., to the NDPC, to national agencies), limiting their time even further. It thus seemed unrealistic to expect they would make a major time commitment to this enterprise, and gaming-simulation design is time-consuming. I did not anticipate any difficulty attaining good attendance at a 2- or 3-day design seminar, but I doubted that any of the staff would have even half-time available during my January visit to devote to design. It was important, however, that they were willing to lend their classes for field-testing and for modifying any games that were to be designed.

Two steps were taken to deal with this situation. First, all faculty members were given a copy of *Designing Games and Simulations* (Greenblat 1987) and were urged to read it before my January return. Second, and more importantly, I was convinced that the success of the project was dependent upon my returning with a colleague, to permit continuous working on the design enterprise while faculty were intermittently available. My proposal that Diana Shannon accompany me, was approved by both the department and by UNDP in New York, and thus two of us arrived in Kumasi in early January.

In terms of materials, the situation was a mixed one. Gaming-simulation design requires databases, case studies, and library materials. While the library holdings in the planning department were found to be very limited in number, I was very impressed with the research studies that had been undertaken and were available to the game designer through the department library. Indeed, many of the topics proposed for design could be addressed quite easily using available information, with little time required to gather supplementary data.

The two areas of greatest need were supplies and a microcomputer. Although I gathered that the situation even a few years ago was considerably worse, there remained a shortage of such items as paper, card stock, tape, plastic bags and cups, push-pins, transparencies, and so on, which were needed for design. I thus requested a small budget so I could assemble a box of such supplies to bring in January. A Macintosh equipped with word processing, graphic arts, and desktop publishing capabilities had been requested from the outset of my talks with UNDP; the utility of this machine became even more obvious during my first trip. Unfortunately, while the request for the supplies budget was granted, the Macintosh request was caught in a series of project-funding problems, and was not met. At the last minute, Diana and I brought computers with us to do the design work, but we were not able to train department faculty to use them in the way we had anticipated.

In my first report I had proposed that a 3-day design workshop be run for staff. This did not prove feasible, given the already heavy commitments they had. Instead, 3 afternoons were devoted to design seminars, and individuals worked with us as their time permitted. By the time we began the seminars, many of them had read quite a bit of the design book, and hence were prepared to discuss the process in a fuller fashion. The seminar was organized in the same fashion as the book, proceeding through the steps of design as outlined there.

The first seminar focused on decisions about specification of objectives and parameters. During meetings held in May, a number of topics had been proposed for the design of new simulations (supposing that no existing ones were found which could be used or modified). The list included a Planning Agency simulation, a Market Planning simulation, a Regional Planning—Resource Needs Assessment simulation, a simulation for Regional Planning—Impact Assessment, an Urban Settlement simulation, and a set of Community Participation simulations. Following my urging that we needed to identify a topic for the continuing efforts, these were discussed and it was agreed that attention would be focused on two: (1) an elaboration of the proposal to design an agency game, but with emphasis to be placed on interagency activities, and (2) a gaming-simulation that could be used to help elicit citizen participation in discussions on problems, goals, and projects. During a follow-up session, it was suggested that the agency game would be useful not only for students, but at the district level and at the national level if it highlighted the sources of inefficiency that currently hamper interagency coordination, in particular, the lack of horizontal communication. Careful specifications were made for this simulation, which eventually became the Planning Agency simulation, PLANet (Table 1). Major efforts were devoted to the design of PLANet when it was suggested by several staff members that the second project could be accomplished satisfactorily through a modification of the AT ISSUE! simulation. The specification of objectives and parameters developed and subsequently used was as in Table 1.

The second seminar focused on the agency simulation, as department members helped us to build the conceptual model, detailing their understanding of the interagency network and the sources of inefficiency. By the time of the third seminar, Diana and I had done a great deal of the third step in design—decisions about the style and form of representation—in intensive work sessions with one another and sometimes with available faculty members. We

Table 1. PLANet: Specification of objectives and parameters

Subject matter:
The gaming-simulation will model the problems of agency and network operation in a situation in which inefficiences are endemic. The sources of such inefficiencies are at the system (network) level, in terms of such factors as overlapping functions, insufficient materials, and so on; at the agency level, where operating procedures are counterproductive; and at the individual level, where such factors as lack of skills, nepotism, and holding of second jobs impair efficient task performance in the work realm.

The gaming-simulation should be sufficiently abstract to be relevant to analysis of a wide range of situations in which a set of organizations, agencies, or ministries have not only their own goals, but are expected to contribute to the achievement of a larger systemic goal.

Purpose:
Through participation in the gaming-simulation, players will experience firsthand both the difficulties of completing agency tasks and the failure to achieve network success. The impediments to agency and network success created by both agency and individual inefficiencies should also alert participants to the manner and extent to which their failures contribute to the failures of others. Analysis of the various factors that create and foster inefficiency in the game should lead to greater ability to recognize such factors in the real-world situation. Analysis of which factors are more subject to change will lead to discussion of how to implement a reduction of real-world agency/network inefficiencies.

Potential participants:
The gaming-simulation will be designed to be played by several types of participants:
— Ministry personnel
— Agency personnel at the regional or district or local level
— Students of planning, management, sociology, etc.

Potential operators:
The gaming-simulation should require little in the way of operator skills. Operators would be faculty members or others in charge of workshops or classes.

Context of use:
The gaming-simulation should be usable in a workshop or classroom context. Hence it will be designed to require approximately 3–4 h of play and discussion/analysis. It is anticipated that such time must be consecutive (i.e., a 3–4 h block will be required). It will be designed to accomodate a group of 25–35 participants, which should be appropriate for many workshops and most classes. If there are a larger number of participants to be accomodated, sequential or parallel sessions would have to be run.

Resources of users:
Since the gaming-simulation is aimed at audiences in developing countries, the resources that can be devoted to its use will often be limited. It should be designed so as not to require computer facilities, and should utilize materials that are relatively easily and inexpensively acquired locally. To the extent possible, materials should be reusable, requiring minimal photocopying or replacement of materials.

had, in addition, begun work on the fourth step: construction of a prototype. This seminar, then, involved faculty members in not only making some decisions, but also in reviewing the decisions and the materials developed on the basis of them.

At each of these sessions, a large bulletin board was used to provide participants with an overview of the entire set of decisions, and of what had been done since the last general meeting. Through these sessions, I believe that a number of staff members gained insight into the design process; because of their lack of time to work with us on a sustained basis, however, they are not at the stage of being able to easily design new materials without guidance.

As indicated above, Diana and I were able to do a great deal on the design of the prototype agency game PLANet. A complete set of preliminary materials was designed and field-tested with students before Diana's departure in mid-February. Modifications based on these field tests are still needed before the materials are in a final form, and an operator's manual remains to be prepared, but everyone agreed that it had considerable potential and quality. Both faculty members at UST and staff of the NDPC believe that it will be a useful tool for classes and workshops, as well as for training sessions with district level agency staff and with ministerial staff at the national level.

My experience elsewhere in Africa suggests that PLANet models problems that are not unique to Ghana, but that it shows problems that appear in varying degrees of severity in many other developing countries. I thus believe that it would be a useful tool in a wide range of contexts. We deliberately designed it in a modular fashion to allow relatively easy alteration for somewhat different audiences. The "big picture" on which agency role-players must work is taken from a planning problem in agricultural economics; modifications that made it focus on another topic such as health care management or higher education management should not be difficult.

Diana also worked with Professor Kofi Tamakloe on the second design project that had been proposed, that is, a tool for eliciting the views of community members at district level meetings. This involved use of the AT ISSUE! concept, but with considerable modification and the ultimate addition of graphic elements.

Success of the Training Project

Both the students and the staff were extremely energetic in their participation in all gaming-simulations, and many of the staff members commented that the students were much more active in the discussions that ensued after play than they usually are. Staff were highly enthusiastic about a number of these gaming-simulations, and particularly spoke of continuing use of BAFA-BAFA, CAPJEFOS, the HEX Game, STRATAGEM, and AT ISSUE! I quite firmly believe that they will do so, in their classes and in the coming workshops they will run. Discussions were also held at NDPC about their using gaming-simulations as a component of their nationwide training programs. A UST faculty member on 3-year assignment to NDPC had been able to attend most of the demonstrations at UST; he developed a firm grasp of them, could run most of them, and was eager to do so. As we left, he was working with the

NDPC training director on modes of integrating the gaming-simulations into the 12 workshops currently being developed.

I believe that there are several factors that account for the success of the first project. First, there was a high level of interest in gaming within the department, and the consultation was supported enthusiastically by both the two prior chairs, and the current chair, Professor Tamakloe. This assured us maximum access to staff, documents, and other resources. Second, there was already a commitment in the department to an active approach to learning. While most classes are presented through a lecture mode, various department publications urge that "doubting, questioning, exploring, enquiring are the essence of good teaching and learning." The commitment to active learning is evidenced in the workshop courses that comprise a significant element of the curriculum at both the undergraduate and graduate levels. These workshops are well-conceived and very dynamic; they exemplify the ideas of learning through direct inquiry and through cooperative team effort—aspects clearly related to the gaming enterprise.

Third, splitting the consultation into two separate parts allowed me to do a needs assessment and then to plan carefully for the subsequent stage, which could not have been done had I made one long visit. It also allowed me to recognize the essential need for an associate if the design phase was to be successfully executed. The fourth factor, then, was the presence of a team of two consultants for the second phase. Diana and I had both worked in a variety of international contexts, and had worked together on numerous occasions before, permitting a smooth division of labor and maximum effectiveness.

The major impediment to our goals was the relative lack of availability of existing materials. Before the first visit and during the period between the two visits, I extensively searched for gaming-simulations that focused on (or at least were usable in) a developing country context, and that dealt with one or more of the following:

— Urban and regional planning
— Management of renewable resources
— Development economics
— Planning unit/agency management and operations
— Community participation in decision making
— Coordination of planning efforts at the local/regional/national levels
— Games that assist participants in skills of finance, budgeting, accounting, record keeping, time management, priority determination, communication, and impact assessment

I located very few. The other problem, fortunately recognized during the first visit and so dealt with in some ways before the second, was the very limited availability of supplies and such services as photocopying. These put constraints on the use of some materials that require extensive "refilling" between runs, and shaped a number of decisions made during the design of PLANet.

I hope that the description of this experience will prove of use to others who seek to transfer knowledge and skills about gaming to other locations.

References

AT ISSUE! In: Duke RD, Greenblat CS (1979) Game-generating games. Sage, Newbury Park, CA

BAFA-BAFA. Shirts G (1973) Simile II, Del Mar (PO Box 910, Del Mar, CA 92014 USA)

CAPJEFOS: A SIMULATION OF VILLAGE LIFE. Greenblat CS et al. (1985) CSG Enterprises, Princeton, NJ (301 N. Harrison St., Suite 156, Princeton, NJ 08540 USA)

COMMONS GAME. Powers R et al. (1982) Oceanside, OR (PO Box 307, Oceanside, OR 97134 USA)

CRITICAL ENCOUNTERS. Greenblat CS (1990) CSG Enterprises, Princeton, NJ (301 N. Harrison St., Suite 156, Princeton, NJ 08540)

FISH BANKS LTD. Meadows D, Fiddiman T, Shannon D (1990) IPSSR, Durham, NH (University of New Hampshire, Durham, NH 03824-3577 USA)

Greenblat CS (1987) Designing Games and Simulations. Sage, Newbury Park, CA

Greenblat CS, Duke RD (1974) Principles and Practices of Gaming-Simulation. Sage, Newbury Park, CA

HEX. Duke RD et al. (1983) Ann Arbor, MI (Multilogue, Inc., 329 Park Lake Ave., Ann Arbor, MI 48103 USA)

RESPONSE: THE AFRICAN AIDS PLANNING GAME. Meadows D, Fiddiman T (undated) IPSSR, Durham, NH (University of New Hampshire, Durham, NH 03824-3577 USA)

STARPOWER. Shirts RG (1969) Simile II, Del Mar, CA (PO Box 910, Del Mar, CA 92014, USA)

STRATAGEM. See Stermon JD, Meadows D (1985) STRATEGEM-2: A microcomputer simulation game of the Kondratiev cycle. Simulation & Games: An International Journal of Theory, Design, and Research, 16(2):174–202

Cathy Stein Greenblat is a professor of sociology at Rutgers University. She has designed many successful simulations, has written several books, including *Designing Games and Simulations* (Sage), and has been an executive board member of NASAGA and SAGSET. She is past president of ISAGA and Editor Emeritus of *Simulation & Gaming: An International Journal of Theory, Design, and Research* (Sage).

The NEW COMMONS GAME

Richard B. Powers[1]

Abstract. One purpose of the NEW COMMONS GAME is to demonstrate how Garrett Hardin's "tragedy of the commons" works. Another purpose illustrates how differences in the power to exploit a commons result in feelings of frustration, alienation, and the desire for revolution on the part of players who begin the game disadvantaged. Finally, the game permits disadvantaged players to ask questions of other players thus illustrating the power and limits of publicity to control the greed of the privileged players. Differences between a college student game and a game played with participants at the ISAGA '91 conference showed that students started with a much higher rate of exploitation than the ISAGA players, almost exhausting the resource, in the first part of the game. However, students reached a more stable cooperative exchange than did the ISAGA participants in the latter part of the game. One reason for this difference may be that a real consequence—points toward their grade—was contingent upon the students' performance in the game, while this was not true for the ISAGA players.

Key words conflict; environmental education; simulation/game; tragedy of the commons

An increasing source of conflict in our world arises because of both real and imagined scarcities of resources. For instance, in the Pacific Northwest of the United States, a bitter fight is raging over whether logging or environmental interests will control what happens to the area's remaining old-growth forests. Examples of global resources which are treated as unmanaged commons are all too plentiful and need not be restated here. There is a desperate need to educate our children about the consequences of continuing to allow the world's resources to be exploited primarily to make a small percentage of the exploiters wealthy.

But how do we change the "exploit-to-exhaustion" mentality most effectively? I suggest that simulations/games have at least two advantages over traditional educational approaches in inducing attitudinal and behavioral changes. The first one is intellectual: players comprehend complex dynamic

---

[1] Oregon Peace Institute, Suite 520, 921 S.W. Morrison, Portland, OR, 97205, US; phones 503-228-8563 (w), 503-842-7247 (h)

systems much more quickly and thoroughly by being active participants in the system and discovering how it works for themselves (Abt 1987, Duke 1974, Greenblat 1988). The second advantage, perhaps as important, is that players' experiences in a simulation foster empathy for persons operating in real world systems. Playing a game which models a systemic dilemma such as Hardin's (1968) "tragedy of the commons" makes students much more appreciative of the pressures and temptations of groups or nations to harvest maximum yields of fish year after year. "Blaming the other" as an explanation is not resorted to so easily after you have watched yourself yield to the trapping characteristics of the commons system and become an exploiter.

The original COMMONS GAME has been played effectively with college undergraduates for over 9 years at Utah State University and has demonstrated considerable effectiveness in promoting understanding of the commons dilemma (Kirts et al. 1991, Powers 1985–1986). A new game, based on the original, was developed which, I believe, increases the potential for learning. In the original game a well-rehearsed facilitator could operate effectively with up to 12 players. The new game uses two-person teams and the increase in players approximates the number of students in a classroom which means that an entire class can play the game at once. Second, the speed with which players experience the consequences of their choices has been increased by restructuring the payoff matrices and related costs: collective choices to cooperate or exploit the commons now have twice the effect on the resource. Third, the new game incorporates a power differential, with some teams having as much as three times the ability of other teams to exploit the resource. The differences in power to exploit a commons model the real world, in which people of the developed nations have the money and technology to exploit a global commons, such as the oceans, while peoples of developing nations are either ignored or hired to help in the exploitation of the resource. This power differential may engender feelings of frustration, alienation, and a desire for revolution on the part of players who begin the game in the disadvantaged position, much as it might in the peoples of the developing nations. Finally, the new version provides for some redress for disadvantaged players. They can expose "super exploiters" by playing a disclosure card which allows them to ask another team a question which must be answered truthfully (at least, the rules specify that a question must be answered truthfully). Thus, a team may ask another how many exploitive choices it has made in the last five trials or how many points it currently has (a revealing question). Will the threat of exposure by the less powerful be sufficient to convince the more powerful that it is in everyone's interest to regulate the use of the commons?

Description of the Game

The NEW COMMONS GAME requires about 90 minutes to play and may be played with as few as six players operating as individuals, or as many as 24 players operating in two-person teams. Teams choose one of five colors each trial with the various colors representing cooperation (red), exploitation (green), withdrawal (yellow), punish exploiters (blue) and reward cooperators (orange). Teams make their choices in private behind a shield so that only the combined choice of the group is known. Communication is limited to conferences held at intervals if the majority votes to hold one (and are typically not allowed until at least 10 and as many as 20 trials have been played). Pay (in points) is determined by a payoff matrix which may increase or decrease depending upon what the group does. If collective play is predominantly cooperative, the payoff matrix gradually improves. If play is predominantly exploitive, the matrix gradually grows worse. The power to exploit the commons is not equal—some teams are given a card with a multiplier number (a 2 or 3) and when the numbered card is played, it multiplies the exploitive choice by that number. Other teams, while not having multiplier cards, may play their disclosure card once during any part of the game. The team being asked the question is instructed to answer truthfully.

Teams play for points, which represent units from a renewable resource such as fish in the ocean. It may help some players to think of points as any item of value to them, such as money. With college students, I use points earned in the game as contributing a small amount towards their course grade. However, one should not tell players that the winning team (or teams) will be given prizes, as some adopters have done. This turns the game into a win-lose game and eliminates one of the most important lessons the game has to offer, namely, that all players can win something (even if some win more than others) and

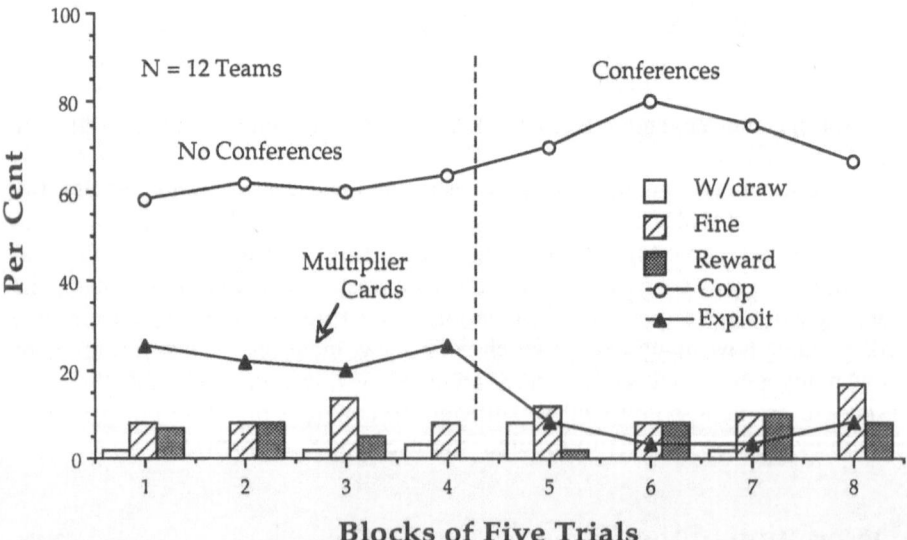

Fig. 1. The percent of the five choices across trials in the game played with participants at the ISAGA'91 conference. Cards which multiplied the power to exploit and cards which permitted questioning of other teams were introduced after trial 10 (*arrow*). Conferences were permitted every four trials at trial 20.

that the greater the cooperation among teams the greater the long-term collective take. The game permits several win-win strategies and players can discover them if they are not forced to "beat the other person."

In addition to choices to cooperate and exploit, a team has three additional choices. Suppose in a group composed of six players, one player choose blue; it eliminates any profit for players choosing to exploit *and* imposes a 25-point fine on them. An orange choice gives an additional 30 points to players choosing to cooperate. Both blue and orange choices cost players five points per person or 30 points for a group of six players. Finally, a yellow choice pays the player choosing it a fixed sum of points (30 points with six players) at any point in the game. The game is designed to be played for about 50 trials with communication not allowed for the first 10–20 trials. The multiplier and disclosure cards are handed out after the players have an opportunity to play a number of trials and the game director is confident that the rules are understood. A third of the players are given a multiplier card with a 3 on one side and a 2 on the other; another third are given multiplier cards with only a 2, and the last third are given one card with the letter D on it (disclosure card). This distribution of power is arbitrary and the game director should experiment with alternative distributons.

A Comparison of Two Games

A recent game was played at ISAGA in Kyoto, with 12, two-person teams made up of individuals from more than eight countries. Most of the players were academics or other professionals rather than students and all but one team had a good command of English. Most were also knowledgeable with respect to environmental issues, and so their results were not expected to be similar to those found for most student groups—and indeed they were not. Figure 1 shows the percent of the five choices made across the 36 trials played; cooperation was chosen by a majority (60% or greater) of the teams throughout the game. The multiplier and disclosure cards were introduced after trial 10 and the increased ability to exploit resulted in a small increase in the number of exploitive choices (fourth block of trials). Exploitation decreased significantly with the introduction of conferences (after tiral 20) and there was also a slight increase in the percent of cooperative choices. It might be thought from examining the graph that this was a highly cooperative group with little dissension among teams. This was not the case, however. While the number of multiple exploitive choices decreased from 11 to 3 from the no-conference to the conference period, there were some exploitive choices in each block of trials throughout the game. An analysis of the groups' discussions during the conferences revealed one reason why the green choice continued to be played. One strategy that emerges in groups evolving towards mutual trust and a stable cooperative exchange is an alternating red/orange play with teams rotating the orange play. This group added a blue to the alternation sequence ensuring that any teams choosing green would be fined. This strategy, however, was only

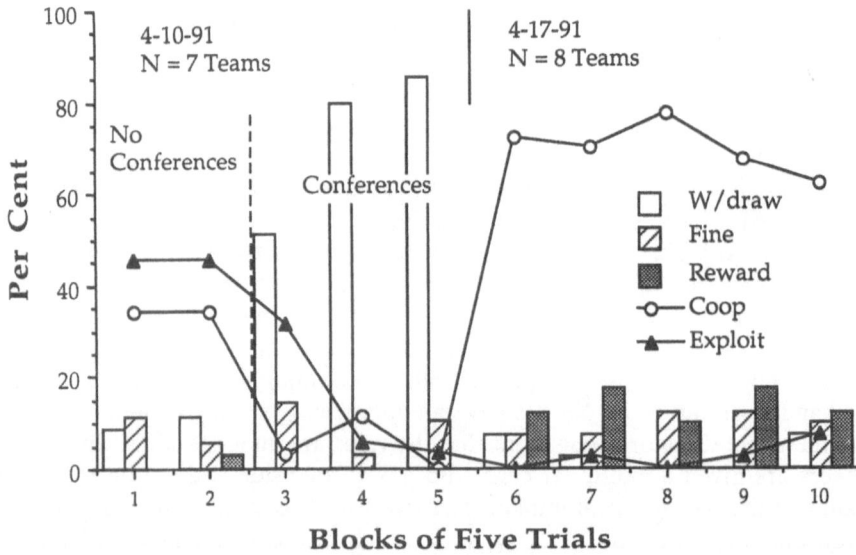

Fig. 2. The percent of the five choices across trials in the game played with Portland State University students. The game was played in two sessions held a week apart and conferences were introduced after trial 10

adopted by about half of the teams with the result that occasionally the rotational sequence fell apart. The unpredictability introduced by the occasional failure of a team to play its scheduled blue or orange choices encouraged some teams to continue trying to score big by choosing green. Towards the latter half of the game, there was a heated exchange between players of teams who were bearing the entire cost of policing and rewarding others and the players in the remaining teams. During this conference, all teams agreed to play the red/orange/blue strategy but when one team played red instead of their scheduled play, cries of "betrayal" arose from several players. After this incident, one team argued that "we should all be free to do what we want"—a sure formula for destruction of the resource.

Only two disclosure cards were used by the four teams who had them. Both questions asked about the use of the blue or fine choice and were not particularly revealing. In conclusion, in spite of a high level of cooperative play throughout the game, this group did not achieve a climate of mutual trust and I was not optimistic that they would ever have reached a stable state of cooperation with all teams participating. I base my judgement on the continued use of fines and the low but consistent exploitive choices made in the latter half of the game. From our college student groups, we have learned that one of the best predictors of a group eventually reaching one of the cooperative solutions and maintaining it, is its decision to stop using the fine option after a history of relying on fines to control exploitation.

During the debriefing a spirited discussion ensued and, in spite of the sophistication and expertise of the players, I felt that several had gained some valuable insights into the workings of a commons and, perhaps, of themselves.

Figure 2 shows data from a commons game played in the Spring '91 quarter at Portland State University over two sessions held a week apart; it is much more representative of the play of college student groups than that in Fig. 1. Exploitation starts out higher than cooperation and the rapid rate at which the matrix decreases induces teams to switch increasingly to the yellow option as both the cooperative and exploitive payoffs at first approach and then drop below that for yellow. The orange choice is rarely played in the first half of the game because players "see no reason to give points to others" at this juncture. Conferences were introduced after trial 10 in the hope that communication would slow down the rate of exploitation. (Multiple cards were not used with this group because of the high exploitation rate). There was a strong appeal made by several teams to "stay with yellow at least for 4 trials" (until the next conference) and exploitation did drop considerably by the fourth and fifth block of trials, but it is not clear how much of the yellow play was due to the opportunity to communicate or to the low payoffs for the cooperative and exploitive choices.

During the 24 trials of the first session, the red/orange alternating strategy was suggested by two or three players, but there was some confusion during conferences because several players talked simultaneously and, at times, only to their immediate neighbors. The last two conferences ended with players agreeing only to adopt the low-risk yellow strategy. The sudden emergence of cooperation in the second session a week later was due to a forceful and persuasive female who informed the group before the game started that she had done some hard thinking and had a strategy that would work. She expounded the red/orange/blue alternation strategy, organized which teams were to do what, and successfully answered the questions and objections of the other teams. After the first two trials in which the group fumbled with this strategy, the red/orange/blue alternation was followed fairly closely for the remainder of the game. There were occasional lapses as when a team forgot its rotational choice, but these were tolerated. During the 27 trials of the second session only five green choices were made and the group had recovered from a dangerously low -5 matrix to a -3 matrix by the end of the game.

One explanation for continued use of fines throughout the game may be related to the presence of one outspoken male student, playing without a partner, who characterized himself as a "lone ranger" and expressed a desire toward the end of the game to "be free to choose whatever he wanted." In an earlier conference, there was a desire expressed to forego the blue choice because it was "costing us and we don't need it." It was probably the lone ranger who kept most of the teams insisting on staying with the red/orange/ blue strategy. Otherwise, I believe, that this group may have eliminated the fines because of the discipline and trust they had developed during the second session.

Comments on the Differences Between the Games

One of the differences between the two games was in the consequences of the payoffs to the players. In the ISAGA game, the points were only game points, while with the college students the points were tied to bonus class points which applied towards their grade. This distinction is important because what the points stand for makes players do different things. In the debriefing after the ISAGA game, considerable discussion occurred between those who wanted to try different things even when it went against a previous agreement and those who argued that agreements should be kept. Those seeking variety argued that cooperation "was boring" while the other side countered with, "So what? Peace is boring, preventive medicine is boring, but that doesn't mean that these objectives are not worthwhile"! Another anomaly occurred in one team playing a sacrificing role (orange) almost from the outset of the game in their belief that this is what a "good citizen" should do in a commons. Students simply do not behave this way because points have real meaning in their lives. Student players consider the costs for playing an orange or blue carefully and do not make these choices lightly. One comment I overhear regularly in student games where a cooperative strategy has been reached and which was also made towards the end of the PSU game was, "I know this is boring, but remember the points count towards our grades. So let's keep going!"

Conclusion

The NEW COMMONS GAME, with two-person teams and with the changes described earlier, produces results that are little different in process from the original game played with individuals. The addition of a power differential in the game raises salient issues having to do with equity and social justice in the use of the world's resources. It may also permit those who are exploited to discover ways of controlling super-exploiters through the effective use of the disclosure cards, although in the ISAGA game they were not so used. I should mention that when I use the game in my classes, I have encouraged students to play the game more than once and many have. Approximately 90% or more of experienced groups reach a stable cooperative exchange; one enormous benefit is the feelings of relief players experience in learning that they can trust their fellows, as well as a sense of accomplishment that they can solve this difficult problem. In a repeat play of the game, players may also learn to ask more revealing questions and learn, in the process, the power of publicity to control runaway greed.

The NEW COMMONS GAME can also be played as a series of games, in which the next generation starts play where the previous one left off. How many generations can play the game before the commons is exhausted? In such a series, the objective is to show teams from all generations that in an unmanaged commons, not only will the present group "have no game" when the resource is inevitably consumed, as Hardin suggested, but that future generations of players will lose out as well because they never had a chance to play the game.

Acknowledgments

I thank Cathy S. Greenblat, Rutgers University, who has generously supplied good measures of feedback, ideas, and insights, all of which have resulted in a better game. Joe Thomas, Stephen Covery Leadership Center, provoked several improvements in the game by his enthusiasm for experiment and by asking good questions.

References

Abt C (1987) Serious games. Viking, New York
Duke RD (1974) Gaming: The future's language. Halsted, New York
Greenblat CS (1988) Designing games and simulations. Sage, Newbury Park, CA
Hardin G (1968) The tragedy of the commons. Science 162:1243–1248
Kirts CA, Tumeo MA, Sinz JM (1991) The COMMONS GAME: Its instructional value when used in a natural resources context. Simulation & Gaming: An International Journal of Theory, Design, and Research 22(1):5–18
Powers RB (1985–1986) The COMMONS GAME: Teaching students about social dilemmas. Journal of Environmental Education 17(1):4–10

Richard B. Powers, emeritus professor of Psychology at Utah State University, has been been a gamer since 1975 and is currently helping the Oregon Peace Institute develop a peace education program. He is a member of the Editorial Board of *Simulation & Gaming: An International Journal of Theory, Design, and Research*. If he had to express his educational philosophy in one sentence, it would be: "There is no reason why serious learning should not also be fun!"

Asian Agriculture: A Tragedy of the Commons in the Making?

Khalid Saeed[1]

Abstract. Patterns of growth in agricultural production existing in Asia and the changing conditions of its agricultural resources are reviewed. A simple system-dynamics model of the underlying generic relationships is then developed and experimented with using computer simulation to illustrate how agricultural resources are being overstrained and how the continuation of the policies causing this may precipitate a widespread decline in agricultural production across the board. Technological solutions to this problem may exist, but these will be ignored unless social and institutional reforms are introduced to create the incentives for adopting sustainable agricultural practices. The general directions for these reforms are outlined.

Key words agriculture; computer simulation; public policy; renewable resources; resource economics; sustainable development; systems; system dynamics.

This paper employs published time-series data from a cross-section of Asian countries to illustrate the pattern of changes taking place in the agricultural resource system of the region. The underlying trends indicate that while the consumption base has expanded across the board through increases in population and income, food production has increased largely as a result of intensive cultivation. Agricultural land resources have mostly stagnated. In some instances, agricultural land under cultivation has increased, but at the cost of a reduction in forest land. The paper also translates the characteristic relationships underlying the interaction between agricultural management policies and the ecological mechanisms of the agricultural resource system into a system dynamics model. Experimentation with this model shows that a sharp decline in agricultural production may be expected across the board unless sustainable agricultural technologies are adopted. However, sustainable agricultural technologies, even when available, will not be put into practice unless the ownership of the resource system is internalized into the agricultural production organization. As long as the technological and economic considerations

[1] Industrial Engineering and Management, Asian Institute of Technology, Box 2754, Bangkok, Thailand; phone (66-2) 524-5681 (w), (66-2) 524-5902 (h), facsimile (66-2) 516-2126; telex 84276 TH

governing agricultural policy remain divorced from environmental information concerning land resources and soil ecology, short-term private gains in production will be sought at the cost of the decay of the resource system that sustains agriculture. Thus, an institutional framework needs to be created, so that appropriate ecological information becomes a regular basis for the economic decisions leading to appropriate technological choices.

Policies and Practices Consuming Agricultural Resources

Agricultural policies and practices currently in use or advocated are based on two almost dichotomous paradigms: neo-classical economic theory and the environmental movement. While economic theory attempts to create policies for maximizing the welfare of society without considering the costs imposed on the environment, the environmental movement has emphasized preserving the environment without considering the pain this may cause to society. The two paradigms do not have a common information base, a situation which seems to have placed environmental issues and economic development agendas in conflict with each other [Saeed 1985].

The environmental issues stand at the losing end in this conflict. They tend to be thrown into the background since information about resource stocks and the ability of the environment to regenerate itself is scarce. Even when available, environment-related information rarely enters economic decisions at the individual or state levels since time elapsing between a societal action and its environmental impact is long and the short-run gains accrued from consuming environmental slack have high utility. This creates irreversible damage in the long run [Simon 1982, Hardin 1986].

Barring a few exceptions, the various statements made about indiscriminate consumption of resources and deterioration of the environment have led to much pontification and few viable policy agendas, possibly due to a compartmentalized treatment of the resource system and the human social organization that is sustained by it. There have been, albeit, several attempts made to explain the decision processes creating the neglect of the resource environment. One of the seminal writings on this subject is by Hardin (1968). Entitled "The Tragedy of Commons," Hardin's essay attempts to understand the logic of the decisions creating individual gains at a cost to the commons. Unfortunately, little if any effort has subsequently been made to translate this understanding into the design of policy instruments that might influence the dysfunctional decision processes.

Ecological constraints may become particularly stringent among the Asian nations representing relatively older civilizations, since long-term existence within their finite resource bases has created rather delicately balanced resource environments with little slack in them as compared with many developed nations, which have been able to exploit virgin territories or to make considerable resource transfers through colonization. Organizational slack, a concept originally advanced by Cyert and March (1976), implies the accumulation of resources that may serve as security against natural shocks. Thus, economic development strategies based on consumption of natural endowments may be quite appropriate when implemented in a land with ample resource stocks. The same strategies may leave people in a vulnerable state, with very low slack, when implemented in a land with a low level of resource stocks. A rather sad example of the dire consequences of development not cognizant of the environmental limitations is that of the Sahel region, where international development efforts ignored the delicately balanced environment creating an enormous tragedy from which the region has not yet recovered [Picardi and Siefert 1976].

Evidence of a Tragedy of the Commons in the Making in Asian Agriculture

Some 300 time series, covering fourteen selected countries representing the Asia and Pacific region over the past three decades, were constructed from published UN sources to serve as a data-base for the analysis.[2] The selected countries were divided into three categories based on per capita income. Australia, Japan, Korea and Singapore were placed in category (A), representing relatively *high* levels of income. Malaysia, Thailand, Philippines and Indonesia were placed in category (B), representing *middle* levels of income. China, India, Nepal, Pakistan, Sri Lanka and Vietnam were placed in category (C), representing relatively *low* levels of income. This classification is consistent with the one proposed by the Asian Development Bank [Okita 1989]. It also adequately covers the variety of the countries in the Asia and Pacific region, in terms of geographic location, form of government and economic conditions. The presence of a particular trend in the selected countries over all three categories provided the basis for the deduction that the trend is pervasive in the region covered by the sample.

Time series plots for the various categories of countries were prepared for population, GDP and GDP per capita to examine growth in the consumption base. The use of agricultural resources was examined through per capita food production index, fertilizer and pesticide application, cultivable land and area under forests. The following observations were made with respect to growth of consumption base and the condition of renewable agricultural resources.

Growth of Consumption Base

Considerable population growth is shown over the three decades covered by the data in all categories, although growth is much higher in the lower-income countries. GDP growth is highest in the middle-income countries, while growth rates in the high- and low-income countries are comparable. Consequently, GDP per capita has grown at comparable rates in the high- and medium-income countries due to moderate population growth in the former and high economic growth in the later. However, high population growth rates and moderate economic growth have led to stagnation in GDP per capita in the low-income countries.

According to the projections of UNCHS (1987), population is expected to continue to rise in all countries well into the twenty-first century, although the rates of projected population growth are negatively correlated with the levels of income—lower income countries experiencing higher and continued rates of total population growth and urbanization. In all cases, the growth in the consumption base originates from two sources, growth in population and expansion in economic activity. It remains to be seen how far the growth in the consumption base can be sustained by the natural resource base and the environmental capacity.

Condition of Renewable Agricultural Resources

Food production per capita, an internal measure of the changes in food availability in each country, exhibits a rising trend in all cases in spite of considerable population growth, while agricultural land per capita shows a declining trend, except in Australia, where it has been possible to maintain it at a steady level. On the other hand, fertilizer and pesticide application has drastically increased in all countries of the sample over the past three decades. This indicates that increases in food production have been obtained largely through an increasing intensity of cultivation and application of chemical fertilizers and pesticides.

Irrespective of the increases in yield, the absolute quantity of cultivable land has not increased much in most of the countries of the sample, except in Australia, where it has been possible to commission large tracts of unused land. In general, where cultivable land did increase, it was at the cost of the forest area, which is already very small in the countries with a stagnant level of land under agriculture. Some jumps again appear in the plotted data, due to variations in the definitions used to delineate the forest area and agricultural land categories.

Unfortunately, deforestation not only reduces valuable timber and fuel wood resources, it is also known to cause soil erosion, water loss, flooding or drought, desertification and silting of irrigation reservoirs, depending on the particular function of a forest in the complex organic relationships existing in the ecological system [Bowonder 1986]. In spite of this knowledge, about half of the area under forests in the developing countries was cleared between 1900 and 1965. At current rates of deforestation, the rest is likely to disappear in 50 years [UN/ESCAP 1986].

The observed trends in data taken from geographically, economically and politically diverse set of countries show that in all cases, increase in agricultural production—a clearly private gain whether pursued by individuals or collectives—has been achieved in the first instance by making intensive use of the land resources viewed as capital inputs rather than an environmental system. It is also evident that expansion in agricultural land has been achieved by consuming forests—another environmental system which is important to the

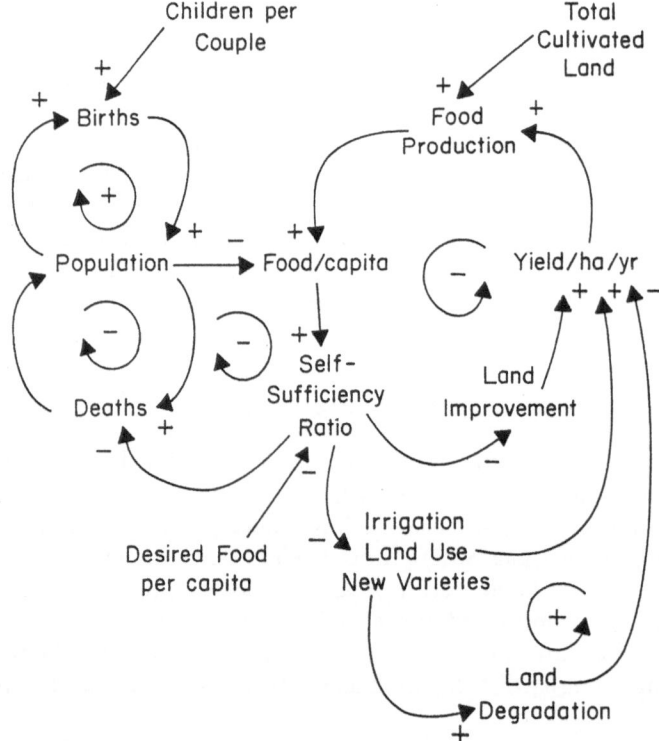

Fig. 1. Important feedback loops in the food production system

maintenance of agricultural land as a sustainable resource, but which is viewed by agricultural individuals and collectives as an unused endowment. Both these processes disregard proper maintenance of the means of production while attempting to increase production in the short run, hence they are the main building blocks of a tragedy of commons in making.

System Relationships in Non-Sustainable Agricultural Development

Excessive use of agricultural land has been known to depreciate soil quality. Soil degradation has occurred in many countries due to erosion, loss of soil-nutrients, loss of texture, water logging and salinity, usually caused by intensive land use [Bowonder 1981]. Centrally-planned as well as market-economy countries have encountered these problems. The market economies tend to externalize private costs to the environmental commons since private decisions often view the environment as a free resource. The centrally-planned economies, likewise, fail to internalize environmental costs because incentives are provided to managers to boost production, not preserve the environment. Furthermore, since the resources placed in the hands of the managers take no account of scarcity value, opportunity cost or real price, the cost of using resources is essentially irrelevant and competition absent. Consequently, there appear wide-spread inefficiencies of production as well as environmental abuse [Chandler 1987].

The food production system of the Asian countries can be characterized by the feedback loops shown in Figure 1. Food fulfills nutritional needs of the population; hence, food sufficiency is related to the average life expectancy. An increase in population expands the food consumption base. Consequently, food consumption is stepped up through intensive land use, high yielding seed varieties and extensive irrigation—all of which degrade land in the long run. Yield may also be increased or sustained through investment in land improvement, which is only resorted to after much damage has already been caused. The model subsumes three subsystems: population, food production and the ecology. The structure of each subsystem is discussed in Bach & Saeed (forthcoming).

The model was parameterized for the ten countries in medium- and low-income categories of the sample. The countries of the high-income category seemed to exercise technological options that were not based purely on pressure. It was, therefore, not considered appropriate to apply the model to those countries. Figure 2 summarizes the results of the simulations of the model with ten parameter sets representing the ambient conditions in the countries of the sample in the middle- and low-income categories.

It is observed that, while food production per capita can be sustained in most cases until the year 2000 or later (possibly from increasing application of capital inputs and intensity of cultivation), land quality may be expected to deteriorate continuously. This may cause sudden declines in output at the turn of the century unless concerted land conservation and reclamation efforts not included in the model have been implemented. The countries most affected are those demonstrating relatively higher population growth rates and more intensive land use. Similar trends obtained with the ten-parameter sets representing the countries of the sample also show that the system is parameter insensitive. Hence, the trends shown must be taken seriously by all countries of the region.

Since the magnitude of deterioration in the land quality index shown in the simulation does not take existing land management programs into account, it indicates only the volume of effort needed to overcome the deteriorating trends. Yet, the results of the simulations are borne out by the experience of many developing countries with large agricultural economies where fertile agricultural lands have already been extensively damaged due to water logging, salinity and erosion of top-soil, even though some of these countries have already instituted extensive land-management programs [Allauddin & Tisdell 1988].

The trends observed in historical data provide persuasive evidence that, over the course of the development process, munificence has been created by consuming the resource environment. The projections made with the help of a model incorporating organic cause-and-effect relationships show that this process has gone far enough, so that it cannot be sustained for long into the future.

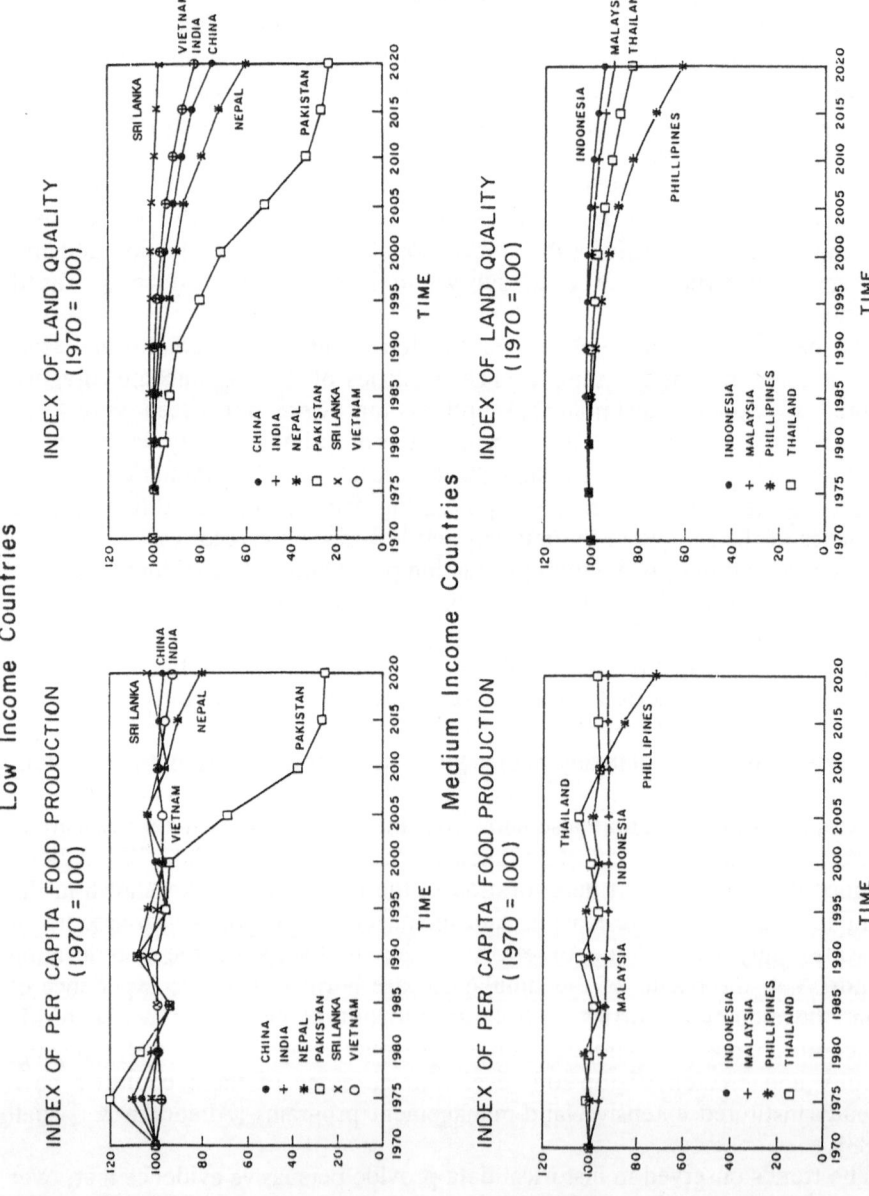

Fig. 2. Projections of food per capita and land quality index based on the model

Generating Sustainable Agricultural Policies and Practices

A decision process cannot be divorced from its organizational context, and it would be impossible to go beyond pontification on issues concerning a commons unless an institutional framework accepting responsibility for the commons is created. Sustainable agricultural technologies in the form of energy-efficient crops and soil-preserving cultivation methods are now widely available. However, there exist neither an adequate awareness of the environmental issues, nor an institutional framework to take responsibility for the environment. Hence, sustainable technologies have a poor chance of being adopted. Social and institutional policy directions must be pursued to create responsible attitudes towards the environment and lead to the adoption of sustainable technologies.

Social Policy Directions

In most countries, agricultural land ownership is concentrated in the hands of only a few people. Hence, cultivators do not have a stake in the preservation of the land they till (often on rental or share-cropping basis). Even when cultivators do own the land they till, it is often in small tracts that must support large self-employed families with a much higher priority on consumption than on preservation. Finally, there may be an awareness of the environmental issues and a commitment to preservation, but there are no institutional mechanisms for mobilizing individual initiatives into the maintenance and preservation activities that call for collective effort. Thus, the following social reforms must precede any sustainable agricultural development initiative:

1. Fiscal instruments and direct intervention must be introduced for transferring absentee-held (both state and private) agricultural resources to the cultivators, so that they have a stake in the preservation of these resources.
2. Extension services should be provided to help cultivators increase their income, so that conservation can take priority over consumption.
3. Infrastructure development must be undertaken in order to increase social mobility in the rural areas making it possible for the surplus labor to move into other production sectors, reducing pressure on land.
4. Fiscal policies should be designed and implemented for creating a crop mix that emphasizes food self-sufficiency on a long term basis reducing the dependency on income generated from export of primary products.
5. Population control should be stepped up in order to contain food demand and urbanization.

Institutional Policy Directions

Current approaches to national planning and policy design often entail fragmented efforts undertaken in the compartmentalized wings of government, which are ultimately assembled into a statement of targets by the central planning organization. This process can rarely identify organizational mechanisms for creating a change—a shortcoming which plans often ignore. Sometimes, impromptu agendas originating at the top also determine the directions of a plan. To create environmentally sound development policies, it is necessary to re-orient planning activity in order to make possible the design of organizational instruments of change along with the preparation of statements of targets. This calls for rethinking the structure of the planning institutions as well as for education in socio-economic planning along the following lines.

1. The information base affecting the policy space in the preparation of plans must be expanded to include geological and environmental information and models of organic relationships which may be experimented with to generate environmentally rational scenarios.
2. An information system must be created to ensure that the expanded information base does not reside in idle compartments but is actively accesible for making private and public decisions.
3. Since environmental issues may clash with powerful corporate and political interests, environmentally conscious institutions must be created in the form of private, non-government and government interest groups. Such groups must also be given legislative protection and a role in public decision making.

Conclusion

This paper has attempted to illustrate how the pervasive indifference towards the conservation of the environment on the part of the public and government institutions are creating dangerous long-term trends depleting agricultural resources in Asia. Such attitudes arise from a view, in both private and public decisions, that emphasizes increasing production in the short run through externalizing costs on the environment whose maintenance is not a responsibility of any of its beneficiaries. The increased availability of production, however, also causes growth in the consumption base, which limits availability, requiring further rises in production. The environmental abuse so inflicted has, by now, created serious deteriorating trends which threaten sustained agriculture in the region.

The policy space for the private and public decisions must clearly be expanded to incorporate the information about stocks and organic relationships in the interaction between human decisions and the environment, if the deteriorating trends are to be reversed. This is not a simple task and would require major organizational reform, both in planning and use of information and in changing existing attitudes.

Notes

1. Research for this paper was partly supported by the United Nations Economic and Social Commission for Asia and Pacific (ESCAP) under a consulting agreement. Earlier drafts have been presented at the Environment and Policy Institute

(EAPI), East-West Center, Honolulu, Hawaii in July 1990, and AIT Workshop on Sustainable Development, Khao Yai, Thailand in April 1991.
2. Data taken from Statistical Yearbooks for Asia and Pacific. 1988, 84, 80, 78, 77, 76, 74, 72 and 70. New York: UN Publications; World Energy Statistics 1950–64. New York: UN Publications; Quarterly Bulletin of Statistics. 1989. FAO; Environmental Data Report. 1987. UNEP; Demographic Yearbook 1978, 1979. New York: UN Publications; World Resources 1987. New York: World Resources Institute; World Population Prospects. 1988, 1989. New York: UN Publications. All monetary data has been converted to US dollar equivalents. Food Production indices based on 1979–81 = 100. Some of the time series required unit conversions and reconciliation of contradicting numbers. In all cases, more recent numbers were preferred over the older data. Some of the missing data cells were computed through interpolation between existing entries.

References

Alauddin M, Tisdell C (1988) Impact of new agricultural technology on the instability of food grain production and yield: Data analysis for Bangladesh and its districts. Journal of Development Economics, 26(7)

Bach NL, Saeed K (forthcoming) Food self-sufficiency in Vietnam: A search for a viable solution. System Dynamics Review

Bowonder B (1981) The myth and reality of high yield varieties in Indian agriculture. Development and Change, 12(2)

Chandler WU (1987) Designing sustainable economies. In Brown, LR et al. State of the World 1987. Worldwatch Institute. W.E. Norton & Company

Cyert RM, March JG (1963) A Behavioral Theory of Firm. Prentice Hall Englewood Cliffs, NJ

FAO (1989) Quarterly Bulletin of Statistics. 1989. FAO, Rome

Hardin G (1968) The Tragedy of Commons. Science, Dec 13, 1968. pp 1243–48

Hardin G (1986) Filters Against Folly. Viking Press, New York

Picardi AC, Siefert W (1976) A tragedy of commons in Sahel. Technology Review. 76(6):1–10

Saeed K (1985) An attempt to determine criteria for sensible rates of use of material resources. Technological Forecasting and Social Change. 28(4):311–323

Simon H (1982) Models of Bounded Rationality. Cambridge, MIT Press, MA

The World Commission on Environment and Development (1987) Our Common Future. Oxford University Press, Oxford, New York

UN/ESCAP (1986) Environmental and Socio-Economic Aspects of deforestation in Asia and Pacific. Proceedings of the Expert Group Meeting. United Nations, Bangkok

UNCHS (1987) Global Report on Human Settlements 1986. Oxford University Press, Habitat Nairobi

UNEP (1987) Environmental Data Report. 1987. UN Publications, New York

United Nations. 1988, 84, 80, 78, 77, 76, 74, 72 and 70. Statistical Yearbooks for Asia and Pacific. New York: UN Publications

United Nations (1950–64) World Energy Statistics. UN Publications, New York

United Nations (1979) Demographic Yearbook 1978. UN Publications, New York

United Nations (1988, 1989) World Population Prospects. UN Publications, New York

United Nations (1988) World Resources 1987. World Resources Institute, New York

Three Simulation/Games and Their Impact on Participants

Ludmilla I. Kryukova[1]

Abstract. Three types of simulation/games on environmental issues are presented, and the effects of these three types on the participants are examined. In the first type of simulation, where participants interacted only with the computer, stress levels were high, but participants became aware of the importance of rules. In the second simulation, interaction took place among participant groups and the atmosphere was more relaxed, but groups tended to make ecologically unsound decisions. In the third simulation, participants had to take personal responsibility for their decisions and arrived at effective environmental and economic policies.

Key words ecology; environment; impact on participants; simulation/gaming

Our approach to simulation/games is based on a game-image conception (Kryukov and Kryukova 1988). The game-image deals with both the impact the model has on the players and the element of realism. In 1988, a session entitled "Simulation Games in Ecology" was conducted in the Department of Biology at Moscow State University. This paper describes the characteristic features of three out of six of the interconnected games demonstrated at that session, as well as the different impacts they had on the players.

STRATEGEM-2

This is a computerized simulation/game of the Kondratyev Cycle, which deals with the economic long wave (Sterman and Meadows 1985). The game focuses on the capital and production sectors. The initial reaction to the game was negative, as the players, mostly biologists and geologists by profession, were threatened by the economic nature of the game. This fear proved to be an obstacle to their comprehension of the introductory lecture. After the lecture, the players (two per computer) read the game rules from the display and discussed them with a partner.

---

[1] Central Economics and Mathematics Institute, Soviet Academy of Sciences, Krasikova Str 32, 117418 Moscow, USS(Russia)

To involve the players in participating more fully in the game, we gave them questionnaires which encouraged them to analyze their decision-making skills and to evaluate their courses of action. The questionnaire revealed that only about 40% of the players felt they had any amount of control over the system. The remaining 60% felt they were at the mercy of outside forces.

Our experience in organizing numerous games has shown that this reaction is normal. The first stage of a game usually produces a strong and unfavorable impression. After the introductory lecture, participants tend to prepare for a game as if it were a traditional seminar. It is only after the game that they realize that any active method demands more knowledge, attention, and personal involvement. Some players did not want to admit their failures, although they were insignificant, and blamed the outcome of the game on the leaders or the game itself, saying that the rules were not fully clarified. Some players did, however, become more self-critical.

Debriefing the game was crucial; it allowed the players to become aware of the advantages and specific features of active-learning methods. The effectiveness of individual decisions and strategies was analyzed, and the essence of the wave cycle was revealed. We were also able to compare the processes in the game with those from the real world. These included such issues as water supply and overdoses of mineral and organic fertilizers. The discussion resulted in participants gaining further insight into the game and establishing analogies with the real world. Fortunately, the initially negative reaction gave way to surprise and a desire for further clarification concerning the game results.

Fish Banks

This game (Meadows et al. 1987; for a review see Crookall 1990) proved to be very captivating for the participants. The rules were explained on the first day. The participants were grouped into teams with names like "Albatross," "Green Bay," and "Southern Cross." In the game, each team makes several decisions, such as how many new ships to order, whether to buy ships at an auction or from other teams, and whether to send their ships to sea, to coastal banks, or to the harbor. Decision results calculated by the computer are given to teams and enable them to make decisions for the next round.

The overall strategy adopted is interesting. As in STRATEGEM-2, participants constantly increased their production capabilities. However, in the first game there was an opportunity for participants to stop the process, while in FISH BANKS the decisions led to a drastic decrease in fish stocks and made normal fish regeneration impossible. This second game, in comparison with the first, was characterized by the participants' interaction in all processes. The newspaper provided a forum where participants could express and read various opinions. It was possible to view the course of the development of the game and situations as a whole. This greatly contributed to the development of personal contacts (which was not foreseen in the original version).

An interesting strategy was chosen by the Green Bay team. As one of the participants put it:

> To prevent damage to the fish banks, our ships did not leave the harbor until the end of the decade. We strictly adhered to the following tactics—we did not do any net fishing and subsisted wholly by selling ships and accruing interest on our money. We developed an intricate economic model, which had a positive effect on the environment at the hands of our competitors, and we reduced the damage from our own activity to a minimum.

This team was also the first to advertise their ships through the newspaper. This stimulated contacts with the "Atlantic" and "Golden Arrow" teams.

Immediately after the game, we had a brief discussion, and then the participants were given a questionnaire. The goal of the questionnaire was to help the participants analyze both their own and their partners' behavior in the game, and to compare it to the real world. A detailed analysis of the game events and of the participants' opinions was carried out by the administrators within a week, and then a special debriefing session was held.

A complete picture of what happened during the simulation could be reconstructed only with the help of the game administrators. In comparing this picture with the participants' impressions, it became clear that some participants were unable to assess the situation, and none of the participants were able to reconstruct the picture in its entirety. As one of the participants noted, reconstructing an overall true picture could only be done at the debriefing session.

In the final discussion, a number of interesting insights emerged. Only two participants regarded the simulation as merely a "game." Those participants whose real jobs were directly related to fisheries remarked that similar situations are typical not only of Western countries, but of the Soviet Union as well. "The game model is applicable to the activity of any enterprise run on a self-supporting basis. The game is a simplified version of real-life situations." Some participants gave contradictory and sometimes evasive answers. There was no sense of realism felt by the participants during the course of the game, but after the end of FISH BANKS they admitted that similar things did happen in real life. It was suggested that, in the game, economics should stimulate ecological policies and penalties should be enforced for ecologically-damaging actions. Although in the long run fish companies found the depletion of fish financially damaging, the participants were of the opinion that this happened too late. Such non-ecological policies should have been stopped at the initial stages.

This discussion revealed that participants were ready to progress on to the next stage of the session. Here, an effort would be made to work out strategies that would be both economically and ecologically sound. Again, some of the participants did their best to lead the discussion away from their personal responsibilities and towards the design of the game. They tended to regard what had happened without any sense of realism, but as a caricature of their ecological ideals. Their injured pride spilled over into finding fault with the game type and structure. It was not that the game was ecologically weak; the trouble seemed to be in the team's interpretation of the game. There was little data upon which to analyze each participant's personal successes and failures, and we could only examine the role of the team as a whole. This allowed individual participants to shirk responsibility and to lay blame on the game itself.

Thus, we needed a game where every participant would make personal decisions and be held accountable for the consequences. We needed a game which (1) provided sanctions against unsound ecological decisions, (2) gave the participants the opportunity to make individual judgements, and (3) enhanced coordination among participants. Another simulation fulfilled these requirements, and became the heart of the third simulation stage.

AT THE LAKE

This Soviet game is a modified version of the COMMONS GAME (Powers et al. 1983). We tailored the game to our audience and modified it to include characteristics of the two previous games. Six to eight people can take part in the game, each of them playing the role of a company manager, with all companies consuming the same natural resource—water from the lake. Decisions are made behind special screens and by choosing cards of different colors. Red indicates nonpurified water disposal, which results in huge profits but spoils the lake for that period. Several red decisions thus lead to rapid lake pollution. Green indicates water purification, which leads to a considerable decrease in profits from its utilization, but does not change the ecological condition of the lake. Other cards include violet (indicating a fine against red decisions), blue (reinforces green decisions), and yellow (production using no water with a modest but guaranteed income). The game is run for 40 months (i.e., 40 game rounds). Every 12 months the lake is purified by spring flows. Every 8 months the participants are allowed to hold a conference. The goal of each participant is to get the highest score. In all decisions, profit depends on the state of the water, and is shown on a graphic model.

The participants knew, from their experiences in the previous games, that they should use natural resources carefully. They also knew from the very beginning what the consequences would be if they chose only red. The temptation, however, was too strong and the majority played red at the beginning of the game, the water in the lake became polluted and the participants' profits decreased. At this moment, a public inspector surfaced, and instead of seeking profits, chose to impose a fine (violet).

At a conference, participants have an opportunity to engage in a face-to-face dialogue. More often than not, they decide on a common green action, but soon become disillusioned since there are always people who continued to pollute the lake water in order to boost their own profits. At later conferences, a more realistic pattern usually emerges, and penalties are used as well. If somebody notices that an agreed policy is being violated, the debate at the next conference will be heated and some participants made to feel awkward.

According to Komarov (1986), there are four types of participants in the game:

1. Lucky individualists: participants who seek victory at any cost and pollute water more often than others. However, they usually manage to avoid the fines (by changing red for green when yellow is played) whenever they have the feeling that the situation demands it.
2. Unlucky individualists: participants who take risks, but often get fined and suffer losses instead of expected profits because they have no feel for the situation.
3. Passive public participants: people who prefer ecologically sound types of technology. In the game, however, they do not try to stop aggressive individualists.
4. Active public inspectors: people who, instead of working and using natural resources themselves, prefer to use sanctions against individualists.

In a second phase of the game, participants are grouped into teams of two or three to work out economically viable projects which guarantee greater profits. The game was played four times with different participants. The participants of the third and fourth run were asked not only to draw their own plans, but to analyze and make critical comments as to projects worked out by the previous teams.

The most idealistic projects relied exclusively on the conscience of the water users and were not supported by any economic sanctions put forward by the individualists. When they were shown by other participants to have pursued their own private interests, they were both upset and embarrassed. More complicated and unrealistic projects were put forth by passive participants. Active public inspectors turned out to be the most practically minded. The special ecology department managed to work out a project that was economically realistic only after the fourth run, that is, after they had analyzed and fully discussed the previous unsuccessful projects. By the end of the session, the participants had become deeply aware of their responsibility to the world and of the consequences of their actions.

Summary

We have dwelt upon three game types. In the first game, teams only interacted with a computer. In the second game, they interacted with each other, with each team's decisions influencing the environment, and thus indirectly affecting all other teams. There are no teams in the third game. Participants influenced the environment on an individual basis and communicated only while holding conferences.

The impact of the game on the participants and the image portrayed greatly depend on the game type. The first game caused stress, but at the same time the participants realized the importance of rules for the game. In the second game, people were more relaxed, which balanced out differences, such as age and social position, but at the same time caused collective control problems, such as how to stimulate sound ecological modes of behavior. In the third game, the participants managed to coordinate their activities and to show their abilities in working out ecologically and economically effective policies. However, the game brought about a certain tension between the participants, since they became aware of their responsibility for the results of their activities.

At the end of the session, most of the participants had more or less detailed plans for continued active environmental protection. Here is what two participants said:

All of us are to some extent responsible for environmental protection problems. It serves no purpose to hold other people accountable for these problems and leave yourself blameless. My next project will be aimed at teaching children sound ecology rules by way of simulation games.

I have been preoccupied of late with trying to save certain regions from a narrow, bureaucratic approach to ecological and economic problems. I'll continue my efforts by advocating positive ecological actions in such a way as to encourage people to influence their leaders to introduce better environmental protection laws.

References

Crookall D (1990) Review of FISH BANKS, LTD. Simulation & Gaming: An International Journal of Theory, Design, and Research 21(2):208–211

Komarov VS (1986) Methodological instructions for AT THE LAKE (in Russian). Institute of Industrial Economy and Ogranization, Soviet Academy of Sciences, Novosibirsk

Kryukov MM, Kryukova LI (1988) Principles of economic reality: Reflexion on business games (in Russian). Sciences, Academy of Moscow

Meadows DL, Fiddaman I, Shannon D (1987) FISH BANKS, LTD: A microcomputer-assisted simulation on the principles of sustainable management of renewable resources. University of New Hampshire, Durham, NH

Powers RB, Duus RE, Norton RS (1983) COMMONS GAME. Utah State University, Logan, Utah

Sterman JD, Meadows D (1985) STRATEGEM-2: A microcomputer simulation game of the kondratiev cycle. Simulation & Games: An International Journal of Theory, Design, and Research 16(1):174–202

Ludmilla I. Kryukova works at the Central Mathematical and Economics Institute of the Russian Academy of Sciences in Moscow. She has been involved with ISAGA for a number of years and has recently become Associate Editor of *Simulation & Gaming: An International Journal of Theory, Design, and Research* (Sage).

Global Change and the Cross-Cultural Transfer of Policy Games

Ferenc L. Toth[1]

Abstract. Studying and managing global change presents new challenges for the scientific and policy-making communities, and they require improved forms of communication at the science-policy interface. This often involves cross-cultural transfer of policy games, strategic simulation exercises, and techniques to address long-term, global problems in their policy context. Over the past 2 years, the author was involved in a United Nations Environment Program (UNEP) project looking at the socioeconomic impacts and policy responses resulting from climate change in Southeast Asia, and used the policy exercise (PE) approach to generate local and national scale policy responses to the potential impacts of climate change in Malaysia and Indonesia. The study in general, and the activities related to the PEs in particular, provided valuable insights into the perception of the global problems, attitude towards long-term issues, environmental risk perception, and the management of uncertainties. The second group of lessons is associated with the cross-cultural transfer of a strategic gaming exercise rooted primarily in the North American and European cultures (war games, operational games, structured workshops) to the Southeast Asian cultures. Dramatic cultural differences have been revealed in the preparation phase and at the PE workshops in the formality/informality of procedures, in the styles of debating, criticizing and challenging each other's ideas, and in the issues related to authority and the hierarchy of people and institutions.

Key words climate change; climate policy; cross-cultural transfer of games; decision making; environmental management; policy exercises; risk management; risk perception

The history of mankind has been characterized by a series of clashes between socioeconomic development and the natural environment. The term global change emerged relatively recently to describe changes in the global biogeochemical systems that involve long time horizons (decades to centuries), large geographical scales (continents to global), and complex interactions among the sources and impacts of those changes (global energy production to

---

[1] Department of Economic Geography, Budapest University of Economics, Fovam ter 8, Budapest, H-1093, Hungary; phone (361)-117-6706 (w); facsimile (361)-117-67-14; telex 224186 mkke h

climate change to crop production). These issues are characterized by significant scientific uncertainties and are the subject of intense policy debates.

Studying and managing global change challenges both the scientific and the policy-making community, and they require improved forms of communication in order to interface. Given the temporal and spatial scales of the problems and possible remedies, there are several cross-cultural transfer and communication processes involved in all aspects of global change. The bulk of new scientific knowledge is created in the developed countries of the North and it is often viewed with suspicion in the South when it suggests policy implications that would prove disadvantageous for them. The management side necessarily involves globally accepted strategies in the forms of international agreement and conventions, but sharing the burdens of such strategies often invites bitter debate.

In the past few years the most widely researched and discussed component of global change was the issue of climate change resulting from the increasing concentrations of radiatively active trace gases in the atmosphere. Two major types of policy-related investigations were undertaken to manage the problem. The first type of study was concerned with prevention strategies: how to reduce the emissions of greenhouse gases (GHGs) in order to stabilize climate change. The second type was aimed at identifying adaptation strategies: what are the most significant impacts of climate change and how might societies adapt to those impacts. The science-policy interface and the cross-cultural communication between more-developed countries (MDCs) and less-developed countries (LDCs) are becoming increasingly important as the need to manage global climate change and its impacts becomes more apparent.

The policy exercise (PE) approach (Brewer 1986, Toth 1988a, 1988b) has been tested and subsequently used in the context of several long-term, large-scale environmental problems in Europe and North America since the mid 1980s. Over the past 2 years, the author was involved in a United Nations Environment Program (UNEP) project looking at the socioeconomic impacts and policy responses resulting from climate change in Southeast Asia (UNEP 1990). PEs were organized to provide a structured interaction between scientists and policy-makers to generate local and national policy responses based on the potential impacts of climate change in Indonesia and Malaysia. This paper is intended to summarize the problems of transferring the PE approach to the Asian culture and the steps required for use in the context of a climate change study.

Perceptions of Climate-Related Risks in Southeast Asia

The UNEP climate change study in Southeast Asia consisted of two major components. Objectives of the scientific component were to identify the most important impact areas and to assess possible impacts of climate change on the environment, natural resources, economy, and social aspects of those areas. The policy component was aimed at developing and analyzing possible strategic policy responses to mitigate adverse impacts of climate change, thereby minimizing environmental, economic, and social disruptions related to those impacts. To understand the difficulties of adapting the PE approach to this particular study and of using it in Southeast Asia, we first needed to understand general attitudes toward global environmental problems and perception of the risks associated with local impacts of global climate change.

The first issue was that of the long-term horizon. GHGs responsible for the changing radiative balance of the atmosphere accumulate slowly and have a long residence time. Most climate studies tend to look several decades or even a century into the future in order to project the climatic implications of various GHG emission scenarios. Impact assessments use these projections as input, hence their results are also several decades in the future. While this fact seems to cause no major problem in perceiving the severity of the potential risks in MDCs, it is apparently a major problem in LDCs.

The long-term projection of future prospects for the economy and society is apparently not alien to the nations involved in the study. In 1990 Malaysia completed an ambitious and well-conceived long-term development plan (Outline Perspective Plan), and the achievements demonstrated the breadth of vision and the depth of wisdom of its creators. At the time of the climate study, a group of people from all walks of life and from various government offices was busy finalizing the Second Outline Perspective Plan, which would provide important policy guidelines and development strategies for the next decade.

The long-term development strategies tend to consider a variety of impacts from the global economy as they might affect the options of the domestic economy. However, it seems to be difficult to include environmental components such as climate among the factors that are (at least partly) driven externally. While it is most obvious for strategic planners to prepare several variants to cope with uncertainties of the external economic factors, it is much more difficult for them to think in terms of several possible climate scenarios. If climate is changing, there is little we can do about it, so why bother?

Besides these perhaps fatalistic views, the perceptions of how serious the climate-associated risks are and the attitudes toward how urgently and what kind of management actions are needed also depend on the prevailing patterns of social management. In Malaysia paddy farmers are fairly dependent on local and national governmental organizations: they provide irrigation water, seeds, and fertilizers for the farmers, and there is even a cash subsidy when farmers sell their produce to authorized government dealers. Under such circumstances, when the agricultural sector is so tightly regulated, all the initiatives, including those to prepare and plan adaptive measures to impacts of climate change, must also come from governmental organizations.

In Thailand the case is almost the opposite. Farmers freely decide what to grow and they benefit from fluctuating market prices. Therefore, they have experience in switching crops and cultivating techniques as the market dictates. They are probably much more self-reliant in choosing their adaptation strategies as well.

Culturally Rooted Concerns about PEs

The project involved a long preparation phase, in which expert teams in the participating countries worked with external consultants to prepare detailed assessments of the potential impacts of climate change on various agricultural crops, river basins, and coastal areas. The impact assessments were intended to serve as primary inputs to the PE workshops where senior policy-makers would review the potential impacts and prepare strategic responses in their own areas of the governmental management scheme. It became apparent in the very early phase of the project that both sides had serious concerns about interacting with the other in such an informal setting as suggested by the original PE protocol. Two main reasons seemed to be behind these concerns, and they were both related to a single cause deeply rooted in the Asian culture.

The first concern was related to the subject matter on both the expert and policy sides. It was clear from the very beginning that the project was expected to generate and analyze the widest possible range of policy response options to climate change impacts, and without the policy-makers' contribution it was simply impossible to achieve this objective. Still, a recurring theme at the first preparatory meetings of the Core teams of experts in each country was whether we should involve senior policy-makers in the project at all. They felt that their present knowledge about the regional and local impacts of climate change and the uncertainties characterizing the rate and magnitude of global climate change did not justify the involvement of the policy community at this point. "Aren't we going too fast when we want to present highly uncertain results to senior policy-makers?" asked a senior member of one national study group at the first meeting.

Senior policy-makers who were approached for potential participation in the PE workshop were initially concerned because of their limited knowledge about the subject matter, the very distant horizons characterizing the climate change issue, and because of their fear that the climate problem (and some of its policy implications) is yet another environmental issue pursued by MDCs that might threaten their economic development. Although they appeared to be ready to discuss long-term development issues where they felt comfortable to use their strategic planning perspectives, their inclination was to consider environmental repercussions of development at the same time scale.

The second concern on both sides originated from the unfamiliarity with the PE procedure. The expert community was afraid of being trapped into questions for which they do not have ready answers. Moreover, even the most senior members of the expert team had only occasional and largely ceremonial personal contacts at the senior level of the policy community which was expected to participate in the PE workshop. The initial fears of the policy-makers were no less significant. They were worried about the roles they were expected to play and had at the beginning no idea how they would be expected to perform. An additional major concern was that what they said at the PE workshop might be taken as an official standpoint and, even worse, might be publicly cited.

This kind of very reserved reaction to an invitation to a PE is quite the opposite of my experience in the West, where scientists are usually eager to get the attention of the policy community and both sides seem to welcome the opportunity to address each other in an informal setting. Others have similar experiences. For example, Clark (1986) "found the best scientists and the best policy people expressing a growing dissatisfaction over their inability to address each other, except through stultifying layers of reports and bureaucracy or in ritualized and guarded public encounters. Carefully designed policy exercises might provide the channel and forum for the communication they seek." This mutual interest in the West is coupled with a variety of increasingly appropriate and comprehensible techniques to present and use uncertain or incomplete scientific knowledge in policy formulation.

These techniques are well-known in Asia as well, but the traditional cultural heritage makes it difficult to use them. The underlying reason for all the concern listed above stems from the fear of "losing face." Loss of face is a difficult concept to understand for Westerners. "It has been described as: making someone feel embarrassed; making someone feel humiliated; causing someone to feel inferior; giving insult to someone" (Craig 1979).

Scientists and experts were afraid of losing face when they presented the results about patterns and local/national impacts of climate change which are not at the level of scientific reliability as other matters on which they usually report. They would rather not say anything, even if it took 10–20 years to achieve the "appropriate" level of scientific evidence, than they would risk losing face by speaking out on such an uncertain issue only to be proved wrong. Policy-makers, on the other hand, were concerned about their loss of face because they were expected to propose strategic responses based on limited and uncertain information (even though the information was the best available) and their policy proposal might turn out to be profoundly wrong later in light of newly acquired scientific knowledge.

To make things more complicated, many participants on both the expert and policy sides had Western education and several years of exposure to Western culture. They nevertheless preserved the basic values and traditions of the Asian culture. The result is that their professional activities are more directly based on their Western education, while their interpersonal relationships preserved the customs of their Asian heritage.

A PE workshop is a very special mix of professional activities (scenario building, policy formulation, and evaluation) and diverse forms of intense personal interactions (expert teams vs. policy teams, mixed small groups assigned to special tasks and rearranged several times in the course of the workshop). Faced with something new, the Asian cultural heritage of loss of face dominated the attitude of the expert and policy communities at the outset. One major challenge for the PE facilitator in the preparation phase was how to relax the concerns about losing face and prove that participation in the PE workshop would be rewarding for both sides.

Relaxing the Concerns in the Preparation Phase

Recognizing these concerns and cultural differences early in the project helped in the design of the preparation phase of the PE. The PE protocol (Toth 1988b) provides a variety of techniques specifically suited for this task: a series of preparatory meetings of the Core group of experts, pre-interviews and briefing material for the policy participants, and the channelling of evolving perspectives and information between the two groups.

The concerns were slowly and gradually relaxed as the project evolved. The key argument was that PEs are specifically designed to address issues involving incomplete scientific knowledge and uncertainties by offering multiple scenarios of "not impossible" futures. This should make clear to the policy participants that what they are dealing with is in the realm of possibility. The objective is not to find the single best management strategy (which would be impossible anyway), but to identify robust strategies that would provide appropriate mitigation regardless of the exact timing and nature of the climate change and its regional impacts.

The expert teams gradually became more self-confident as they worked through the impact assessments and recognized the importance of the study by learning the magnitude of the possible problems (the severe local and regional impacts of climate change). The teams used computer simulation models and other advanced impact assessment techniques which increased their self-confidence in the plausibility of the results. Preparatory meetings of the Core group provided the opportunity to share their results with each other and to rehearse their presentations. Knowns and unknowns were sorted out and it became clear that input from the potential users of their results (the policy community) would be extremely valuable for their own future work.

The policy community had similar fears at the beginning. Their knowledge about climate change and the potential local impacts was not surprisingly very limited and they did not feel comfortable engaging in professional discussions on a topic about which they knew so little. The pre-interviews were therefore partly turned into private briefings for the policy-makers whom we wanted to invite to the workshops. The facilitator conducting the interviews first summarized the basic problems and processes of global climate change. The intimate atmosphere and the privacy of these briefings practically eliminated the danger of losing face by the senior policy people which could have resulted from acknowledging ignorance, especially in the presence of subordinates.

The next part of the briefing in the pre-interviews was an overview of the preliminary results produced by the expert teams concerned with regional and national impacts of climate change. As the presentation proceeded from the first-order biophysical impacts to higher-order economic and social implications, the interest of the policy-makers increased greatly. The real breakthrough came when they realized that their own long-term development programs (rural development, elimination of poverty, modernization of agriculture, land use planning in coastal areas, flood protection schemes, and water management in river basins) were all affected by the impacts of climate change. From this point on, they were eager to hear more from the experts even though it was made clear to them that uncertainties about the impacts were significantly higher than in other cases involving expert advice.

Parallel to making progress on the subject matter (impact assessments by the experts, briefings and defining the policy context with the help of policy-makers), the informality of personal interactions at the PE workshop increased. Rules, roles, and procedures were established by using elements of the PE protocol that were felt most appropriate by the invited policy-makers. Here I observed that Asian cultures are much more formal and much more sensitive to superior-subordinate relationships than Western cultures. Therefore, the procedure involved more formal contributions to the workshops (short presentations, short "speeches") from both the expert and policy sides, as opposed to the quick rotation of short contributions (questions, short answers, new questions) common to workshops in Western culture. The very open and sometimes heated confrontations of views and opinions, which characterize PE workshops in the West, were turned into a milder and cooler style of discussion.

From a Western point of view, these procedural changes may have reduced the efficiency of the process, but they had one clear advantage: they worked. Plunging into the exercise with Western style rules, procedures, and facilitation would have produced a disaster: participants losing interest quickly and withdrawing politely after the first few hours.

Conclusions

Despite the strong initial reservations towards a PE type of interaction between experts and policy-makers, I successfully conducted PE workshops in Indonesia and Malaysia. The presentations and discussions at the workshops greatly enhanced the quality of results of the project as a whole. The PE workshops proved to be a unique opportunity for experts to present their findings and explain all the current shortcomings of the data, analytical techniques, and results; and for policy-makers to learn about a global environmental risk that will have to be taken into account as it will inevitably affect outcomes of their long-term development programs.

Despite the overall success of the PEs in the project, there were problems and difficulties as well. Probably the biggest drawback at the workshops resulted from the fact that there was only a partial overlap of policy people who were briefed and pre-interviewed in the preparation phase and those who participated in the workshops. There was a striking difference in the attitudes and performances of the two groups. Those participants who went through the pre-interviews were much more productive and cooperative. They were familiar with the subject matter and understood the objectives of the project and their role in it. Once again, the lesson is that the time and money invested in the PE pre-interviews will produce a handsome return at the workshop.

My experience with the Southeast Asian climate change project clearly demonstrated the PE approach is useful in addressing long-term, large-scale, and complex problems characterized by significant scientific uncertainties but demanding the attention of strategic planners and policy-makers. By understanding the local cultural traditions, it was possible to design the PE procedures so that it followed the basic concepts of the original protocol rooted in the Western culture, but was sufficiently flexible to accommodate organizational, behavioral, and interpersonal aspects of the Asian culture.

References

Brewer GD (1986) Methods for synthesis: Policy exercises. In: Clark WC, Munn RE (eds) Sustainable development of the biosphere. Cambridge University Press, Cambridge, pp 455–473

Clark WC (1986) Sustainable development of the biosphere: Themes for a research program. In: Clark WC, Munn RE (eds) Sustainable development of the biosphere. Cambridge University Press, Cambridge

Craig J (1979) Culture shock! Singapore and Malaysia. Times Books International, Singapore

Toth FL (1988a) Policy exercises: Objectives and design elements. Simulation & Games: An International Journal of Theory, Design, and Research 19(3):235–255

Toth FL (1988b) Policy exercises: Procedures and implementation. Simulation & Games: An International Journal of Theory, Design, and Research 19(3):256–276

UNEP (1990) The potential impact of climate change in Southeast Asia. Interim report. UNEP, Nairobi, Kenya

Ferenc L. Toth is an economist and policy analyst interested in problems of natural resource and environmental management. His activities in simulation/gaming started in 1983 when he was involved in the development of STRATEGEM-1, a computer-based management training game. Later he designed, tested, and applied the policy exercises approach to provide a structured communication forum between scientists and policymakers who study and manage long-term, large-scale, and complex environmental problems.

Policy Exercises on Global Environmental Problems

Laurent Mermet[1]

Abstract. A network of scientists at and in connection with IIASA has been working for several years on the development of Policy Exercises (PEs) to explore the options available for acting on global environmental issues. A number of experimental workshops have shown that the methodology—which uses gaming techniques in a context of scientific research and of policy-making—has a high potential. It has made significant progress, but some basic problems remain, both in design and debriefing. The paper presents some pragmatic options for overcoming them in future developments of PEs. But these difficulties also point to deeper theoretical issues which will have to be addressed fully if PEs are to reach their main objective, that is, to simulate the dynamics of global environmental problems.

Key words environment; gaming; methodology; Policy Exercises; policy making; research; simulation

Can we capture the complexity of global environmental issues in a gaming/ simulation workshop? As Meadows clearly demonstrated during the conference, the answer is certainly "yes" in an education or training context, where the purpose is to help participants understand the paradigms of long-term and global thinking. But Policy Exercises have a different ambition—that of applying gaming/simulation as a tool for scientific research and policy making. After 4 years of experiments on the methodology, it has become apparent that this different context imposes a whole new set of challenges, some of which would previously have been difficult to anticipate. It is time to step back, look at the more basic difficulties in the endeavour, and reflect upon the main options for further development.

Learning from Experience

To deal successfully with global environmental problems, it is necessary to achieve both an in-depth understanding of their social and institutional aspects

[1] AScA, 69 rue des Rigoles, 75020 Paris, France; phone 33 1 43 66 88 98; fax 33 1 43 66 85 58

216

and a constructive synthesis of the various disciplines of natural science involved. The methodological challenges involved in this proposition brought Brewer (1986) and Clark (1986) to suggest the development of Policy Exercises (PE). The agenda they put forward was basically to adapt and apply free-form gaming to research and policy making in the field of global environmental problems. An international network of scientists, at and around the Institute of Applied Systems Analysis (IIASA, Laxenburg, Austria) began to work out a more precise base for the methodology (Toth 1988, Underwood 1988, Underwood and Duke 1987). The basic concept they retained rested on the following features:

— The core of the exercise is a gaming/simulation workshop.
— Participants include decision makers as "players," and scientists as a "control-team."
— The control team presents a hypothetical situation at time T.
— Each player, in role, proposes a set of actions.
— The control-team assesses the impact of all actions in the round, and presents the new, resulting situation at $T + \Delta T$.
— Through a number of such rounds, the exercise builds up a "future history" and ends with a debriefing session.

Clearly, this concept aims at combining the strengths of improved meeting processes, scenario-writing methods, and an actual simulation of the complex systems involved, with both human participants and computer support if necessary.

Quite a number of experimental exercises were run from 1987 to 1990 (for an overview, see Mermet 1992 and Duinker et al. 1989). They used many variants of the basic design, differing in topic, group size (2–30), duration (3 hours to 3 days), and design (from very flexible to more rigid rules).

Scientists involved in running the experiments met with external experts for a review workshop at the Institute of Environmental Studies, Toronto, in October 1990. They confirmed that the methodology does have high potential, in particular for:

— Integration, through an explicit and dynamic treatment of the time dimension, of many disciplinary perspectives on global environmental problems problems
— Creating a stimulating and well-structured communication between scientists and policy-makers

The experiments have come up with solutions concerning many of the difficulties involved in running such complex gaming/simulations. Continued exchanges between the designers of the experiments has led to a form of collective learning. Despite much progress with PE "technology", some central problems still remain. If PEs are no longer just a concept, neither are they a mature methodology. Both the preparation and the debriefing are still difficult to master, and PEs remain a fairly unpredictable and risky experience.

The preparation of PEs is apt to become a very large undertaking, somewhat disproportionate with its outcome, that is, a workshop lasting from 1–3 days. When good material has come out of a workshop, it has required a disproportionate amount of work, and it has been difficult to identify to what extent, and in which way, it could be identified as a specific product of the exercise.

Pragmatic Options

To overcome these limitations, one pragmatic solution is to modify the general design of PEs. Four options seem to be available.

In the course of preparing experimental PEs, it became obvious very soon that some topics were more "gameable" than others. Examples were geographic or thematic subsystems of global environmental problems, such as the management of a regional watershed under conditions of climate change (Mermet 1992), the long-term evolution of a rural region, or the future strategies of one corporation in the wood industry (Duinker et al. 1989). The strategy of focusing on gameable situations can help identify and approach interesting aspects of global environmental problems. This already gives it a high potential for further applications in the field. However, it cannot be considered a suitable way of grasping these global problems themselves. Lest we fall into an instrumental bias (looking under a street lamp for a key lost elsewhere in the dark), PEs will have to go beyond this solution.

A second way to dodge the difficulties inherent in the PE concept stems from the fact that it is the element of actual simulation (of the consequences of decisions on the evolution of the global system) which generates the worst contradictions between the intricacies and uncertainties of our initial understanding and the demands of workable design. Why not invest more in other components of PEs, by fully exploiting the potential of qualitative scenarios, as well as paying greater attention to rich, multi-perspective discussions between participants, and thus minimizing the role of actual simulations in the workshop design? Experiments have shown that this does make the exercise more workable. Writing scenarios directly provides a tangible product, and the more freewheeling discussions are easier for participants to accept.

In such designs, the remaining element of simulation is only valued for its contribution to the meeting process. However, the satisfactions obtained through this formula should not conceal the underlying inability to simulate more decisively the global problems approached. Simulating these in a credible way is a statement of the kind of understanding we are aiming at, even if we are unable as yet to deliver it. The success of a more qualitative exercise should pave the way for an increasingly firmer treatment of its topics. It cannot, however, be a substitute for it.

A third possibility is to go beyond the reduced scale of a gaming workshop, and mobilize more participants, more information, more time, and more means. Instead of a 4-week effort, culminating in a 2-day workshop, for instance, one could organize a 2-year effort, in a network arrangement, including six 2-day interactive workshops or the use of teleconferencing facilities. This possibility is certainly very promising because it can produce more elaborate material, allows time to process more information, provides an opportunity to hammer out difficulties and to explore sidetracks that could ruin a workshop, and makes easier the collaboration of participants with different research rhythms and personal time-constraints.

However, it leaves two major pitfalls open. First, more time and more resources make the undertaking vulnerable to precisely the same pitfall that caused so many difficulties with global computer models—the futility of expecting that massive treatment of information could take the place of new and appropriate theoretical insights on global problems. Second, the bigger the exercise becomes, the closer it would have to be to a classical inter-disciplinary research proiect. Choosing this orientation may lead us back to a problem which is still basic in all PEs: how to ensure the sound conceptual and practical insertion of a gaming event in the processes of science and policy making?

The first phases in designing a gaming-simulation need a clear formulation of the problem to be addressed, an adequate description of the system within which management action takes place, and the selection of a set of elements and relations in the system sufficient to account for its main features and dynamics. On the topic of global environmental problems, a satisfactory treatment of this first phase would be a fair achievement, even if the simulation itself never took place!

Clearly, all protocols suggested for PEs have insisted on the importance of the preparation phase (Toth 1988). But the point here is that it should, and could, be made to produce more than just material for a workshop. The experiments so far have shown that trying to meet the requirements for running a simulation provides a stimulation, framework, and test for the selection and integration of knowledge into a coherent picture. What is needed now is a workable way to extract documented and usable information from all the trial, error, and conceptual elaboration which occur in the preparation phase. This is certainly a promising orientation, but again, it does not completely address the challenge inbedded in the PE concept.

Central Problems

All four options above can help alleviate the difficulties of designing and running PEs. However, each of them tends to avoid a central core of difficulties, which seem to be inherent in the initial agenda of PEs. They also lead to varying types of designs. As a result, the identity of the PE as a methodology is at risk if we simply follow pragmatic options for further development.

It is certainly necessary for PE designs to move away from pure gaming/simulation. But to harness the whole potential of the concept, this tendency will have to be balanced by an awareness that the central aim is to simulate the dynamics of global environmental systems. For this, we have to step back from the pragmatic challenge of designing exercises. Three basic problems then become apparent.

Science and public decision making are central in the process of managing global environmental problems. But in both arenas, PEs have not yet found a very well–defined and legitimate niche. From a scientist's point of view, their conceptual status remains unclear. The links between theory, data, and assumptions, which exist in a computer model for instance, are lacking in a "future history." Also, the products tend to be difficult to publish—a major source of institutional obstacles. From a decision-making perspective, PEs are valued as thought-provoking meetings, but they consume large resources for a small number of participants, with little relation to the actual process of decision making and, so far, with few solutions for a wider diffusion of results.

The second main problem rests at the heart of gaming as a methodology. To organize a simulation is to offer a model of reality. However, the model and its simplification of reality must rest upon an adequate theory. As Greenblat (1988) plainly states: "Unless you take the time required to understand the system, you cannot simulate it, even in highly abstract form." But if you have to understand before you design a game, the idea of introducing gaming as a means of understanding the system leads to a vicious circle. This throws some light on one of the main obstacles encountered in experimental exercises: when trying to design simulations for high level scientists and decision makers, the margin of maneuvre is very narrow. What the designers know is also known to most participants and tends to appear trivial to them. And what designers only understand vaguely tends to be the basis for confusing workshops.

The third central problem lies in the aims of PEs. The initial proponents of the PE set very high expectations. Among other benefits, the methodology was expected to enable new research results on global environmental problems to be transferred directly to policy makers, to integrate the findings of "hard" and "soft" science and also the benefits of the personal experience of high-level decision makers, to prepare informed decisions for high-level decision makers, and to identify and study new aspects of complex systems which govern global problems and their management. In view of these exceedingly high expectations, an actual PE is a sobering experience: the quantity of information which can be treated in one workshop is strictly limited; if all kinds of components can presumably be included in an exercise, trying to take them all in at the same time can only be a source of confusion; and preparing a good simulation with a very imperfect understanding of how a system works is a risky and frustrating experience. But these ambitious aims have inspired a strong and sustained motivation to both the scientists and decision makers involved.

In short, the prospective PE was burdened with all the changes that appeared to be necessary in our field of global environmental problems. This gives a good sense of the frustration that the community of specialists of global environmental problems felt a few years ago at the perceived inability of traditional research and decision-making procedures to give adequate treatment to these problems.

New Challenges

Although much progress is currently being made in that direction, the needs that generated the agenda for the PEs have not disappeared. PEs are one element of a whole set of approaches that need to be developed. We need to improve our understanding and theories in two main directions.

A more direct and detailed analysis of the procedures used in science and in policy making in the field of global environmental problems is an urgent necessity. Our difficulties in clearly defining the appropriate niches for the PE is only one symptom of a blind area in our awareness which may hinder our efforts to solve global problems. This is only apparently trivial. Rayner (1991) analyses some of the obstacles which stand in the way of formulating an explicit diagnostic and innovation in this field.

The design of a PE demands that we transform the vague concept of "managing global environmental problems" into explicit and structured decision-making situations. As soon as we try to do so, we realize the inadequacy of our understanding. As long as gaming serves training purposes, it may be sufficient to rely on the toolbox of soft-system analysis and of game-theory as base for ad hoc "conceptual modelling." However, in a research and decision-making context, it is necessary to rely on adequate, state-of-the-art understanding of the actual dynamics of very complex situations. The design of PEs is then no substitute for new theoretical perspectives on the management of global environmental problems. This may well be the most interesting contribution to research made by the experimental PEs so far. From the coherence and clarity of understanding that it requires, the design of a game reveals both the missing links of our understanding and the general form we would like it to take.

The PE design may be the product, not because it could answer our questions on very complex problems, but because it can help formulate and organize them better. Furthermore, this observation leads to the hypothesis that a theory based on game concepts may have a high potential in addressing these problems. Beyond game theory, we have few ready-made theoretical foundations for using game-based concepts for an analysis and synthesis of very complex problems (Mermet 1987). A new effort to lay deeper theoretical foundations to gaming may be necessary if it is to be applied to very complex issues, and with research and policy making (as opposed to training) as its main purpose.

Conclusion

The PE methodology can be useful in giving us intuitive access, and potentially a testing ground, toward important theoretical issues in the management of global systems. These issues could be very easily overlooked when using a more conventional methodology. Its development has a significant contribution to make to the diversification and improvement of the procedures by which hard science, policy science, and decision makers can work together towards the resolution of global environmental problems. Finally, experiments with PEs raise important questions about the use of gaming for research purposes and as an approach to complex problems.

Acknowledgments

The author's work on Policy Exercises has been funded by IIASA, and by the Groupe de Prospective of the French Ministry of the Environment.

References

Brewer GD (1986) Methods for synthesis: Policy exercises. In: Clark WC, Munn RE (eds) Sustainable development of the biosphere. Cambridge University Press, Cambridge

Clark WC (1986) Sustainable development of the biosphere: Themes for a research program. In: Clark WC, Munn RE (eds) Sustainable development of the biosphere. Cambridge University Press, Cambridge

Duinker P, Nilsson S, Toth F (1989) Policy exercises in the forest study of the biosphere project: A methodological review. Working document

Greenblat C (1988) Designing games and simulations: An illustrated handbook. Sage, Newbury Park, CA

Mermet L (1987) Game analysis—An analytical framework to bridge the practitioner-researcher gap in negotiation research. IIASA Working Paper 87-084

Mermet L (1992) Les exercices de simulation prospective: Une méthode pour étudier des politiques en situation de complexité, d'incertitude, de long terme. Natures, Sciences, Société, 1

Rayner S (1991) Think globally, act locally: Harnessing diverse expertise for managing the global environment. Actes du colloque: Les experts sont formels; controverses scientifiques et décisions politiques dans le domaine de l'environnement, tenu à Arc et Senans

Toth F (1988) Policy Exercises: Objectives and design elements. Simulations & Games: An International Journal of Theory, Design, and Research (19)3:235–255

Toth F (1988) Policy Exercises: Procedures and implementation. Simulations & Games: An International Journal of Theory, Design, and Research (19)3:256–276

Underwood SE (1988) The policy exercise: Cooperative learning for long-run policy assessment. In: Crookall D, Klabbers JHG, Coote A, et al. (eds) Simulation and gaming in education and training. Pergamon Press, Oxford

Underwood SE, Duke RD (1987) Decisions at the top: Gaming as an aid to formulating policy options. In: Crookall D, Greenblat CS, Coote A (eds) Simulation-gaming in the late 1980's. Pergamon Press, Oxford

Laurent Mermet's background is in ecology. However, skepticism about our ability to manage our environment properly has led him to a doctorate in management science. He presently works, both as a scientist and as a consultant, on decision-making processes, long-term strategies, and conflict resolution issues in the field of environment.

An Open Simulation-Game with a TV Studio as a Tool for Long-Term Policy Formation

Kiyoshi Arai[1]

Abstract. In order for people to reach a consensus concerning the future of a region, the author proposes THE MAYOR'S ELECTION, an open simulation-game with a set of facilities in a studio. A preliminary trial run was performed in February, 1991 at Kinki University, Kyushu Campus. In the scenario, a region is divided into three areas, each of which has a candidate for the mayor of the region. Each candidate persuades interest groups to get ballots. The groups include city council members, shop owners, constructors, medical doctors, women's groups, and farmers. The players exchange their ideas, discuss the policies of the region, and communicate with each other via written messages and face-to-face discussions. A TV newscaster provides information about the messages as regular and special news programs, based on information collected by the news media group. The game can be developed as a policy exercise tool to form long-term, regional policies.

Key words   election; frame game; general picture; long-term policy; policy exercise; regional planning

It is vital to incorporate the learning process into the planning process to make citizen participation more effective and valuable. The author proposes an open simulation-game as a policy exercise tool for long-term regional policy formation. A set of facilities in a studio helps people to discuss regional problems, required policies, and action programs for the future of a region.

A trial run was performed in February, 1991 at Kinki University, Kyushu Campus, located in Iizuka City, Fukuoka. This paper deals with the general framework, structure, expected effectiveness, and implications of the game obtained by performing the trial run.

Design Concept

A frame game is known as an effective tool which deals with a long-term policy. Because nobody has a clear picture of the future, the gradual refine-

---

[1] Department of Industrial Engineering and Management, Kinki University, 11-6 Kayanomori, Iizuka, 820 Japan; phones 948-22-5655 (w), 92-661-4382 (h); facsimile 948-23-0536; e-mail f77541 @ Kyu-cc.cc.Kyushu-u.ac.jp

ment of policies is required. Participants can create details of the future if they are provided with an appropriate gaming framework. A frame game can be an open simulation in two senses: it becomes a richer experience through replay and it can influence decision making in the real world. In a replay, a frame game can be developed into a more structured game. The more proficient the players become, the richer the game. This means that players accumulating knowledge in the game world will share this information with other participants and bring this knowledge back to policy making in the real world. A frame game as a policy exercise tool should function as a Delfi exercise.

It is crucial for players to know what others are thinking and doing in the real world in order to formulate their strategies. Specific communication methods are required which make it possible for someone to collect, analyze, and deliver information about what happens in the game world. Computer network systems and written message exchange systems are suitable as a level-one communication media between individual players. Mass media systems like TV news are suitable as second-level communication media for real-time feedback purposes.

A game scenario is important because it determines the game's attractiveness and effectiveness. THE MAYOR'S ELECTION adopts a scenario in which a region is divided into three areas, each of which has a candidate for the mayor of the region. The three candidates have to concentrate on other areas from which to get ballots, as well as their own areas. This scenario makes players think about competition and cooperation among the three areas.

Structure of a Trial Run

The simulated region—Chikuho—is divided into 25 local governments which have difficulties in planning. It is hard for people to share a general picture of the region. Each local government has its own interests, and does not cooperate with the others. Although the Chikuho region actually consists of 25 municipalities, the region is, in the scenario, divided into three areas. Each area has one candidate who tries to be elected as the mayor of the region.

The timetable of the trial run is shown in Table 1. The first six rounds each consist of four phases: (1) strategic meetings with each candidate and local conferences in each area, (2) political transactions via written messages, (3) inter-area conferences in each social group, and (4) regular TV news. In the seventh round, the mayor is elected by vote.

Roles of players are shown in Table 2. Players are categorized into four groups: the news media group, candidate groups, social groups, and general voters. The three areas have the same number of players. The news media group consists of three kinds of people: a newscaster, news staff, and messengers. They check, collect, and analyze messages exchanged between players, and announce as regular and special news what is happening in the game world. Their role is between the controllers and other players. Each candidate group has four members: a candidate, a chair of the city council, a chair of

Table 1. Timetable of the trial run

Total timetable

9:00–9:20	Orientation
9:20–10:10	1st round
10:10–11:00	2nd round
11:00–11:50	3rd round
11:50–12:30	Lunch
12:30–13:20	4th round
13:20–14:10	5th round
14:10–15:00	6th round
15:00–15:05	Break
15:05–16:05	7th round
16:05–16:15	Speech by the elected mayor
16:15–16:30	Break
17:00–19:00	Social meeting

Timetable in the 1st turn

9:20–9:28	Strategic meeting with each candidate and local conference in each area
9:28–9:53	Political transaction via written-on-paper messages
9:53–10:00	Inter-area conference in each social group
10:00–10:10	Regular TV news

Timetable in the 7th turn

15:00–15:45	Speeches by three candidates
15:45–16:00	Electing the mayor by vote

Table 2. Roles of players

Newscaster	• Announces regular and special news
Newsdesk staff	• Make news drafts by analyzing messages exchanged
Messengers	• Check and deliver messages
Candidates	• Aim at being elected, announcing their promises, trying to break ties of competitors, and making them certain in their own camp
City councils Chambers of commerce Constructors' associations	• Help candidates who run from their local areas because of their interests, but sometimes betray them in the final stage by making coalitions with other candidates
Medical associations Agricultural cooperatives Women's associations Amateur sports associations Labor unions	• May make coalitions with other people, if they are beneficial to their social interests
General voters	• Are independent of political transaction, evaluating candidates mainly from TV news. May question candidates

Table 3. Message management

Contents of messages
- Sender, receiver, and time should be written clearly
- Secret political transactions should be put in parentheses to avoid disclosing them to the public via TV news
- Scandals, money and privileges without any policies are prohibited
- Replying to messages should include such items as when the original message was received, who sent it, and what was the subject
- A ballpoint pen is recommended because four carbon copies are made for one message

How to exchange messages
- A sender keeps the original message and passes three carbon copies to a messenger. Then the messenger checks if necessary items are written in the message and delivers the first copy to the news media group, the second to the receiver, and the third to the audience

How to make use of media
- Reveal policies with new ideas because new ideas are often announced via TV news
- Discourage the unity of other groups by showing policies which are attractive to each member
- A secret political transaction is not open to the public if the statement is put between parentheses

the chamber of commerce, and a chair of the constructors' association. The candidate and the three other members discuss strategies to get as many ballots as possible. Sometimes, some members betray other members in the final stage of the game. The category of social groups consists to several separate roles: medical associations, agricultural cooperatives, women's associations, and labor unions. They are loosely coupled with each other, but make coalitions if it seems beneficial to their interests. General voters are independent of the political transitions, and evaluate candidates mainly from TV news. Open questions to candidates are possible.

There are two general voters in each area and every player has one ballot except the media group. There are 33 ballots in this game, and the newscaster is the tie breaker.

Messages are exchanged with carbon copies. Players are expected to adopt suitable information exchange strategies. Instructions for players about how to manage messages is described in Table 3.

The studio layout is shown in Fig. 1. It consists of a classroom for 300 persons. There is a screen between the news media group and other players so that players cannot directly see the news media group. The audience, on the other hand, can see all the players.

Results of the Trial Run

Participants were as follows: 150 people, including 33 players, 22 media people, 12 controllers, and about 80 people in the audience. Players were from

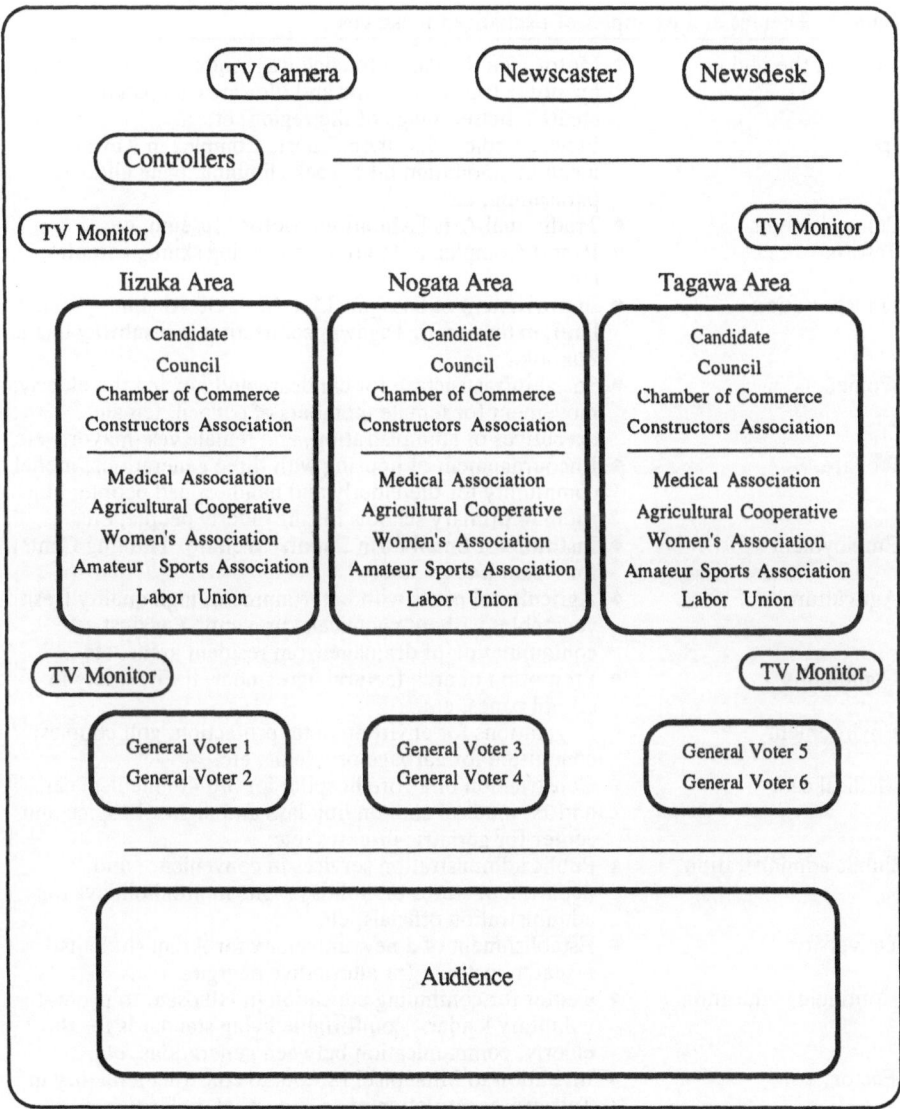

Fig. 1. Studio layout

various occupational backgrounds, such as shop owners, farmers, local government officials, workers from associations, and educators. All were volunteers from the Chikuho Seminar of Kinki University and the Chamber of Youth. The media group, with the assistance of students, consisted of a professional newscaster from RKB Mainichi, a professional reporter from NHK, newspaper people from Asahi, Mainichi and Yomiuri, and a media hardware specialist. This group played their roles effectively. Controllers consisted of similar mem-

Table 4. Themes and examples of exchanged messages

Image of the region	• Motor Island Kyushu to encourage industries in Kyushu by motor factories, Fruits and Flowers Campaign to create a better image of the region, etc.
Sports	• Flower Ladies Marathon. Sports Complex in Kawara, including mountain bike, rock climbing, hang gliding, parasailing, etc.
Cultural events	• Traditional Arts Exhibition, Motor Museum, etc.
Resorts	• Resort Complex in Hikosan, including skiing, camping, etc.
Transportation	• Improvement of Sasaguri Line (i.e., electric line, parallel line), extension to Tagawa, connection to a subway line in Fukuoka, etc.
Women	• Social infrastructure for childcare and nursing the elderly; movement for female members of council, female executives of administration, and female vice-mayors, etc.
Welfare	• Encouragement of housing with three generations, model community for the elderly and handicapped people, lunch room in primary schools for the elderly people, etc.
Employment	• Institute for Small Firm Labors' Welfare, Training Center for Asian Labors, etc.
Agriculture	• Agricultural parks with entertainment, high quality fresh vegetables without chemicals, prevention against contamination of drainage from resident areas, etc.
Community	• Promotion of architectural agreement, traffic safety in school zones, etc.
Environment	• Regulations for environmental protection, golf courses, committee for garbage problems, etc.
Medical care	• Construction of a core hospital for 3rd/4th medical care and 1st medical care on holidays and nights, hospice and center for geriatric diseases, etc.
Public administration	• Public administration services in convenience and department stores on holidays, one month holidays for administration officials, etc.
University	• Establishment of a new university for Asian students, research institute for alternative energies, etc.
Continuing education	• Center for continuing education in Hikosan, to promote voluntary leaders, comfortable living standards for the elderly, communication between generations, etc.
Factory	• Invitation to Mitsubishi Motors to construct a factory in Tagawa, construction of an industrial park for factories with relation to Toyota Motors, etc.
Commerce	• New image for the region, etc.
Laws for reconstruction	• Construction of better residences, negative opinion against laws for reconstruction, etc.
Information networks	• 24-h CATV networks, information center for agriculture, etc.

bers. The audience attended the trial run because they had read articles announced in different newspapers.

During the simulation, 466 messages were exchanged, with each person receiving an average of 14 messages. The themes and examples of messages are shown in Table 4.

Participants were satisfied with the game because they enjoyed themselves and learned a great deal about regional policies. They recognized that there was a lot to learn about the region itself. It was difficult for the players to deal with all the information from the messages, face-to-face conferences, and TV news. They found difficulty in writing all the messages. Policies were not intensively evaluated in the gaming process, and they will be evaluated before the next run. Real-time feedback via TV news was very effective because the news media group was organized by professionals.

Communication via computer networks can resolve two different levels of problems. It can accumulate knowledge in evaluating policies and it can make message exchanges much easier. New ideas and policies will be evaluated and accumulated in the database as a menu of policies after the gaming is performed. In the next run, players can choose any number of policies from the menu, revise them in formulating their messages, and send them to other players. All players can share their policy menu in the database.

The roles of players can be changed according to the situation. The media group should offer commentary programs, as well as reports on what happens in the game world. At several stages of the gaming process, it is useful to administer questionnaires to gain information on priorities or weights on policies through networking. THE MAYOR'S ELECTION is a self-developing frame game, whose aim is to accumulate new ideas and viewpoints on regional policies through playing the game repeatedly.

Conclusion

The February 1991 session was successful as a first trial run. It also drew public interest, for example a TV station reported on the session in the evening news. The game may be developed as a more structured frame game for policy exercises when suitable communication tools are developed, such as on-line systems or machine readable off-line systems, and when specialists in each field are involved to evaluate policies.

Acknowledgment

The author would like to express special thanks to Hiroharu Seki who suggested TV news as a powerful tool for feedback. The timetable of gaming follows that of the GLOBAL SIMULATION designed by Seki, which is performed annually at Ritsumeikan University.

Kiyoshi Arai specializes in computerized simulations to evaluate voting systems, and has recently been trying to introduce gaming aspects into his work. His background is in physics, economics, and engineering. He has been working in the field of social engineering for 10 years. He is a member of the Editorial Board of *Simulation & Gaming: An International Journal of Theory, Design, and Research*.

A Computer-Supported Meeting Environment for Policy Exercises

Rüdiger B. Wysk[1]

Abstract. Complex and long-range issues present significant challenges to policy-making methods such as computerized models and expert panels. Policy exercises are recognized as a more open and appropriate way to synthesize information, tools, and methods in policy-making. From brainstorming to questionnaires, from group dictionary to stakeholder identification, more tools and methods are now automated and/or supported. This article, after reviewing policy exercises, describes a COmputer-Supported Meeting Environment (COSME) framework and evaluates the use of that information technology to support policy exercises. A possible extension of policy exercises to electronic meeting systems with global reach is mentioned.

Key words analysis; communication; computer-supported meetings; decision room; electronic meeting systems; global reach; group systems; policy exercises

Policy exercises are one of the most important developments in the area of policy making involving complex and long range issues such as environmental conservation. As an emerging method, the policy exercise is facing several challenges, such as documentation at the group level (Mermet 1990, Toth 1988a). Although documentation has been a task traditionally handled by computers and recently covered by features of information technology for work groups, no system seems to have been suggested to tackle such problems in the policy exercise arena.

This article provides a description of the policy exercise process and then presents a recent development in the area of workgroup information systems, the COmputer-Supported Meeting Environment (COSME). It is an environment in the sense that several technologies, tools, and software packages are made available to a group of people who use them in an integrated way with the help of a local area network (LAN) in a decision room. The following descriptions of the policy exercise process and of the COSME provide the foundation for showing how a COSME can be used to answer several of the challenges that policy exercises are facing.

---

[1] Computer Information Systems, College of Business Administration, Northern Arizona University, PO Box 15066, Flagstaff, AZ 86011-5066, US; phone 602-523-7370; facsimile 602-523-7331; e-mail wysk@nauvax.bitnet

Policy Exercises

The policy exercise approach was identified by Brewer (1986) as a method to synthesize and assess scientific information for use in policy making. Policy exercises were then seen as a hybrid approach that would integrate the benefits of existing computerized models and expert panels within a gaming context. Historical and/or future scenarios would be generated by scientists and then presented to scientists and policy makers who, through discussions, would recommend policies for the next round of the scenario-based exercise or game (Clark 1986).

Scientists, policy makers, and support staff work through several stages to carry out a policy exercise. Toth (1988b) identified three main phases for a policy exercise: preparation, workshop, and evaluation. A sample policy exercise could consist of one preparation meeting, several parallel workshops in a couple of days, and one final evaluation meeting (see, e.g., Mermet 1990).

Difficulties exist in running a policy simulation based on a computer model because of the steep learning curve and the long runs. In general, too much procedural sophistication can be counterproductive (Mermet 1990). Another of the main difficulties is related to documenting and analyzing the recordings, paperboard sheets, and notes produced. The analysis is complicated even further because multiple stakeholders from different disciplines need to be considered. The documentation effort also limits the size of the groups. It only allows for one discussion at the same time (Mermet 1990, Toth 1988a).

A Computer-Supported Meeting Environment

The advantage of using a COSME has to do with the main bottleneck of policy exercises: for operational and documentation reasons, only one discussion can be carried at the same time in a long sequential process. A COSME allows parallel processing of issues raised in a meeting and, at the same time, automatically handles the documentation. At one site, because of this meeting structure, a comparison of man-hours expended resulted in 56% savings attributable to the use of a COSME (Nunamaker et al. 1989).

But there are other reasons for exploring the potential of a COSME for policy exercises. Planning under uncertain conditions requires flexible communication and richness of information to reduce equivocality, and data to reduce uncertainty. COSMEs allow both for the flexibility and richness of face-to-face communication and for the structure of an information system.

The face-to-face communications possible in a COSME also support the informality found in contexts that lead to the appropriate introduction of information technologies (Cooper and Zmud 1990, Johnson 1990). Because they also allow for an informal, face-to-face communication channel, COSMEs are more appropriate for complex rather than simple tasks (Bui and Sivasankaran 1990).

Table 1. A framework for the computer-supported meeting environment. Adapted from Chen and Nunamaker (1989), DeSanctis and Gallupe (1987), and Martz (1991)

Level 3	Metaplanning			
	Scheduling available tools			
	Extended PLEXSYS Session Planner			
	Scheduling tools to be generated			
	Extended METAPLEX			
Level 2	Analysis			
	Stakeholder Identification			
	Group Matrix			
Level 1	Communication			
	Issue(s) Generation	Organization	Evaluation	Exploration

The type of task for which COSMEs are used most is planning (Straub and Beauclair 1987). From the planning perspective, one of the main features of a COSME is the backbone string of tools used in this environment—issue(s) generation, organization, evaluation, and exploration. An issue-based COSME facilitates the communication between experts and non-experts (Mason and Mitroff 1981).

The following description of a COSME is based on the environment developed and used by the University of Arizona for about 5 years (Martz 1991, Nunamaker et al. 1991). It is a particular case of the Electronic Meeting Systems (EMS) environment in that it considers both the computer network and the informal face-to-face channels in a decision room. This description of a COSME follows the three levels of design suggested for all types of EMSs by DeSanctis and Gallupe (1987).

Level 1 is composed of a set of software tools and presentation technologies that support communication. Level 2 adds more analytical tools to the set. Level 3 is seen as a support to the structuring of the process that the group is going through. All three levels are shown in Table 1.

Level 1 tools generate, organize, evaluate, and explore issues/ideas. Level 2 tools, such as stakeholder identification and group matrix, allow for different kinds of analyses. The former graphically compares how stakeholders impact and are impacted by a policy based on the stakeholders' assumptions (Martz 1991, Mason and Mitroff 1981, Nunamaker et al. 1991). The group matrix displays two lists (row and column) in a way that the cells indicate the consensus among the group about the relationships mapped (Martz 1991, Nunamaker et al. 1991).

Level 3 supports metaplanning—planning the planning process. The planning of the process can be done a priori in defining the sequence of tools to be used. If the tools are available, a question and answer session planner could support that decision (Wysk 1990). System-monitored, level-1 group-process parameters can trigger a reactivation of level 3's session planner for a review of the pre-established sequence of tools (DeSanctis and Gallupe 1987).

This framework (Table 1) not only integrates different technologies to support a semi-structured approach (communication plus analysis) to reduce equivocality and uncertainty (Agarwal and Tanniru 1989), but also covers the main problems we are trying to eliminate in policy exercise processes— sequential processing of the discussions, documentation, and analysis.

Level 1 faces the communication and documentation problems. Level 2 addresses the comparative analysis and integration problems. Both levels 1 and 2, understood as one tool kit, allow for flexible arrangements of the various tools for different policy processes (Nunamaker et al. 1991).

A COSME for Policy Exercises

To show how a COSME could be used for policy exercises we need to map the appropriate tools into the existing process (Martz 1991). Practice has shown that mapping of COSME tools into an existing process leads to higher satisfaction when the method in place only needs to be automated, not changed (Bullen and Bennett 1990). The basis for this process mapping "scenario" is the policy exercise detailed by Mermet (1990). The software tools are available in the market under the GroupSystems label (Martz 1991).

Starting the *preparation* phase/session, electronic brainstorming followed by idea organization and policy formation could be used to develop a conceptual framework for the exercise and the problem statement. After the definition of the key disciplines, key policy stakeholders could be identified through the use of stakeholder identification. Group writing could help with scenario writing (see Table 2).

In the *workshop* phase, each specific workshop could be seen as a series of COSME sessions. A group could be broken up into workshops using the electronic brainstorming tool (Dennis et al. 1991). The groups could have access to policy formation and/or group writing to formulate policies and to update scenarios. Discussions could be held with the use of electronic brainstorming followed by another cycle of policy formation and group writing. Access to computer models, external data-bases, stakeholder identification, and group matrix would always be available on the network, as would the voting tool.

The *evaluation* phase would benefit not only from the documentation collected on the COSME, but also from the electronic questionnaire. A sequence of electronic brainstorming, idea organization, and topic commenter, with or without voting, could support the process in this phase. Throughout the exercise the use of the group dictionary would generate a common language between all the domains involved in the exercise. With the help of a COSME, the whole policy exercise could not only be documented but also structured for research purposes.

The key for a successful meeting is the facilitation. The same is even more true for a COSME (McGoff and Ambrose 1991). Mermet (1990) mentions that the preparation for a computer modeling workshop is a very demanding pro-

Table 2. COSME tools for a policy exercise. Adapted from Martz (1991), Mermet (1990) and Toth (1988b)

Policy exercise phase/session	COSME tools	Policy exercise use
Preparation	Electronic brainstorming	Conceptual framework problem
	Idea organization	statement
	Policy formation	
	Stakeholder identification	Identifying stakeholders
	Group writing	Scenario writing
	Group dictionary	Common language
	Session manager and planner	Support facilitation
Workshop	Electronic brainstorming	Group brake-up
	Policy formation	Formulate policies, update scenarios
	Group writing	
	Electronic brainstorming	Discussions
	Policy formation	
	Group writing	
	Group dictionary	Common language
	Computer models	Available access
	External databases	
	Other networks	
	Stakeholder identification	
	Group matrix	
	Voting	
	Session manager and planner	Support facilitation
Evaluation	Documentation	Documentation
	Electronic questionnaire	Evaluation process
	Electronic brainstorming	
	Idea organization	
	Topic commenter	
	Voting	
	Group dictionary	Common language
	Session manager and planner	Support facilitation

cess, requiring a steep learning curve. After defining the design of an exercise, the only skill required to use a COSME is typing. The design of the exercise and the coordination and support of each workshop, however, place a heavier burden on the facilitator. The session manager and session planner tools support the facilitator in her/his role.

Given its toolkit structure, a COSME can be used for several groups with different tasks within the same room. Research with COSME has shown that, for the same task, a single large group generated more ideas of better quality than several smaller groups or all the same individuals without interacting. This is in contrast to several findings with non-computer-supported studies (Dennis et al. 1991). For an overview of gains and losses in group processes using COSME see Nunamaker et al. (1991). From the above it is possible to conclude that COSME maps promisingly into policy exercises.

The sequential workshop option (A and B in Table 3) assumes that the whole group goes through the same process, A in the same room and B not

Table 3. Information technology for policy exercises.

C o m m u n i c a t i o n	c h a n n e l s			
		COSME: Face-to-face and computer	A	C
		EMS: Computer only	B	D
			Sequential	Parallel
			Workshop phase structure	

necessarily in the same room. The parallel workshop structure (C and D in Table 3) expects the group to be broken up into several subgroups which then go through different processes. C may require different rooms because of possible noise in the face-to-face channel. While D does not necessarily require separate rooms for each subgroup, it may allow for people in different rooms and buildings to participate in a meeting at the same time.

The simultaneous availability of the COSME tool kit for different groups or tasks also makes other alternatives within the EMS framework possible (Nunamaker et al. 1991). The alternatives within the EMS framework lose, however, the informal face-to-face communication channel, a possible source of group member satisfaction (Jessup and Tansik 1991) and of higher outcomes for members with integrative bargaining orientation (Sheffield 1989). On the other hand, the EMS options could take the exercises to a bigger number of participants in multiple individual or group sites and also extend them to a global dimension.

Conclusion

The descriptions of the policy exercise process and of a COSME lead us to the presentation of an overview "scenario" of how a COSME could be used for policy exercises. The challenges facing policy exercises find several answers in the use of a COSME. A COSME can also become a powerful research instrument to evaluate process gains and time savings. It could tell us which approach is more appropriate and how information technology could be used to broaden the scope of policy exercises to a global level.

References

Agarwal R, Tanniru M (1989) Technological support for decision making in the presence of uncertainty and equivocality. In: Proceedings of the 10th International Conference on Information Systems. Boston, MA, Dec 4–6, pp 19–30

Brewer GD (1986) Methods for synthesis: Policy exercises. In: Clark WC, Munn RE (eds) Sustainable development of the biosphere. Cambridge University Press, Cambridge and New York

Bui T, Sivasankaran TR (1990) Relation between GDSS use and group task complexity: An experimental study. In: Proceedings of the 23rd Annual Hawaii International Conference on System Sciences. Kailua-Kona, HI, Jan 2–5, vol. III pp 69–78

Bullen CV, Bennett JC (1990) Learning from user experience with groupware. Proceedings of the Conference on Computer-Supported Cooperative Work—CSCW'90. Los Angeles, CA, Oct 7–10, pp 291–302

Chen M, Nunamaker JF (1989) Metaplex: An integrated environment for organization and information systems development. Proceedings of the 10th International Conference on Information Systems. Boston, MA, Dec 4–6, pp 141–151

Clark WC (1986) Sustainable development of the biosphere: Themes for a research program. In: Clark WC, Munn RE (eds) Sustainable development of the biosphere. Cambridge University Press, Cambridge and New York

Cooper RB, Zmud RW (1990) Information technology implementation research: A technological diffusion approach. Management Science 36(2):123–139

Dennis AR, Valacich JS, Nunamaker JF (1991) Group, sub-group, and nominal group idea generation in an electronic meeting environment. Proceedings of the 24th Annual Hawaii International Conference on System Sciences. Kailua-Kona, HI, Jan 8–11, vol. III pp 573–579

DeSanctis G, Gallupe, BR (1987) A foundation for the study of group decision support systems. Management Science 33(5):589–609

Jessup LM, Tansik DA (1991) Decision making in an automated environment: The effects of anonymity and proximity with a group decision support system. Decision Sciences 22(2):266–279

Johnson JD (1990) Effects of communicative factors on participation in innovations. Journal of Business Communications 27(1):7–23

Martz WB (1991) GroupSystems 4.0: An electronic meeting system. Proceedings of the 24th Annual Hawaii International Conference on System Sciences. Kailua-Kona, HI, Jan 8–11, vol. III pp 799–804

Mason RO, Mitroff II (1981) Challenging strategic planning assumptions. Wiley, New York

McGoff CJ, Ambrose L (1991) Empirical information from the field: A practitioner's view of using GDSS in business. Proceedings of the 24th Annual Hawaii International Conference on System Sciences. Kailua-Kona, HI, Jan 8–11, vol. III pp 805–811

Mermet L (1990) Policy exercises in the IIASA european case study. Proceedings of the International Simulation and Gaming Association's 20th International Conference—ISAGA'90. Durham, NH, pp 172–199

Nunamaker JF, Vogel DR, Heminger A, Martz WB, Grohowski R, McGoff C (1989) Experiences at IBM with group support systems: A field study. Decision Support Systems 5(2):183–197

Nunamaker JF, Dennis AR, Valacich JS, Vogel DR, George JF (1991) Electronic meeting systems to support group work. Communications of the ACM 34(7):40–61

Sheffield J (1989) The effects of bargaining orientation and communication medium on negotiations in the bilateral monopoly task: A comparison of decision room and computer conferencing communication media. CHI'89 Conference Proceedings. Austin, TX, Apr 30–May 4, pp 43–48

Straub D, Beauclair R (1987) GDSS technology in practice (Working Paper 88-3). University of Minnesota, MISRC, Minneapolis, MN

Toth FL (1988a) Policy exercises: Objectives and design elements. Simulation & Games: An International Journal of Theory, Design, and Research 19(3):235–255

Toth FL (1988b) Policy exercises: Procedures and Implementation. Simulation & Games: An International Journal of Theory, Design, and Research 19(3):256–276

Wysk RB (1990) Expert systems in the context of decision-support-related interventions. Proceedings of the ACM International Conference on Trends and Directions in Expert Systems, Orlando, FL, Oct 31–Nov 2, pp 475–490

Rüdiger Wysk is an Assistant Professor in Computer Information Systems at the College of Business Administration at Northern Arizona University. His research interests include computer-supported meeting environments and expert systems for planning and strategic management.

Municipal Planning Room for Policy Exercises

Yasufumi Igarashi, Toshiyuki Kaneda, and Yoshinobu Kumata[1]

Abstract. The informatization of planning tasks has the potential of upgrading municipal authorities. Our study tries to create a Planning Information Room for municipalities— a highly informatized working environment equipped with various information technologies. Aiming at designing and realizing the Planning Information Room by planners themselves, an experimental project has been set up through cooperation between Utsunomiya city office and the authors. In the Utsunomiya experiment, a policy exercise (PE) as an application of simulation/gaming has been tried as a user-participation approach to designing and realizing the Planning Information Room. This paper focuses mainly on the issues of (1) our planning concept of a Planning Information Room and (2) the policy exercise for activating informatization of planning administration at the Utsunomiya city office.

Key words informatization; municipality; park location; Planning Information Room; policy exercise; system realization; Utsunomiya

In the past 30 years, many municipal governments have tried to install computerized information systems to upgrade the efficiency of information processing. Infact, many of the systems which are being substituted for routine human work have been successful. However, there is still much improvement needed in order to support non-routine work such as planning tasks.

Planning is closely linking not only with realistic decision making but also with processing ideas exchanged among politicians, citizens, administrative practitioners, and others. Therefore planning tasks are important and the informatization of planning tasks have the potential of enhancing the work of municipal authorities.

Recent innovations in computers enable us to solve many technical problems in developing effective planning-support systems. The informatization of planning tasks shifts the focus from technical to organizational problems. This is because introducing a new information system forces users to change the way they work.

---

[1] Department of Social Engineering, Tokyo Institute of Technology, 2-12-1 Òokayma, Meguro-ku, Tokyo, 152 Japan; phone 3-3726-1111 ext. 3191; facsimile 3-3729-1131

Our study tries to create a highly informatized working environment for people engaging in planning within the environment of a complex of rooms equipped with information systems necessary for the unique conditions of the municipality in question. This is the Planning Information Room (PIR). The room is equipped with various basic technologies such as network and audio-visual systems. Moreover, the room is designed to be available from the early phase of system-building and to be flexible with regards to maintenance.

Our study also uses the policy exercise (PE), which is an application of simulation/gaming as a participation-oriented design approach. Through the PE, the authors are seeking and sharing the activity images and realizing the PIR together with the users. This paper explains both our PE approach and the features of the PIR designed and realized through the PE.

The Planning Information Room

Turning Point of In-house Informatization Policies of Japanese Municipalities

Since the first computer was introduced in 1960, Japanese municipal governments have been depending on their in-house information systems more closely. Japanese municipal governments have so far basically succeeded in pushing forward with their implicit informatization policies.

Computerized information systems that are designed to substitute for human routine tasks have grown. These systems work with many service tasks, such as an on-line tax service and citizen record management. However, for the non-routine tasks such as planning and forecasting, only a few examples have survived (Ministry of Home Affairs 1990).

Since the emergence of PCs in the 1980s, the revolution of municipal computing has become worldwide. Today, the management of computers (including mains, minis, WSs, and PCs) is shifting drastically from centralized forms to decentralized ones, particularly with the explosion in use of PCs in the 1990s (King and Kraemer 1991). The informatization policy focusing on "Centralized, Few, and Big" is now collapsing in Japanese municipalities. It is technically possible for every municipality to give a PC to each employee. The "Decentralized, Many, and Small" policy can be conceived, but there still remains organizational and human resource problems involving computer literacy, technical support for PCs, and the multi-vendors', network management principles.

Many of the municipalities need organization-wide development projects for in-house informatization to provide both a mass-introduction of PCs and the construction of integrated networks.

Two Informatization Strategies for Introducing Planning-Support Systems

This study focuses our development target of in-house informatization towards comprehensive municipal planning. Comprehensive plans are administrative guidelines. However, more than 99% of Japanese city governments have their own comprehensive plans. Each comprehensive plan is centered in the planning administration system of each municipality because of the transsectional wide comprehensiveness and an adequate planning time (usually 5–15 years).

Using the data of the Taskforce on the Comprehensive Planning Research of JAPA (1990), typical characteristics of planning tasks have been found. The planning tasks need not only realistic decisions but also futuristic creations among transsectionally wide administrative practitioners, as well as the citizens. This process can be described as a highly complicated parallel-knowledge-processing model. The informatization of the comprehensive planning tasks can be expected to contribute to the enhancement of the organizational decision process.

There are two introduction strategies of in-house information systems of munici-palities: the top-down strategy and the bottom-up strategy. Both strategies are complementary to each other; thus, successful informatization requires the integration of both. However, only the top-down strategy glitters sometimes with technology impact and high funding [see, for example, UIS projects initiated by the Ministry of Construction (1982)], and there are still few studies to refer to the latter bottom-up strategy. Our PIR project is linked to both strategies. Moreover, the authors prepare the PE as a design approach

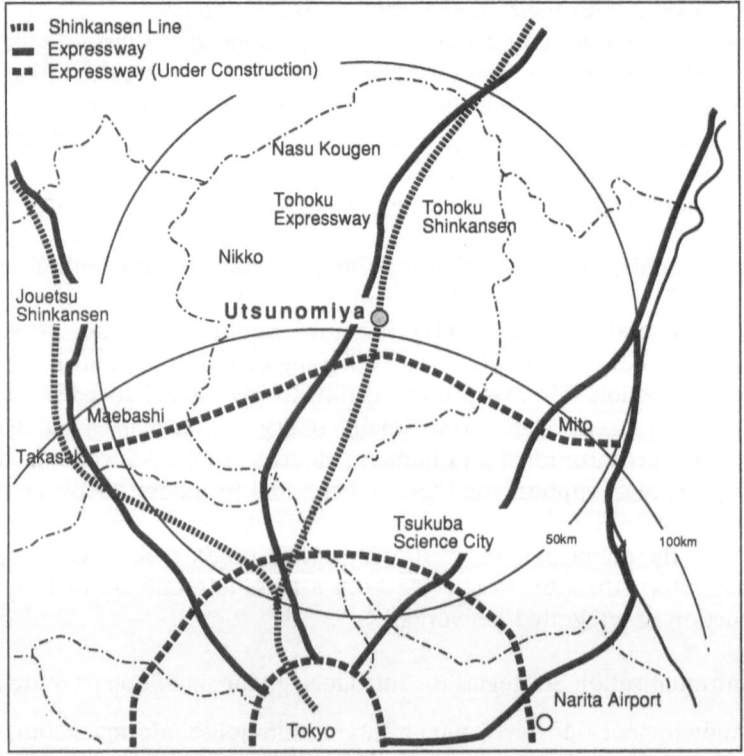

Fig. 1. Location of Utsunomiya City

to the PIR through the bottom-up strategy. This study aims at designing and realizing a PIR by applying the PE method.

Planning Information Room Designed Through PEs

The PIR is a physical and informational environment equipped with various information systems. Planning tasks includes a complex variety of independent, cooperative, physical, informational, and judgmental actions. Our PIR functions as a working environment for such tasks. The room also functions as a basis for experimental projects for organizational informatization.

The requirements of the PIR can be summarized as follows: total support for intelligent cooperative work, easy to use for beginners, ease of reconstruct and extend, and ease for maintenance. Our study deals with the realization of the PIR at the Utsunomiya city office. It is thus designed with cooperation between the city officials and us through PE.

In a PE an actual situation which needs some policy formation is modelled. Then, the practitioners attend to the role-playing in the modelled situation, in that they can be allowed some trial and error. Lastly, they devise and choose the best policy based on their exercises. Arai (1990) has already reported a PE which consists of a creation of a future image of the Chikuho region involving many citizens. Our PE has a double meaning. The first meaning is a PE concerning designing and realizing of the Planning Information Room, and the second meaning is a PE on the actual making of a policy (in our case, a park construction project) in the PIR. The room itself functions as a field for PEs.

Planning Information Room Project for Utsunomiya City

Profile of Utsunomiya City

Utsunomiya is a city in Tochigi prefecture about 100 km north of Tokyo. Since the Edo era, Utsunomiya has been developing steadily as the commercial center of the northern part of the Kanto plain. Now Utsunomiya has about 0.4 million citizens and is the 40th largest municipality, a typical medium-sized city in Japan. Utsunomiya has the potential to grow even further, considering the effects of the Tohoku-Shinkansen and the Technopolis project of MITI (Fig. 1).

A planning organization in the city office, called the planning division, has three task teams: a long-range planning team, a statistical research team, and a planning support team. The PIR project is organized under the collaboration of the planning support team and the authors.

Attitude of the Utsunomiya City Government Concerning Informatization

The Utsunomiya city government has promoted informatization of the administrative work, and has made an attempt to facilitate the use of information systems for planning activities. The authors have cooperated with the Utsunomiya city government as a project of the Japan Administration and

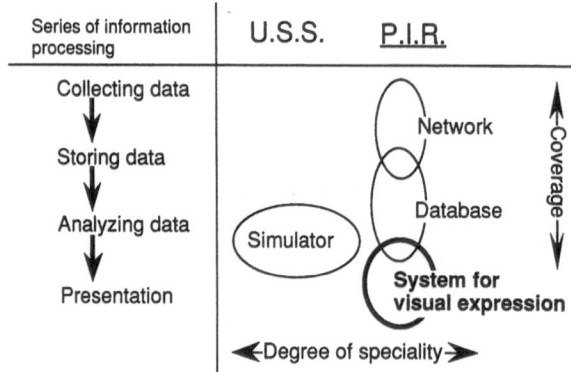

Fig. 2. The stages of information processing and system coverage

Planning Association, and have developed the city simulator named Utusnomiya Symbolic Simulator (USS) (Kaneda et al. 1990). This simulator is based upon a system dynamics model and is intended as a tool for discussing policies through examining the balance between financial planning and physical planning. The USS system has developed a tool for considering the feasibility of long-range planning, and this is utilized by the statistical research and planning support teams. However, the system requires semi-professional knowledge for handling the city model, which means that the number of users of this system is limited.

Therefore, increasing the number and type of user is the next goal in the implementation phase of the simulation development. The PIR project was started in order to achieve this subject.

The Planning Information Room at Utsunomiya

Comprehensive planning is essentially information processing. The processes are collecting, storing, and analyzing large amounts of information for developing plans. The series of information-processing steps for the administration and for the support systems are shown in Fig. 2.

USS mainly covers the analyzing phase of the information-processing stage. Although the simulator has its own database and display systems, it was not sufficient to impart the knowledge acquired from the result of simulation. As comprehensive planning involves cooperative work among various sectors, knowledge conveyance is the key factor dominating the performance of the planning. Therefore, the current concern is to support activities in the presentation phase. This time the PIR system covers the phase of presentation first. The collecting and storing of data phases support the function of presentation.

PEs for Activating the Use of Information System

The coverage of PIR can be identified by the necessity of planning. However, the way to use the system in practice is decided by the officers. In order

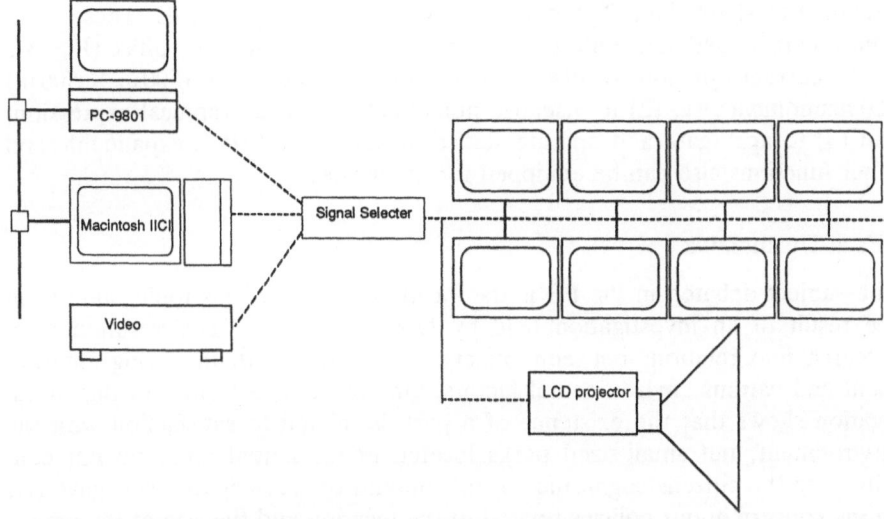

Fig. 3. System components of the PIR

to decide the PIR activity, the design process of this system has to involve officers who are concerned with the activities of planning—this is part of the particiapatory approach of PEs. For its purpose, the fundamental requirements of PEs are that (1) the functions are easy to understand for officers and (2) the subject reated in the exercise is suited to their work. These are the viewpoints for planning the PE.

Objectives of the Planning Information Room

The objective of the PIR is to facilitate top management level meetings in the Utsunomiya city office. The PIR has the purpose of boosting the effectiveness of information exchange through the visual expression of such things as graphical figures and geographic information. The current functions include a database to store statistical data of Utsunomiya and to display mesh maps and figures from the database according to the discussion in meetings. The fundamental requirements are the ability to deal with requests to create these figures on the spot.

Hardware and Functions of the PIR

The PIR consists of three subsystems: a network, a database, and a system to visualize information. The number of users is between 10 and 15. Monitors to display output from personal computers or workstations include 14-inch CRT displays and a 100-inch projection-type display. This system enables us to share and discuss an image created by the computers. Computers for the data-base functions and for creating graphical figures are connected by a (local area network) to share data. The data format is based upon dBASE. These components are settled in a conference room of the Utsunomiya city office (Fig. 3).

The current functions of PIR are (1) to display mesh maps (500 m × 500 m) of Utsunomiya city, (2) to visualize numerical data as a graphical expression, and (3) to aggregate and operate statistical data. The PIR is expandable and other functions also can be equipped for other needs.

PEs at Utsunomiya

The subject debated in the PE is the location of parks. This topic came from the result of an investigation held by Utsunomiya city. This resulted from research into relations between citizens' satisfaction with the living environment and various environmental factors. One of the conclusions of this investigation shows that the existence of a park is related to satisfaction with the environment, but small-sized parks located in residential areas do not contribute to the citizens cognizance of the amount of green space. This gave rise to the concern about policies related to the location and the size of the park.

Participants. This PE was held as an activity of a study group to facilitate information systems for planning. Members of this group are practical class officers from several sections related to long-range planning. Thus, participants in the PE have positive opinions on information systems for planning.

Structure of the policy exercise. The aim of this policy exercise is to figure out the image of policy formation meetings using the PIR. The participants are required to remember the aim and are requested to:

1. Present the personal opinion of what a park should be
2. Argue the policy relating to parks (philosophy, location, and facilities) as a public administrator
3. Think about how a meeting should be organized and conducted
4. Make a scenario of a meeting of policy making in the PIR. Through these tasks, we aimed to draw and share the image of activities in the PIR.

Experiments with the PE

The first PE was held in August, 1991. Twelve participants attended the meeting from various divisions such as city planning finance, secretary for the Mayor, general affairs, task management, and planning. It took about 3 h to debate several topics including the park location project. Further meetings are planned for once a month.

Conclusion

The PIR is conceptualized as a new in-house informatization measure for municipalities. The essential characteristics of the PIR is flexibility to extend its components through user initiatives. An experimental project has been conducted with cooperation between the Utsunomiya city office and the authors.

In the Ustunomiya experiment, the authors and the officers of Utsunomiya municipality also tried a PE as a user-participation approach for designing and realizing the PIR.

The success in the Utsunomiya city office shows that this approach is transferable to other medium- or small-sized municipalities.

References

Arai K (1990) Learning support systems for creating a future imageol a region. In: Proceedings of SCOPE 2000 of ISAGA/NASAGA 1990, Durham, NH

Kaneda T, Yagaguchi N, Igarashi Y, Kumata Y (1990) Utsunomiya Symbolic Simulator (USS) as a quantitative simulator for practical use. SCOPE 2000 of ISAGA/NASAGA 1990, Durham, NH, USA

King J, Kraemer K (1991) Patterns of success in municipal information systems: Lesson from US experience. Informatization and the Public Sector 1(1)

Ministry of Construction (1982) Urban information database (in Japanese). Keibun, Tokyo

Ministry of Home Affairs (1990) Almanac on municipal computing use (in Japanese). Ministry of Home Affairs, Tokyo

Taskforce on Comprehensive Planning Research of JAPA (1990) Newsletter on comprehensive planning research no. 1 (in Japanese). JAPA, Tokyo

Yasufumi Igarashi has a Master's of Engineering from T.I.T. and is now teaching computer simulations and statistics as an assistant professor at Kumata Laboratory. His main interest is knowledge-based simulation.

Toshiyuki Kaneda has a Doctor of Engineering from T.I.T. He now deals with teaching and research activities as an assistant at the department of Social Engineering in T.I.T.. His recent interests are in organizational development methodology for Japanese municipal governments in the information age.

Yoshinobu Kumata, professor of planning theory, has been teaching a course on gaming and simulation at the planning department of T.I.T.. He is still active in developing gaming models as effective tools for university education, on-the-job training, and methods for evaluating urban development plans and programs.

Space Influences on Earth's Ecological and Economic Systems

Norihisa Kaneda, Yoshio Ishikawa, Tatsuo Motohashi[1],
Yoshiki Yamagiwa[2], and Kyoichi Kuriki[3]

Abstract. The remarkable growth of the world's economy makes us feel its growth is unlimited. Economic expansion, however, induces many kinds of unfavorable influences on the environment: the exhaustion of natural resources, the destruction of nature, and a shortage of food. Finally we cannot know the limits of growth. One of the remedies for these ill effects is to promote the utilization of space around the earth. A lunar base and solar power satellites (SPSs) are projected developments. The influence of these developments on the terrestrial systems is assessed here by means of a simulation. The method of "System Dynamics" developed by J.W. Forrester was used to analyze the evolution of the systems considered. Our model involves three subsystems: the earth, the lunar base, and the SPS. The following results are obtained from the present simulation:

1. The use of lunar resources will help us reduce the consumption of terrestrial resources and improve the economic situation on earth.
2. SPS will enable transfer of electric power to earth after an initial stage, which provides a solution to a potential energy crisis. Moreover, we are prevented from Polluting the air, thereby ensuring economic growth.

Key words Earth; ecology; economy; energy; moon; simulation; space development; SPS; system dynamics

At present we are faced with many serious problems, such as the exhaustion of natural resources, the destruction of environments, and a shortage of food, coupled with growth in population and economy. We suggest that it may be possible to solve these serious problems by developing the space around the earth. It is thus important to know which is the better scenario.

In this report, the development of lunar resources and the construction of a space power satellite (SPS) system are chosen as examples of space develop-

[1] Department of Aerospace Engineering, College of Science and Technology, Nihon University, 7-24-1 Narashinodai, Funabashi, Chiba, 274 Japan; phone 0474-66-1111; facsimile 0474-67-9569
[2] College of Engineering, Shizuoka University, 3-5-1 Johoku, Hamamatsu, Shizuoka, 432 Japan; phone 053-471-1171; facsimile 053-472-0251
[3] Institute of Space and Astronautical Science, 3-1-1 Yoshinodai, Sagamihara, Kanagawa, 229 Japan; phone 0472-51-3911

246

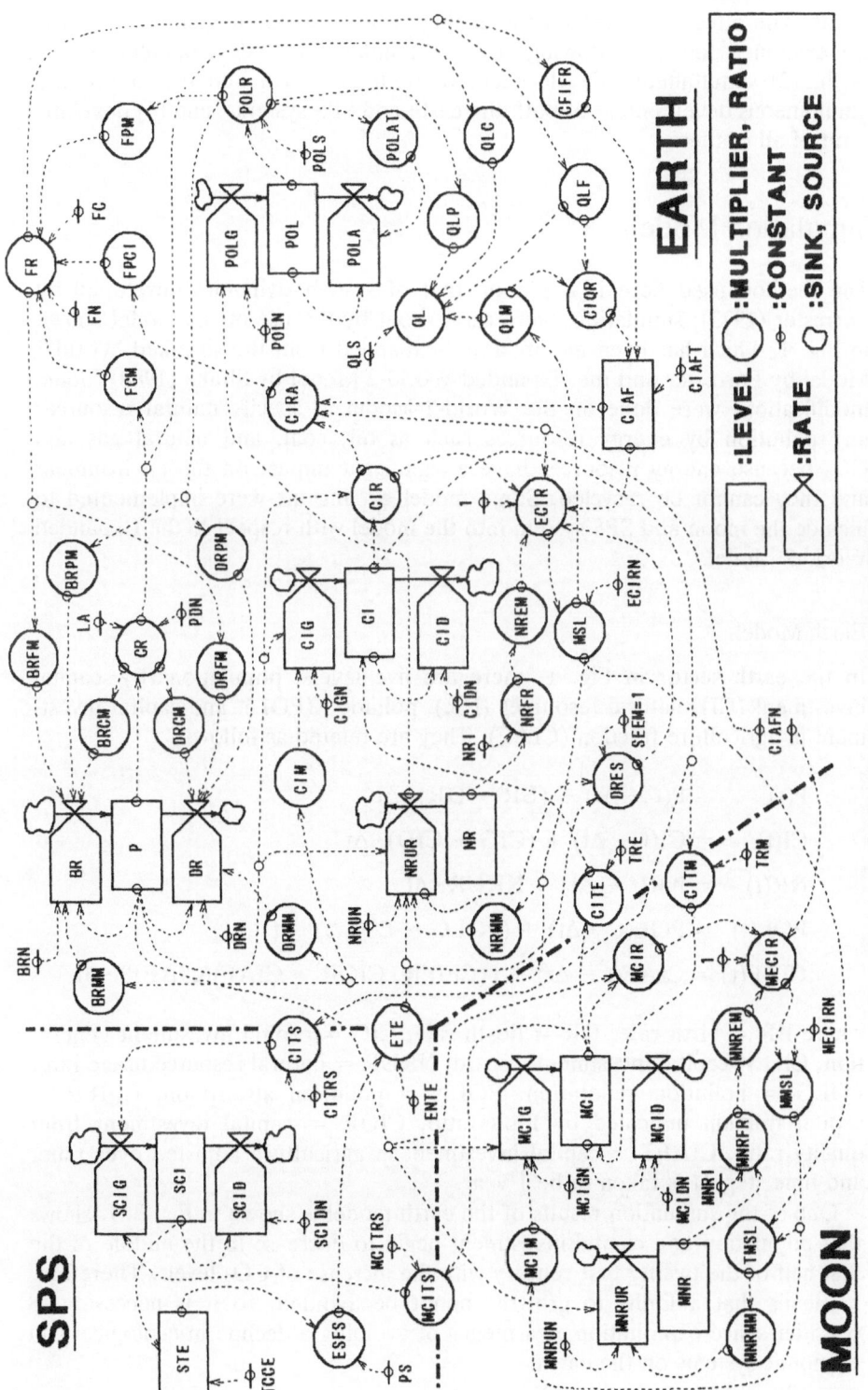

Fig. 1. Earth-moon-SPS diagram. See Appendix for abbreviations

ments. The effects of these on the ecological and economic systems on earth are examined for the following cases: (1) independent development of the earth, (2) simultaneous development of both the earth and the moon, (3) simultaneous development of both the earth and SPS systems, and (4) development of all of these.

Simulation Models

The method used here is based on that of system dynamics developed by Forrester (1973); simulations were carried out by the use of the model shown in Fig. 1, which has been modified and expanded from the so-called World-2 Model by Forrester and the Expanded World-2 Model by Shultz (1988). Some modifications were done on the World-2 Model to specify natural resources and pollution by energy resources such as oil, coal, and natural gas and CO_2 because energy resources have a significant impact on the environment and they cannot be recycled. Some model expansions were implemented to include the moon and SPS system into the model with respect to the Expanded World-2 Model.

Earth Model

In the earth sector in Fig. 1, there are five levels: population (P), capital investment (CI), natural resources (NR), pollution (POL), and capital investment in agriculture fraction (CIAF). They are related as follows;

$$P(t) = P(t - \Delta t) + (BR - DR) \cdot \Delta t$$

$$CI(t) = CI(t - \Delta t) + (CIG - CID) \cdot \Delta t$$

$$NR(t) = NR(t - \Delta t) - NRUR \cdot \Delta t$$

$$POL(t) = POL(t - \Delta t) + (POLG - POLA) \cdot \Delta t$$

$$CIAF(t) = CIAF(t - \Delta t) + ((CFIFR \cdot CIQR - CIAF)/CIAFT) \cdot \Delta t$$

where BR = birth rate, DR = death rate, CIG = capital investment generation, CID = capital investment discard, NRUR = natural resource usage rate, POLG = pollution generation, POLA = pollution absorption, CFIFR = capital fraction indicated by food ratio, CIQR = capital investment from quality ratio, CIAFT = capital investment in agriculture adjusted over time, and time step Δt is taken to be 1 year.

One of the simulation results of the earth model is shown in Fig. 2. It shows that population and capital investment begin to decrease in the middle of the first half of the twenty-first century with the increase of CO_2 levels. Therefore, it means that a limits to growth cannot be avoided, so it is necessary to consider space exploitation as a means of avoiding a decline in ecological and economic systems on the earth.

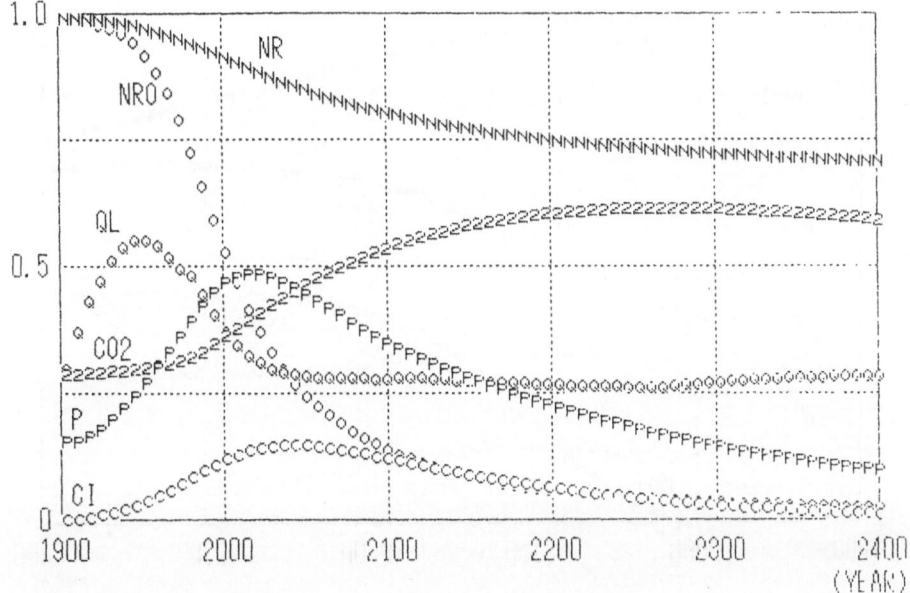

Fig. 2. World model (present model). Axis ordinates: Population (*P*), 1×10^{10} people; capital investment (*C*), 5×10^{10} capital units; natural resources (*N*), 3.24×10^{13} barrels; oil resources (*O*), 2×10^{12} barrels; CO_2 levels (*2*), 1,000 ppm; quality of life (*Q*), 2

Fig. 3. Earth-moon model

Earth-Moon Model

The moon sector of the model shown in Fig. 1 consists of two levels: energy resources (MNR) and capital investment (MCI). These are similar to the same parts of the earth sector of the model, and they are connected with the earth sector through the exchange of capital investments, as shown in Fig. 3.

In this case, capital investment from the earth to the moon (CITM) will be started from SYEAR (2000) when moon development begins and capital investment from the moon to the earth (CITE) is returned every year from MYEAR. This assumes that the capital investment fraction to the moon (CIFM) is equal to 0.3, which means that MYEAR is the year when accumulation of capital of the moon amounts to 30% of that of the earth. [3] He can be thought of as an energy resource on the moon (MNR), from which electric

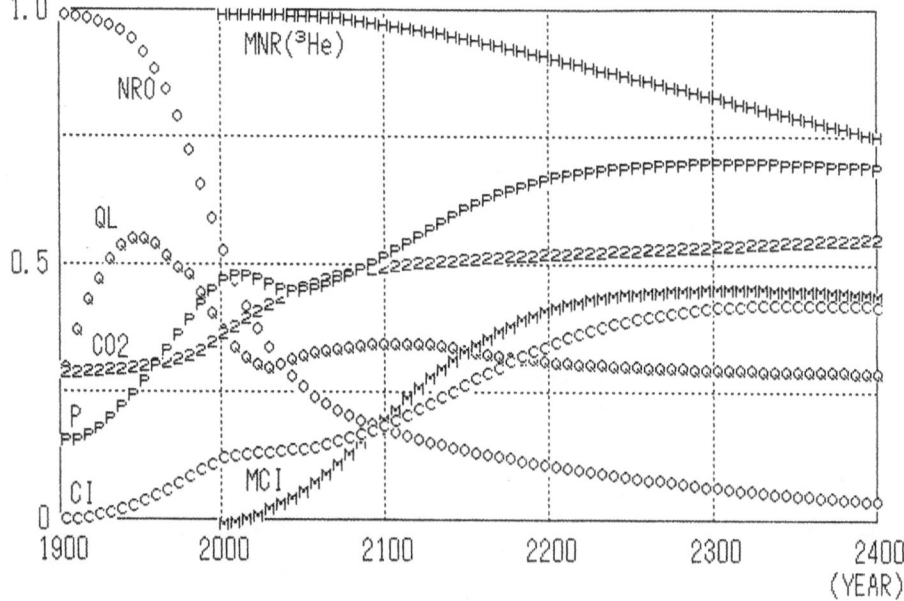

Fig. 4. Earth-moon model. Axis ordinates: Population (*P*), 1 × 10^{10} people; capital investment (*C*), 5 × 10^{10} capital units; oil resources (*O*), 2 × 10^{12} barrels; moon capital investment (*M*), 5 × 10^{10} capital units; moon natural resources (*H*), 1.1 × 10^{9} kg; CO_2 level (*2*), 1,000 ppm; quality of life (*Q*), 2. Parameters: SYEAR, 2000; CIFM, 0.3; MCIDN, 0.025; TRM, 0.004; TRE, 0.02; MYEAR, 2026

energy will be generated by nuclear fusion. The total amount of ^{3}He on the moon is estimated at 1.1 × 10^9 kg. Adopting the value of 0.025 for the capital investment, normal discard on the moon (MCIDN) results in 40 years' life span of machines and devices on the moon.

The result in Fig. 4 shows that population can increase generally until the middle of AD 2300, except a small decrease in the first half of the twenty-first century caused by a time lag of the effects of moon developments and that CO_2 levels will be lower than that of the earth model. As described above, it seems possible to avoid the crisis by using ^{3}He on the moon as an energy resource.

Earth-SPS Model

The SPS system has an electric power generation ability of 5 GW per unit, which can be supplied to the earth in microwaves of 2,450 MHz. According to a NASA/DOE report (Koomanoff and Riches 1980), the production cost for the first SPS system, which has a generation ability of 5 GW, was estimated at US$20,500 per kW and for the following SPS systems, between US$3,100 and US$16,700 per kW. In this case, the value of US$3,500 per kW is adopted.

The relationship between the earth and SPS is shown in Fig. 5. The capital investment to construct SPS systems from the earth (CITS) will be started in

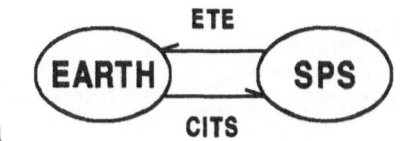

Fig. 5. Earth-SPS model

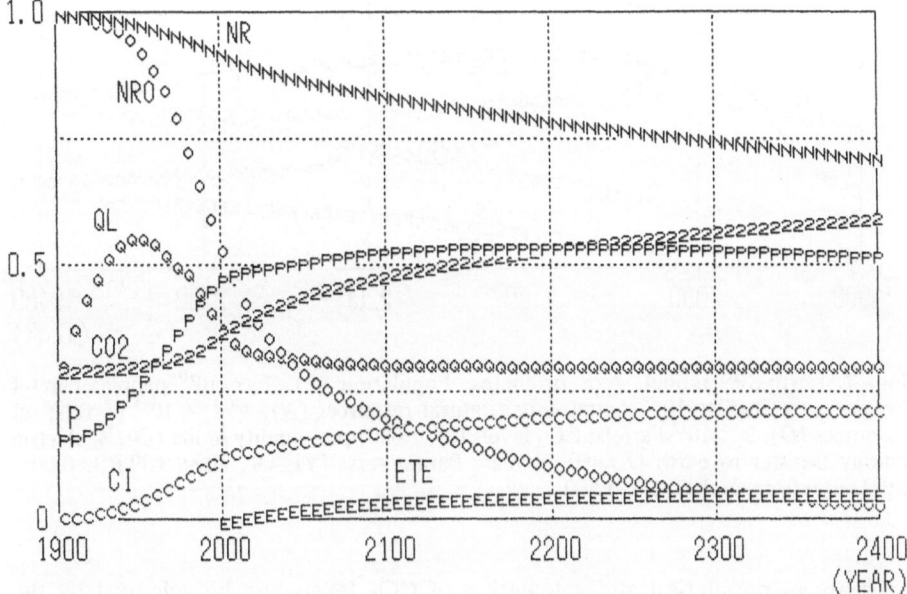

Fig. 6. Earth-SPS model. Axis ordinates: Population (P), 1×10^{10} people; capital investment (C), 5×10^{10} capital units; natural resources (N), 3.24×10^{13} barrels; oil resources (O), 2×10^{12} barrels; CO_2 levels (2), 1,000 ppm; quality of life (Q), 2; electric energy transfer to earth (E), 50 TWyear. Parameters: IYEAR, 2000; CITRS, 0.004; SCIDN, 0.025; DOLE, US$3,500

IYEAR (2000) and continued every year. The SPS systems will supply electric energy (ETE) to the earth to the tune of 5 GW to substitute for the fossil fuels consumed on the earth.

The simulation results are shown in Figs. 6 and 7. Figure 6, in which the construction cost of SPS systems is the same as that of the developmental cost on the moon, shows that population can increase and CO_2 level can be suppressed until the year 2200. But after 2200, the population cannot be maintained because of the lack of electric power generation capacity of the SPS systems caused by a significant depreciation in the SPS system (SCIDN), an increase in CO_2 level, and a decrease in the accumulation of the earth's capital (C).

Moreover, the capital investment transfer ratio from the earth to SPS (CITRS) in Fig. 7, is twice as much as that in Fig. 6, which shows that an

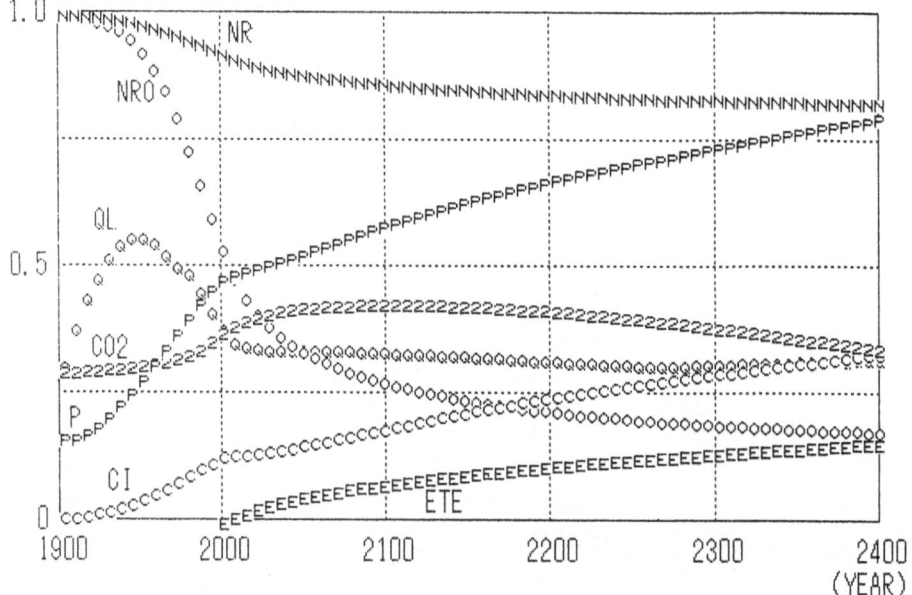

Fig. 7. Earth-SPS model. Axis ordinates: Population (P), 1×10^{10} people; capital investment (C), 5×10^{10} capital units; natural resources (N), 3.24×10^{13} barrels; oil resources (O), 2×10^{12} barrels; CO_2 levels (2), 1,000 ppm; quality of life (Q), 2; electric energy transfer to earth (E), 50 TWyear. Parameters: IYEAR, 2000; CITRS, 0.008; SCIDN, 0.025; DOLE, US$3,500

increase in population and a lowering of CO_2 levels can be achieved by the adoption of the value of SCIDN. This becomes clear in comparing Fig. 6 with Fig. 2. In short, it is clear that considerable capital investments for the construction of SPS systems are necessary to make enough profit in the case of single construction of SPS systems.

Earth-Moon-SPS Model

The relationship of the earth, the moon, the SPSs will proceed in the following three steps (Fig. 8). In the first step, capital investment from the earth to the moon and SPS will be started at the same time (in AD 2000) and the energy supply from SPS to the earth will be increased by 5 GW each time an SPS is constructed. In the next step, capital investment to the moon will be stopped and the capital investment needed to construct an SPS from the moon (MCITS) will be started in order to take over that from the earth at MYEAR, when accumulation of the capital of the moon amounts to 20% of that of the earth. In the final step, all of the construction costs of the SPSs will be covered with the capital investment from the moon. The construction costs of the SPSs from the moon (DOLM) is assumed to be US$2,800 per kW because the gravity of the moon is one sixth of that of the earth.

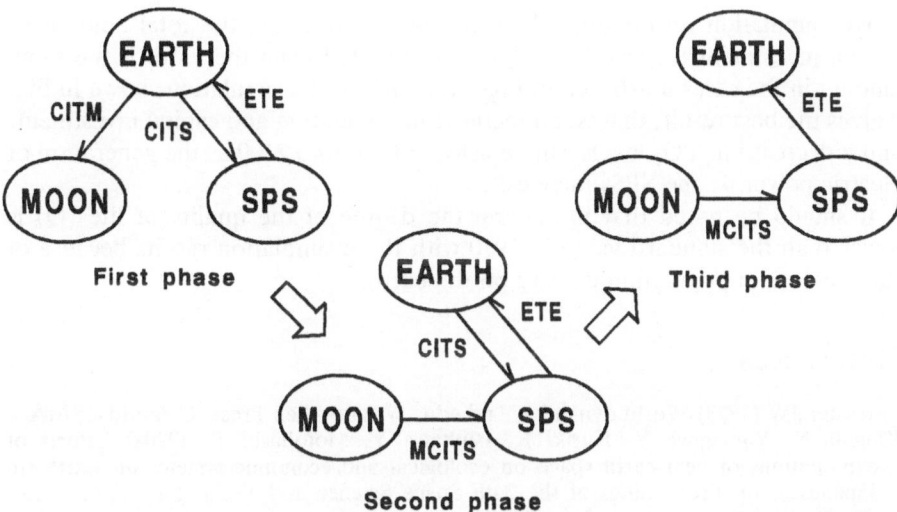

Fig. 8. Earth-moon-SPS model

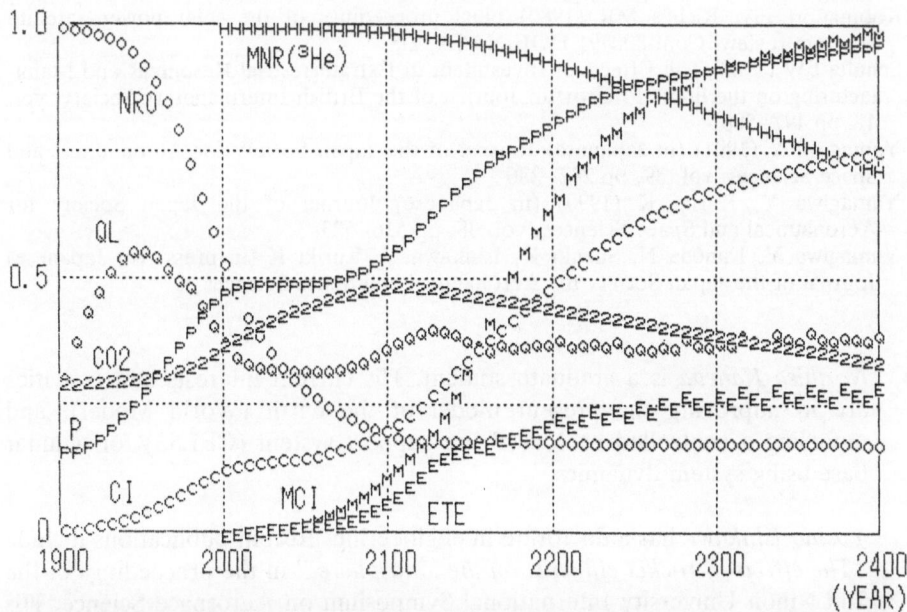

Fig. 9. Earth-moon-SPS model. Axis ordinates: Population (P), 1×10^{10} people; capital investment (C), 5×10^{10} capital units; quality of life (Q), 2; oil resources (O), 2×10^{12} barrels; moon capital investment (M), 5×10^{10} capital units; moon natural resources (H), 1.1×10^{9} kg; CO_2 levels (2), 1,000 ppm; electric energy transfer to earth (E), 50 TWyear. Parameters: SYEAR, 2000; IYEAR, 2000; CIFM, 0.2; MCIDN, 0.025; TRM, 0.001; MCITRS, 0.02; DOLM, US\$2,800; CITRS, 0.003; SCIDN, 0.025; DOLE, US\$3,500; MYEAR, 2042

The simulation results are shown in Fig. 9. Although the total amount of capital investment for space development invested from the earth is the same amount in the cases illustrated in Figs. 4, 6, and 9, the simulation shown in Fig. 9 gives the best result; that is, an increase in population and capital investment, and a decrease in CO_2 levels can be achieved after AD 2100 as the generation of electric power by the SPSs increases.

It should be noted that in general the degree of the quality of life (Q) is lower than the standard value in 1970 with these simulation results because of the increase in population density on the earth.

References

Forrester JW (1973) World dynamics, 2nd edn. Wright-Allen Press, Cambridge, MA

Kaneda N, Yamagiwa Y, Kuriki K, Ishikawa Y, Motohashi T (1990a) Effects of exploitations of near-earth space on ecological and economic systems on Earth (in Japanese). In: Proceedings of the 34th Space Science and Technology Conference, Oct 29–31 1990, Japan, pp 288–289

Kaneda N, Suzuki K, Ishikawa Y, Motohashi T, Yamagiwa Y, Kuriki K (1990b) Influences of developments of space around earth on earth's ecological and economic systems (in Japanese). In: Proceedings of The Tenth ISAS Space Energy Symposium, Feb 14–15 1991, Kanagawa, pp 37–43

Koomanoff FA, Riches MR (1980) Final proceedings of the solar power satellite program review, Conf-800491 DOE/NASA, US

Schultz FW (1988) The Effects of Investment in Extraterrestrial Resources and Manufacturing on the limits to Growth. Journal of the British Interplanetary Society, vol. 41, pp 497–508

Yamagiwa Y (1991) (in Japanese) Journal of the Japan Society for Aeronautical and Space Sciences, vol. 39, pp 282–286

Yamagiwa Y, Kuriki K (1990) (in Japanese) Journal of the Japan Society for Aeronautical and Space Sciences, vol. 38, pp 516–523

Yamagiwa Y, Kaneda N, Suzuki K, Ishikawa Y, Kuriki K (in press) (in Japanese) Journal of the Japan Society for Aeronautical and Space Sciences

Norihisa Kaneda is a graduate student. His current interests and activities are in improving the present model of the earth (World Model), and modeling a controlled ecological life support system (CELSS) for a lunar base using system dynamics.

Yoshio Ishikawa has a doctorate in engineering. Recent publications include "*The effects of rocket effluents on the ionosphere*," in the proceedings of the 2nd Nihon University International Symposium on Aerospace Science. His current interest is in system dynamics simulation of a controlled ecological life support system (CELSS).

Tatsuo Motohashi has a doctorate in engineering and has recently published "*Instabilities around an isolated roughness element*," in the proceedings of the American Society of Mechanical Engineers and Japan Society of Mechanical Engineers Joint Conference 1991. He is interested in turbulence structure and laminar to turbulent transition. A favorite short quote is "Science is nothing without generalizations. Detached and ill-assorted facts are only raw material and, in the absence of a theoretical solvent, have but little nutritive value."

Yoshiki Yamagiwa has a doctorate in engineering and has published "*A model of solar power satellite for world dynamics simulation,*" in the proceedings of the 1991 Solar World Congress, August 1991, as well as "*An Evaluation Model of the Moon Resource Exploitation by the System Dynamics Simulation—Improved Points and Parameter Analysis,*" in the Journal of the Japan Society for Aeronautical and Space Sciences. Current interests and activities include application of system dynamics to the evalution on space development, and study of space propulsion and traveling methods in the future.

Kyoichi Kuriki obtained a doctorate in engineering from the University of Tokyo in 1963. Current interests are electric propulsion, microgravity, lunar bases, and space frontier systems. He is an aerospace committee member of the Science Council of Japan, as well as an American Institute of Aeronautics and Astronautics Technical Committee Member of Electric Propulsion and a Lunar Base Committee Member of the International Academy of Astronautics. He has also been co-investigator of space experiments with particle accelerators (SEPAC) for the Spacelab 1 Mission (1976–83) and is currently leader of the SFU (Space Flyer Unit) Project (1986).

Appendix

List of Abbreviations

SYEAR : Start year for moon development
IYEAR : Start year for investment for SPS
P   : Population
CI   : Capital investment
NR   : Natural resources
NRO   : Oil resources
POL   : Pollution = ppm of CO_2
CIAF   : Capital investment in agriculture fraction
QL   : Quality of life
FR   : Food ratio
MCI   : Moon capital investment
MCIDN : MCI discard normal
MNR   : Moon natural resources
CIFM   : Capital investment fraction to the moon
MYEAR : Year when MCI amounts to some fraction of CIFM
MCIX   : MCI at MYEAR
CITM   : CI transfer to moon [CITM = CI × TRM]
TRM   : Transfer ratio to moon

CITE   : MCI transfer to earth [CITE = (MCI − MCIX) × TRE]
TRE   : Transfer ratio to earth
SCI   : SPS capital investment
SCIDN : SCI discard normal
STE   : SPS total energy
CITS   : CI transfer to SPS [CITS = CI × CITRS]
CITRS   : CI transfer ratio to SPS
MCITS : MCI transfer to SPS [MCITS = (MCI − MCIX) × MCITRS]
MCITRS: MCI transfer ratio to SPS
DRES   : Dependent ratio of earth energy to SPS
DOLE   : US$/kW from earth for constructing SPS
DOLM : US$/kW from moon for constructing SPS
ETE   : Electric energy transfer to earth

Section 4
Economics and Business

Global Changes in Business and Economics

Richard D. Teach[1]

Abstract. The winds of change are blowing on the earth. While there are unseen forces at work, many are known. The biggest change is the recognition that it is the market place and not command economics that drives economic activity and growth. As global markets grow, there will be room for many entrepreneurial endeavors. One can expect to see some equalization of economic well-being across the economically-advanced nations, and more countries being admitted to this club. Simulation and gaming will play a major role in training the managers and planners of the future.

Key words economic changes; global economics; the role of gaming; teaching market place economics

It may be trite to say, but the only two sure things about the future are that (1) it will occur and (2) it will be different. In the 1890s, a bill was introduced into the lower house of the United States Legislature to close the Patent Office. "There would soon be no need for it as there was nothing left to invent. All of the possible useful and meaningful inventions had already taken place!" The bill did not pass.

If one reads the prognostications of only a few years ago and compares the forecasts with current reality, they have been little better than the science fiction writers of the 1930s and 40s. The future is coming but we do not know what it will bring. At the ISAGA meeting in Weimar, how many were able to predict the fall of the "Wall" and the movement of Eastern Europe to new governments, or the current breakup of Yugoslavia? Even 1 year ago today, who could foretell the Gulf War? Even if one cannot predict the future, there are several happenings which will shape this future we all will share.

[1] School of Management, Ivan Allen College of Management, Policy and International Affairs, Georgia Institute of Technology, Atlanta, GA 30332-0520 US; phone 404-894-43-55; facsimile 404-894-60-30; e-Mail RT17@Hydra.GaTech.edu

Worldwide Recognition of the Market Place as the Driving Force Behind Economic Activity

The primary change in the world's economic philosophy has been the almost universal adoption of the consumer sovereignty or the market-place precepts of Adam Smith. The market place with its inefficiencies, redundancies, shortages, surpluses, and miscalculations will shape economic activities. However, there are some important parts of this philosophy which all of us seem to forget from time to time. The part about the "hidden hand" must not be overlooked. The power and danger of either self-made cartels or government-sponsored and/or protected cartels are enormous. One cannot protect the automobile worker in Europe, the electronics worker in the US, or the rice farmer in Japan without economic consequences to and from each affected party.

However, governments must remember to temper the efficiency of a market economy with compassion for those who may not have all the skills to survive the harsh realities of the free market in the short run.

One of the phenomena of global marketing is that it creates niche markets. As firms such as Coca Cola, Sony, and Siemens, with worldwide product recognition, market their products to the world's buyers, cultural differences and local customs are not used to a great extent in the marketing appeals and promotions. These local needs may be too small or change too fast for multinational firms to pursue, but they are perfect markets for small, quick-acting, entrepreneurial firms to exploit.

The time it takes from the birth of a concept or idea to its realization in products available to the general public is getting shorter and shorter. For example, the general adoption of the telephone in the United States took 30 years. Bell Telephone was founded in 1887, and by 1916 the telephone was in general use in the US. The adoption of television in the US was much faster; it took only about 15 years. In 1946, 24 licenses were granted for television stations and by 1960, well over 85% of US households had at least one television set. Faster still is the current trend of microcomputers. Apple Computer introduced the Apple II in 1978, and by 1985 micros were standard fare in American high schools and upscale homes. This spread of new technology took only about 8 years. The spread of future technologies is expected to be even more rapid.

As a Proportion of Gross National Product, Military Output Will Decrease While Civilian Output Will Grow

With a decrease in the level of confrontation between East and West, and hopefully within the Middle East, the military proportion of the gross national product of the West as represented by the US and NATO, and the East as exemplified by the USSR and the now non-existent Warsaw Pact, will fall. This should free up both capital and human resources for the fulfillment of human wants and needs and for consumer satisfaction. The military took first priority in R&D expenditures and "consumed" the best engineers. Given the proportion of GNP spent on defense, was it any wonder that the US lost ground in its internal economic development? The nation's entire secondary educational system could have been reworked with only the cost of one new weapons system. What occurred in the Soviet Union is even a more severe example.

As the resource-consuming industrial-military complex is wound down and these resources are shifted to civilian and public goods, a substantial economic boost should occur. The civilian spinoffs from defense spending have been overrated. This shift will, however, create an enormous need for retraining. Simulation and gaming will prove to be excellent tools to assist in this retraining.

Leveling of the Economic Growth Rates Among the World's Economic Powers

Economic growth rates are somewhat analogous to learning curves. That is, efficiency and growth rates vary with the logarithm of output, not in a linear fashion. As the educational and social bases of the world's greatest economies get larger, their differential growth rates will cause the economies to become more equal. There will be short-term bursts by one country over another as competitive advantages exist for periods of time but, by and large, the economic well-being of citizens of the economically advanced nations of the world will slowly, but inevitably, achieve parity. There are few reasons that the economic benefit, or real income, derived from working in an automobile factory would vary from country to country for long periods of time. The same can be said of managing a retail facility, providing services, or even teaching in a university.

The Changing Role of the US

In 1945, the only heavily industrialized economies to come out of World War II without serious damage were the US and Canada. This dominance in the post-war period could have been characterized as almost imperial, if not arrogant. The US then became the world's policeman, spending a large part of its technological research capability, engineering manpower, and technology development on new and improved military hardware. (The same general comments can be said of the Soviet Union.) This period is coming to culmination now as the war resulting from the Iraqi invasion of Kuwait has ended. In the future, I expect to see the foreign policy position of the US to be more self-centered and less global. The US will devote more of its resources to its own domestic economic well-being and allow the rest of the world to pick up the slack. Note that, currently, Japan spends a higher portion of its GNP on foreign aid than does the US.

The Place of Simulation and Gaming in the Changing Global Economies

Many of the world's newly emerging economies have little recent experience with the market place as the method of apportioning economic wealth. The move from central planning with its centralized decision making has been seen by many as a shift to a more responsive society. However, a market driven economy does not guarantee a tyranny free society. An ISAGA member from the "East" spoke to me about the quick change of philosophy as the German Democratic Republic was absorbed into the Federal Republic of Germany in 1990. He said "The same individuals who strongly espoused the socialist way a year ago, encouraging one to work for the good of their country and socialism, are now quick to praise the market-based economy and they do so with the vigor of a nineteenth century capitalist."

Teaching and training individuals how to move from a centrally-planned economy to one determined by day-to-day events in the market place is not a trivial chore. The experience of 75 years existence after the Russian Revolution has changed capitalism. It has given it heart and compassion.

Those who will manage these new economic entities need to learn quickly. They will learn, not by the book, but by experiencing management through simulation and gaming. The training of these new managers (managers shifting from centrally planned to market-oriented economies and managers shifting from defense-oriented industries to domestic consumption industries) will be the challenge of ISAGA and others whose task is education and training. The continued economic growth and the well-being of the world's population will depend upon how well these new managers are trained.

Richard Teach has delivered academic papers and lectures throughout the United States, and in Canada, England, France, Italy, The Netherlands, Germany, and Japan. He has published articles in *Management Science, Operational Research Quarterly, Journal of Marketing Research, Journal of Marketing, Academy of Management Review, Simulation & Gaming*, and others. His research was awarded the "Best Simulation Research Paper" at the 1990 ABSEL conference. He is an Associate Editor of *Simulation & Gaming: An International Journal of Theory, Design, and Research.*

An Econometric Simulation Model: The Case of FUGI/MS

Masuo Aiso[1], Akira Onishi[1], Fumiko Kimura[2], Masayasu Atsumi[1], Toshiaki Imoto[1], and Yuji Tokiwa[3]

Abstract. FUGI/MS is a collection of computer software to support the building of the FUGI Global Model and its use for simulation. FUGI/MS is under prototype development towards an expert system. The key point of software design is to represent and interpret the model-builder's requirements correctly and consistently. The requirements as knowledges representation are classified into two categories: model specifications and constraints. The model specifications are formal descriptions of a target model. The constraints are the knowledge on model-building procedures which make the model workable and justify its functioning. The information technologies of visualization, micro-mainframe link, and global computer networks will help promote the research tasks.

Key words econometric methods; expert system; knowledge base; large scale; micro-mainframe link; networking; simulation; specifications; visual presentation

One of the most effective methods of analyzing the ever-increasing international nature of economic and social activities is to construct an econometric simulation model on a computer. The use of a model can both increase understanding of the current situation and help to evaluate policy alternatives for future development. The benefit of econometric modeling can be realized through the extensive use of a computer for designing, building, and validating the model.

The procedures of building and applying the model require repeated steps comprising professional tasks. There exists a number of potential problem domains where information technologies, such as knowledge-based expert sys-

[1] Department of Information Systems Science, Faculty of Engineering, Soka University, 1-236 Tangi-cho, Hachioji Tokyo, 192 Japan; phone 426-91-2211; facsimile 426-91-9311; e-mail AISO@JPNSOKA2.BITNET
[2] Institute for Systems Science, Soka University, 1-236 Tangi-cho, Hachioji, Tokyo, 192 Japan; phone 426-91-9430; facsimile 426-91-9431 e-mail FKIMURA@JPNSOKA2.BITNET
[3] IBM Japan, 19-21 Nihonbashi, Hakozaki-cho, Chuo-ku Tokyo, 103 Japan; phone 3-3808-9252; facsimile 3-3664-4839; e-mail IBMSE@JPNSOKA2.BITNET

tems, visualization, micro-mainframe computer links, and global computer networking, could play key roles.

Description of the FUGI/MS

The FUGI Modeling System (FUGI/MS) is a collection of computer software used for the construction and application of the FUGI Model in a simulation. There are four major programming modules: a database module, an estimation module, a simulation module, and a presentation module. In this paper, the current version of the program is described along with some issues on designing and developing a large-scale econometric simulation system. Figure 1 indicates an overall function schematic of FUGI/MS.

The database module accommodates three major types of data files: original, country, and region files. The original files store data from different information sources, such as the UN, the OECD, and the IMF. The original files are aggregated by country to form a country file. The region file is drawn from the country file. The estimation module uses a series of statistical tools to operate on the data. The simulation module is used for validating solutions of the model. The equations are grouped into three blocks: a reduced form block, a simultaneous block, and a successive calculation block.

Several techniques of presenting the results of the simulation are provided. Major examples are the world maps which present simulation results in color with a zooming facility, business graphics such as histograms, bar charts, and text displays through which we can see the model structures and other documents.

Design Considerations

The FUGI Model is an example of a large-scale simulation model composed of mathematical expressions created by the use of econometric procedures. Large-scale systems' complexity was studied and the following objectives were set.

The system should be an expert system exploiting the technologies of artificial intelligence which are aimed at automated and integrated model building (see, e.g., Atsumi 1991). Two kinds of knowledge bases can be identified: model specification and model-building procedures. The current version of FUGI/MS has been implemented using the following model specification:

1. Flexible databases in creation, maintenance, and retrieval
2. Parameter estimation employing expert judgement as well as hypotheses imposed by statistical methods and techniques
3. Verification of the model solution
4. Goal-oriented simulations

The results of the simulation are presented graphically, with vision allowing for easy understanding. Vision means to see and understand the contents of

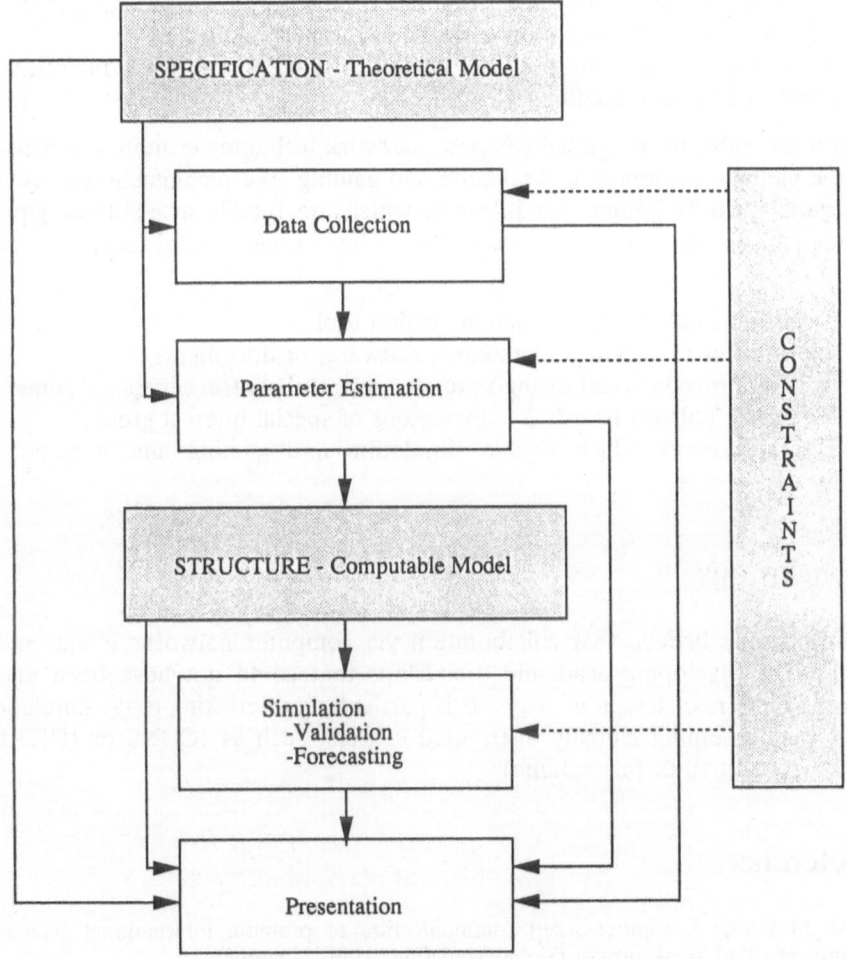

Fig. 1. Overall structure of FUGI/MS

the simulation both intuitively and synthetically. The technology of scientific visualization is becoming more important and complex in information processing with the advancement of supercomputers and graphic workstations. It is said that an important use of vision in manufacturing is defect detection (Helms 1990). Likewise, the use of vision in econometric simulation systems is effective in tuning the model and its computation as well as in presenting the final results for information dissemination.

In large-scale econometric simulation research, we cannot rely solely on personal computers because they cannot always provide capabilities such as high-speed computation or resources such as large storage. The capabilities of a large mainframe computer can be added to the flexibility of personal computers; this is called micro-mainframe linkage. This idea will be realized through the following approaches (Kravitz 1986).

1. Access to data created by host programs from a personal computer program
2. Access to the communication network of the host computer
3. A method of extending personal computer data storage using mainframe-based disk storage facilities

It is advisable to use global computer network technologies more extensively in the global community of simulation and gaming. We recommend the use of networks such as Bitnet and Internet, which are readily available as global academic computer networks (Aiso 1990). They provide useful functions, such as:

1. Electronic mail as a daily communication tool
2. File transfer to exchange databases, software, or documents
3. Remote procedure call or log-in to use powerful central server machines
4. Electronic bulletin boards for discussions of special interest groups
5. Teleconferences which enable simulation and gaming among remotely located people.

Conclusions

Many people believe that collaboration via computer networks is extremely helpful in developing academic friendships that could not have been made through letters or telephone calls. It is particularly interesting to do simulation and gaming among globally distributed centers, such as ICONS or IDEALS (see Crookall et al, this volume).

References

Aiso M (1990) Computer-based communication to promote international academic projects. 2nd Workshop of Global Modeling, Tbilisi, October
Atsumi M (1991) A knowledge-based econometric modelling support system based on hypothesis-based reasoning with case-based hypothesis generation and modification. International Symposium on Economic Modelling, University of London, July
Helms RM (1990) Introduction to image technology. IBM Systems Journal 29(3)
Kravitz LK et al. (1986) Workstations and mainframe computers working together. IBM Systems Journal 25(1)

Managing a Post-Communist Economy: A Gaming/Simulation Study

Miroslaw Dlugosz and Elzbieta Naumienko[1]

Abstract. Successful transformation from a command-and-control into a market economy, whether economic, political or social, has to include changes in attitudes, development of new skills, and adaptation to new circumstances. Ex-communist legislators, policy-makers, or managers have to respond flexibly to fluid, open-system circumstances never before experienced. The process of retraining results in increased intellectual awareness of new management standards. However, it tends to bring about only little change in management practice. To explain this phenomenon as well as some other key issues concerning business transformation, a game-based study was undertaken last year. In this paper we discuss examples of management perception of different challenges when change is accomplished in the simulated environment and matched with verbal statements concerning exactly the same problems. The heritage of management culture and fears concerning business transformation are also presented.

Key words barrier of routinized perception; business transformation; gaming/ simulation; management culture; research laboratory

The cultural orientation of any organization reflects the complex interaction of the values, attitudes, and behaviors displayed by its members. Individuals express cultural norms through behavior considered appropriate in the organization for a given situation. Senior managers need to monitor and influence continually changing patterns of individual and group behavior in order that the organization will grow, develop, and if necessary become transformed (Adler 1991). This conceptual feedback model, although theoretically correct, has been hard to confirm in the post-communist management environment. Evidence shows (Naumienko and Dlugosz 1989, Lawrence and Vlachoutsicos 1990) that because of the absence of strong incentives to adapt to change, the majority of decision makers who developed their skills in a command-and-control economy are neither prepared for economic reform nor able to capitalize on the new economic conditions afforded them.

---

[1] Faculty of Management, University of Warsaw, 02-678 Warsaw, Szturmowa St. 3, Poland; phones (48/22) 47 19 81 (w); facsimile (48/22) 216 000

Business Transformation: The Necessity for Change in Management Culture

Business transformation, however, requires values and skills radically different from previously appreciated ones. For the last 2 years numerous efforts have been undertaken to promote management standard changes in Poland. New legal and economic rules have been introduced, many companies have entered the process of ownership transformation, and assistance of Western consultants and lecturers for a wide range of companies has been provided. Evidence from the World Bank, the British Council, and other international organizations shows that company representatives who have participated in courses on Western management standards were usually very successful in passing final tests. This suggests that the necessary condition for change was accomplished. However, many Western partners, potential investors, or new owners continued to complain about the inability of retrained domestic partners (managers, employees) to act according to the new, market-oriented rules.

Gaming/Simulation as a Research Laboratory

Research Objectives

These repeated complaints, directed to the Ministry of Industry and Commerce as well as to the Ministry of Privatization, encouraged us to undertake a study on behavior patterns versus the ideas and intentions of different interest groups within organizations.

Research Sample

The study was undertaken in 1990–1991 as an integral part of numerous game-based training seminars given for industrial decision-makers by the Polish Parliament (Committee of the National Economy and Industrial Policy), the Confederation of Polish Employers, the Management Development Centre of the Ministry of Industry, the Polish Institute of Management, and the Warsaw University School of Management. Six hundred company members, representing different interest groups in their organizations participated in the study. They were recruited from businesses all over the country. Half of them (executives 9%, managers 42%) represented commonly recognized decision power within their organization, the rest (worker councils 18%, Solidarity 24%, other labor unions 7%) were the major political forces in state-owned companies (Fig. 1). The test group (10% of the whole sample) was drawn from senior and graduate students of the School of Management of Warsaw University.

Research Tools

The research method was chosen to maximize three dimensions: the ability to generalize from the sample, the control and precision with which to evaluate

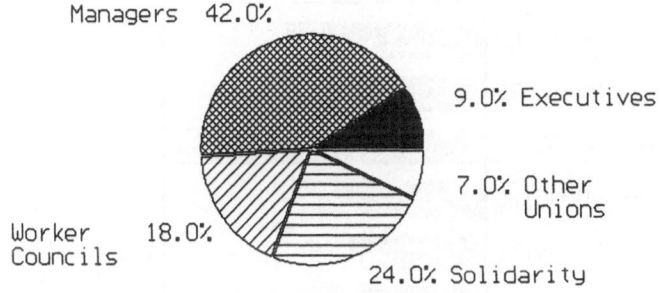

Total sample of 600 industrial
decision-makers

Fig. 1. Distribution of industrial research sample according to interest group from a total sample of 600 industrial decision makers

the behaviors, and the realism of the setting in which the actors behave (McGrath 1982). Gaming/simulation was considered to be especially appropriate here (Keys and Wolfe 1990).

In this study the "research laboratory" was based on two computer-supported organizational games covering different aspects of managing micro and macro organizations. These were the OwnerTransformLE game (Naumienko and Dlugosz 1990), a gaming exercise on the logistics of privatization of a typical enterprise, and STRATEGEM, simulation on national economy development (Meadows 1984). These complex gaming/simulations permitted the researchers to monitor the decision-making processes within highly realistic contexts and in a turbulent business environment (Klabbers 1990, Dlugosz 1990).

Although complex, the two games used in the study represented well-defined problems (Simon and Newell 1972) with clear, numerically defined criteria for winning. The gaming sessions were conducted according to recognized standards (Greenblat and Duke 1981).

During each research session the strategies and game results were carefully recorded. Close observations (often videotaped) of behavior and decision-making patterns were conducted. Prior to the gaming sessions the researchers collected from the players questionnaire-based data on players' opinion and attitudes concerning the same issues as in the gaming/simulations.

Impact of Management Culture on Managing Change in Organizations

Even though the participants were placed in almost "sterile" conditions for the release of their creativity in decision making, the results of the simulations were far from rational. Irrationality in strategy formulation was repeatedly observed in various research groups and during different gaming/simulations.

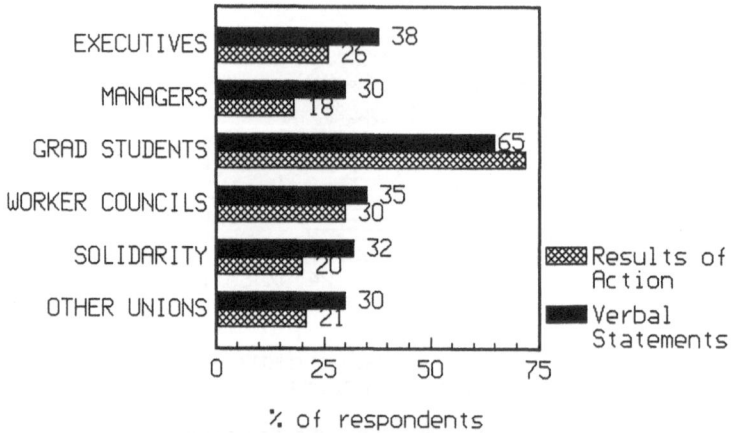

Fig. 2. Focus on economic criteria in decision making

To explain these irrationalities, a series of experiments were conducted, involving step-by-step elimination of the cognitive (information), motivational, and competence limits. However, these alterations did not change the results considerably. The quality of decisions was still poor.

Systematic (and often videotaped) observations of team behavior and discussions with game participants shed some light on the nature of the problem. It became clear that the participants were transferring the patterns of their real-life behavior into the game environment. Further investigation showed that this particular phenomenon had an axiological rather than a psychology-based explanation. This persistence of habit, even in situations where it is no longer functional, may therefore be considered a fourth structural rationality constraint (in addition to the previously mentioned cognitive, motivational, and competence limits). We have called this *the barrier of routinized perception* (Naumienko and Dlugosz 1989, Dlugosz 1990, in press).

Let us concentrate on the four most striking examples of behavior patterns and decision strategies developed by the decision makers under study by comparing them with their verbal statements prior to gaming sessions.

Perception of Organization Goals and Objectives

In managing the simulated enterprises, participants tended to concentrate on capacity utilization rather than on economic aspects of manufacturing. The majority paid almost no or very little attention to profitability analysis when accepting orders from customers (Fig. 2). Instead of looking for the least costly or most profitable options, they continued to define their goals in terms of accepting and fulfilling all incoming orders (even those whose costs exceeded their expected revenues), and in terms of maintaining a high index of capacity utilization (regardless of the costs involved). Although this pattern of behavior conflicted with an economically based criterion for winning (known prior to

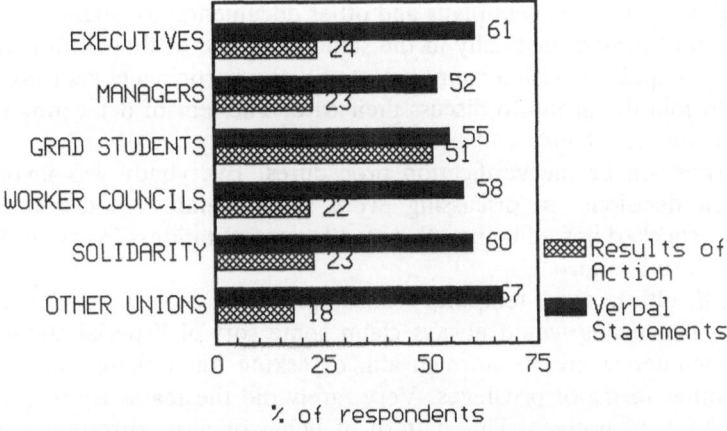

Fig. 3. Focus on quality issues in decision making

the gaming sessions), the participants continued to explain their behavior by referring to their long experience in their own particular organizations. It is necessary to emphasize here that for decades the principal goal of command-and-control firms was to "produce as much as possible, keep the machines going."

Although verbally declared to be important, the problem of quality of production did not play a critical role in creating decision strategies (Fig. 3). Game participants coming from real organizations preferred quantity to quality, although there were significant incentives in the simulation to encourage emphasis on quality and to maintain a high market share. These behavior patterns powerfully reflected players' experience of working for years in a producer-oriented professional environment.

The Simulated Company and Its Environment

Observations were made of the way participants interacted as team members. In each of the games, a set of accessories was available that represented different resources and decision-support tools of varying complexity. Distinct game roles were assigned to different participants or groups of participants. All these accessories constituted the operational frame. The participants, however, had complete freedom to create their own organizational structure with their own matching internal information system supported by appropriate tools and techniques. Although given this opportunity, only very rarely did the teams manage to create their own tailor-made organization structure, complete with delegation of responsibilities and an adequate decision-support system.

Team members were mainly inward looking; very rarely were they market led. Delegation of responsibilities, either internal or external, existed only on paper. This particular feature was especially noticeable when the company or the state had to negotiate credits. Instead of delegating financial representatives (with prepared business plans and other documents) to negotiate with the bank officers (located physically in the same room but at their own table with their own computer decision-support system), the participants kept asking the bankers to join the group to discuss their case. The fear of delegating responsibility on the one hand, and of taking responsibility on the other, resulted in overexpansion of the verification procedures. Everybody was involved in making all decisions, in discussing every single detail. All decisions were frequently checked before being submitted for computation—in case of failure, nobody was to blame.

Any difficulties, even temporary ones, provoked a very specific type of behavior. The teams would always claim some sort of "special status," like getting low-interest credits, foreign aid, or asking for a change in the game rules or other forms of privileges. Very rarely did the teams try to overcome difficulties by themselves. This pattern of hehavior also corresponds directly with players' real-life situations. For many years the companies operated within so-called soft budget constraints. Moreover, regardless of the results of operation, they could relatively easily get exemption from the obligatory rules imposed by the government. The ease with which this special status could be claimed (without any obvious reason), instead of creating an effective strategy within a given space of action, therefore became the axiological feature of decision makers. It was then transferred from the professional reality to the simulated environment.

Managing the National Economy

Regardless of the type of organization (micro, macro) the participants rarely succeeded in creating a strategic vision of their organizations. Instead of defining their strategic goals and aims of operation, they concentrated on day-to-day activities. This short-term orientation often resulted in players making decisions which were suitable in the short term, but which would probably be ineffective in the long run. The lack of strategic vision often resulted in putting such organizations in a state of permanent disequilibrium, which resulted in massive wastage of resources and a lowered level of human satisfaction.

When given investment opportunities, the participants tended to choose options associated with expanding production capacity, even if their existing capacity was being only partially utilized. Only a few players invested in the more efficient utilization of existing capital or tried to increase labor productivity by investing in management education or by development of other social services. At the national economy level, investment priorities were directed towards heavy industry rather than towards technologically advanced and energy-efficient industries like electronics, services such as education or health, and environmental protection or energy-efficiency solutions (Fig. 4a,b). Ecological considerations were at the lowest level of investment priorities.

Any study of statistical data on the Polish government's investments in these areas over the years shows very similar trends. This comparison suggests that ignorance of global issues (those easily recognized and defined by Westerners)

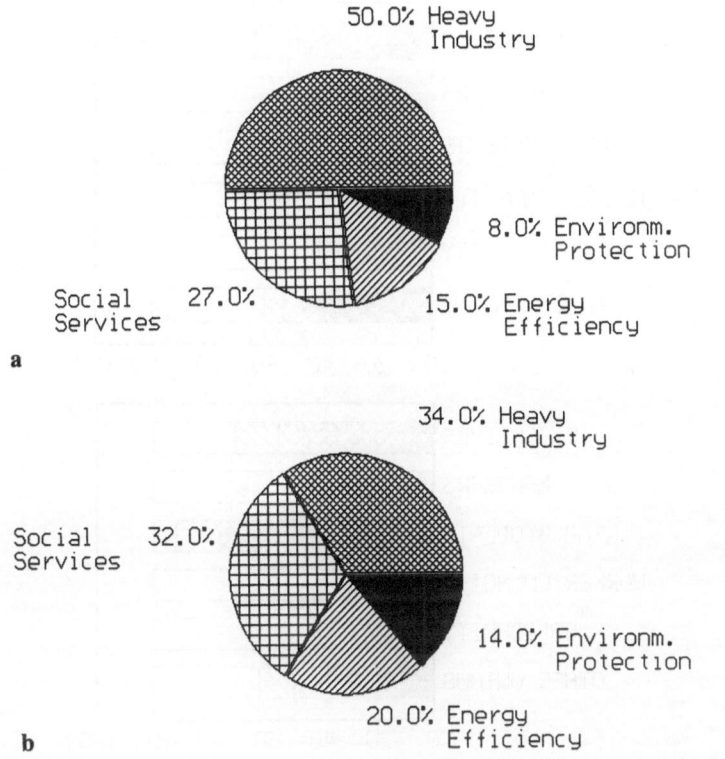

Fig. 4a,b. Investment preferences of industrial decision makers (**a**) and students (**b**)

is a structural feature of our system. To confirm this notion one could compare attitudes towards the same problems shown by participants coming from countries with a well-established tradition of investing in these "soft" areas.

Another result of the absence of strategic vision, especially on the macro economic level, was the tendency of either not to take foreign credits, or, if taken, to use them for operational purposes, namely to maximize short-term interests by satisfying comsumption needs. Only in a few cases were the credits used as a first, initial "investment push" to stimulate the economy, although this strategy led to a very high return after a longer period of time. The fear of being in debt was overpowering. The high level of awareness here, coming from real-life experience of being citizens of a highly indebted country, often resulted in extreme strategies to avoid falling into debt. In many cases teams finished gaming sessions with significant savings, even though the value of their bank account had no influence on the criteria for winning, and this was always discussed in detail prior to each gaming session.

These behavior patterns strongly correspond with the country's reality. Foreign credits taken in the early 1970s were mostly used for consumption purposes. Since then the country has not been able to repay those credits and the accumulated interest, although each year a large portion of export income

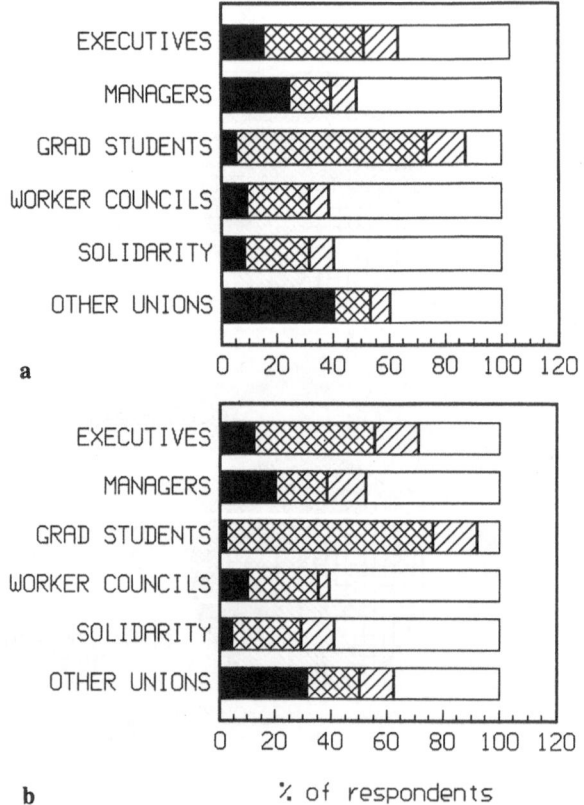

Fig. 5a,b. Owner preferences of different interest groups. **a** results of action. **b** verbal declarations. The four ownership options were employee (*open area*), private domestic (*single hatching*), joint venture with foreign capital (*double hatching*), and state (*shaded area*)

is directed just for these purposes. Poland's politicians, economists, and mass-media representatives repeatedly argue that the low standard of living of society is a direct result of foreign credits taken years ago. But they do not mention that the majority of those original credits were spent on consumption, not on investment.

Perception of Systems Transformation

The major problem to be solved when playing the OwnerTransformLE game was to find the most effective way to transform the company into a private venture. Having four options available (state ownership, joint venture with foreign capital, private venture with domestic capital, employee ownership), different interest groups repeatedly choose patterns characteristic of their political orientation (Fig. 5a). The majority (40% of executives and other labor unions, and up to 60% of solidarity and worker councils) chose a system of

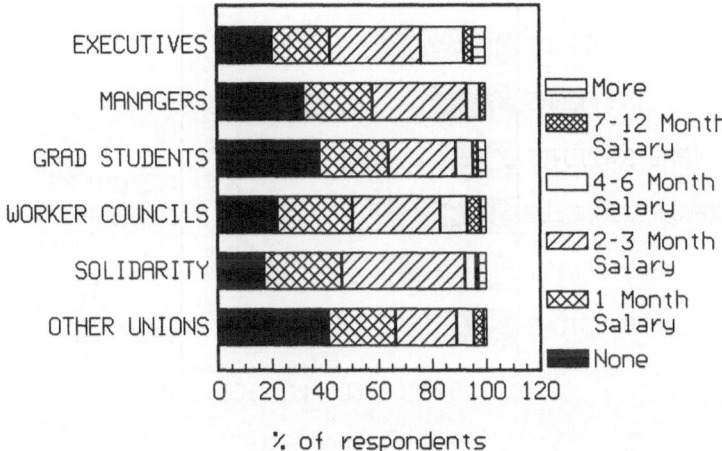

Fig. 6. How many monthly salaries would you spend on shares?

employee ownership. Other options, although not so popular, had strong followers and fanatic opponents. If the private venture option was chosen, both executives and managers preferred a joint stock company with foreign capital to a domestic company (executives, 36% against 12%; managers, 15% against 9%). State ownership was preferred most by other labor unions (40%) and least by Solidarity (8%) and the worker councils (9%). Student preferences concentrated mostly on joint venture (68%) against private domestic (14%), employee ownership (13%) and state ownership (5%).

The results of the simulation almost perfectly matched players' pre-game declarations (Fig. 5b). The distribution of choices corresponded with the hopes and fears of the former and current political leaders in Poland. In the case of the students it became clear that they considered transformation an opportunity for their fuller professional development. They perceived the process of change in terms of professional and financial improvement rather than political self-realization.

The concept of stocks issued by privatized companies and stock market operations also became an issue in the simulation. Behavior patterns observed when players purchased securities showed that more than half of the participants in each interest group were anxious to buy shares in their own company. This positive attitude towards their own shares varied from 51% of other labor unions to 63% of managers, 70% of worker councils, 69% of Solidarity, 70% of executives, and 51% of students. This indicates that the general tendency was very strong, especially if compared to a preference for other shares offered on the market, where only one third of the players on average were willing to take the risk. Considering the amount of money allotted for shares (Fig. 6) the general tendency was to spend 2–3 monthly salaries. Less than 2% of all participants wanted to spend more than a yearly salary on securities. Other labor unions (43%) were the most and executives

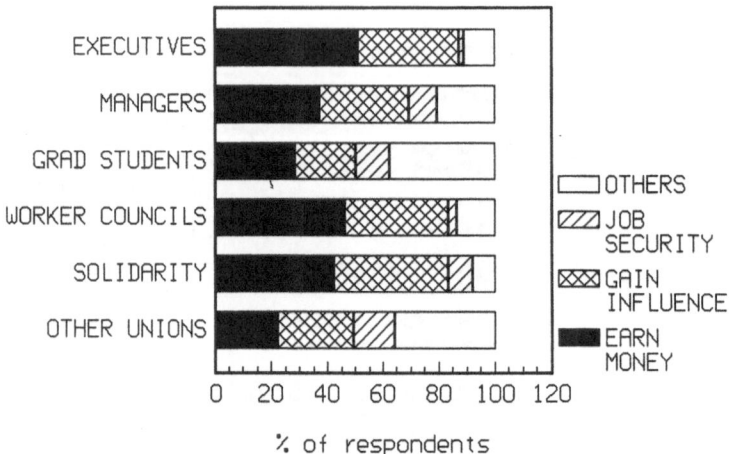

% of respondents

Fig. 7. Why would you buy shares in your own company?

(18%) were the least resistant to buying shares. Here again the verbal declarations expressed in the questionnaire closely matched the behavior pattern observed during gaming sessions.

In order to understand players' behavior concerning securities, additional questionnaire-based data were collected. When asked why they would buy shares of their own company and having four response options available (to earn money, gain influence, for job security, others), approximately one third of all game participants, regardless of their orientation and positions, responded "to earn money" (Fig. 7). But the motive of gaining influence was even more important; approximately 40% for all interest groups except for other labor unions (only 21%). This latter group came out to be mostly undecided and scoring low towards the principal motives (option other, 43%).

Another interesting feature observed was the attitude of different interest groups towards getting dividends in the form of company shares rather than money. The executives were the most (68%), and other labor unions the least, 39% approving of this particular option. The average was 45% for and 55% against non-cash dividends. The attitude observed here reflects the general tendency of the Poles still to be distrustful of non-cash operations.

It is necessary to emphasize here that most of the game participants had never dealt with the concept of securities before. Therefore their behavior strongly corresponded with their overall understanding of this new approach to money management.

Implications

For years management theorists have argued that organizations are beyond the influence of culture and that they are only determined by technology and task.

Today it is generally accepted that work is not simply a mechanistic out-growth of either technology or task, and that at every level, culture profoundly influences organizational behavior (Adler 1991).

The behavior patterns observed during numerous gaming sessions strongly supported this latter interpretation. It appeared that the cultural background of the players created perceptional barriers to their decision making within the simulation. This observation corresponds to those of Dlugosz (1990, in press). There was a discrepancy between players' operational behavior and their intellectual awareness of new possibilities in management style. This tendency to fall back on culturally oriented command-and-control behavior may be hard to understand and even much harder to accept by those brought up and educated in entirely different systems. It may also become a significant barrier of communication between managers of our region and those from Western economies.

It was further observed that players who had not been caught up in a command-and-control management style showed none of the discrepancies between verbal statements and action that have been described above. These players included students with no management experience and real-world decision makers who were accustomed to change. However, this observation should not lead to the conclusion that decision makers having the axiological background of a command-and-control economy cannot adapt successfully to new, market-oriented behavior.

Poland and all other post-communist countries have been and will be exposed to Western aid in the form of technology and knowledge transfer. The standards, methods, and procedures to be transferred are those successful in the countries of their origin. It does not, however, mean that they will automatically be successful or easily adopted in the countries of communist heritage. The results of our research show that in the case of management training and development, a study on sensitivity of the target population of decision makers to issues and standards to be transferred is a necessary prior condition to the transfer process itself. Examples presented here indicate that in some subject areas, all groups of managers will be willing to take this new knowledge for granted, and in some other areas, the most we can achieve using traditional methods is to increase intellectual awareness of new standards without radical change in management style.

Separation between "knowing" and "doing" leads us to another very important conclusion. The results of our research call for reorientation towards action rather than intellectual learning. Referring to the Confucian proverb "I hear and I forget, I see and I remember, I do and I understand," an active learning approach to management education and development should be adopted. This may result in much higher transferability to emerging Western cultures.

References

Adler N (1991) International dimensions of organizational behavior. PWS-Kent, Boston, MA

Dlugosz M (1990) Decision games in organization research and improvement (in Polish). Polish Economic Publishers, Warsaw

Dlugosz M (in press) Post-communist executives and new management freedoms: A gaming/simulation study. Scandinavian Journal of Management

Greenblat C, Duke R (1981) Principles and practices of gaming-simulation. Sage, Newburg Park, CA

Keys B, Wolfe J (1990) The role of management games and simulations in education and research. Journal of Management 16(2):307–336

Klabbers J (1990) Problem-setting through gaming. 7th International Conference of WACRA, University of Twente, The Netherlands

Lawrence PR, Vlachoutsicos CA (1990) Managerial patterns: Differences and commonalities. Harvard Business School Press, Harvard, MA

McGrath JE (1982) Dillematics: The study of research choices and dillemas. In: MacGrath JE, Martin J, Kulka RA (eds) Judgement calls in research. Sage, Newbury Park

Meadows D (1984) User's manual for STRATEGEM-1. RPC, Dartmouth College

Maumienko E, Dlugosz M (1989) Using management games to test Polish managers' preparation for the economic reform. ISAGA '89, Weimar, Germany

Naumienko E, Dlugosz M (1990) Gaming/simulation and the challenges of Eastern Europe. Opening Plenary Session of ISAGA/NASAGA '90, Durham NH

Naumienko E, Dlugosz M (1991) Exploring new management freedoms in a post-communist economy: Gaming approach. WACRA '91, Berlin

Simon HA, Newell A (1972) Heuristic problem solving: The next advance in operation research. Operations Research 4:152–163

Elzbieta Naumienko PhD is a director of the Post-Graduate Executive Program at Warsaw University School of Management, and a member of the Board of Directors of ISAGA (International Simulation and Gaming Association). She is an independent management consultant (and a member of the editorial board of the "Organization Review," a professional monthly journal for managers and management educators. She is also vice-president of the Foundation for Constructive Change and a director of the "Gaming for Poland" project. She has coauthored 7 computer-supported management games, for UNIDO, WHO, and domestic organizations, and authored (coauthored) numerous articles in the field of gaming/simulation.

Miroslaw Dlugosz PhD is a faculty member of the School of Management, University of Warsaw, personal advisor to the Minister of Industry and Commerce, former secretary of the Parliamentary Committee of the National Economy and Industrial Policy, and director of the Parliamentary Monitoring Task Force. He is a general director of the Polish Institute of Management and a deputy editor of the "Management Science" quarterly of the Polish Scientific Publishers and the University of Warsaw. He has coauthored 6 computer-supported management games and authored (coauthored) over 80 articles and four books in the field of decision making and gaming/simulation.

Strategic Decision Making in Business Gaming

Arata Ichikawa[1], Minoru Mukuda[2], and Hideo Inaba[3]

Abstract. Almost every decision-making player in business games in the classroom environment is interested in the final financial status, such as the profit of their company, rather than the attainment of the strategic business goals which their professors expect them to implement through their decision making. Although profit is the most important measure for decision makers in real business, it should be noted that they will find the available options and select from them according to their business strategy. Thus, consistency in decision making could be equally as important as profit to measure the managerial ability of players. In using business games in classroom settings, we should stress that decisions be based on the chosen business strategy, in which case the focus on profit maximization can decrease. In this paper we will show that the goals of business games can be both profit maximizing and decision consistency.

Key words business game; decision consistency; performance measurement; profit; strategic decision; teaching-learning system

In most business games, a decision maker is part of a group of a small number of people and plays the role of manager in an enterprise. The number of groups is between five and nine in most business games. Usually players make decisions repeatedly every round for the period of play. Each round they choose decision options either to maximize their profit over the periods of the game or to avoid their company going into bankruptcy (on performance measurement, see Teach 1990).

Our experiment was with gaming participants who were learning strategic decision making based on that assumption. It was conducted using a qualitative analysis based on Factor Analysis (Okuno and Yamada 1978). There is so much numerical data, such as financial statements, available from other gaming-simulations that we have a tendency to fall into the orientation of

[1] College of Economics, Ryutsu Keizai University, Ryugasaki, Ibaraki, 301 Japan; phone 297-64-0001; facsimile 297-64-0011
[2] Department of Business Administration and Communication, Bunri College, Sayama, Saitama, 350-13 Japan; phone 429-52-1211; facsimile 429-54-7733
[3] Department of Industrial Information, Komatsu College, Komatsu, Ishikawa, 923 Japan; phone 761-44-3500; facsimile 761-44-3506

financial analysis. However, we used the financial data to analyze the decision making by participants in gaming.

Introduction

The experiment consisted of four phases for each gaming run. In the first phase, members of each team put down their managerial plans, including the strategic objectives for the company. In the second phase, players chose decision options and made managerial decisions for each period of the gaming simulation. In the third phase, all the decisions of the game were interpreted in qualitative statements using Factor Analysis (see, e.g., Takeuchi and Tsukuda 1990). Finally, we analyzed their decisions by comparing the managerial plans with the corresponding qualitative statements.

We found that one of the eight teams had made their decisions through the run of game in a manner completely consistent with the strategic plans they had put down at the beginning of the run. In addition, the strategic goals chosen by the players and their style of decision making were those typically found in many Japanese enterprises. If a similar correlation between the nationality of the gaming participants and the characteristics of typical enterprises of their countries could be established, then strategies gathered internationally through business games involving decision makers having culturally different values would be useful for mutual understanding in international trade.

The Business Game

A business game session was organized for this research. A model for the game was programmed from the Systems Dynamics approach and was translated into a computer language. The game run lasted only 1 day due to scheduling difficulties of the participants in the game. The participants were chosen from college students taking subjects related to industrial management in order to minimize differences in their knowledge level.

Table 1. Strategies of all companies

Team	Planned strategies (key words)
A	Good products, higher price, high quality oriented
B	Brand image oriented, relatively lower price, investing in public relations and advertisements
C	Profit making, research and development, high quality products
D	Meeting consumer needs, lower price
E	High quality products, "big business is beautiful"
F	Better quality products, mass production, high welfare employment
G	High quality products, research and development, better production management
H	Brand image oriented, first class products, "small business is beautiful"

We organized eight groups of six or seven participants for the gaming companies. Members of each group chose their management roles: president, sales manager, production manager, financial manager, personnel manager, planning manager, and consulting advisor.

At the beginning of the game, all the companies planned and wrote down their business strategies. Table 1 shows the representative key words extracted from all the strategies. From their key words, we can infer that the students had an image similar to that of Japanese companies that are involved in the production of high quality goods (see, e.g., Davidson 1984).

Decision Collection

The game run consisted of eight rounds, each simulating 1 year of real time. For each round, role-players analyzed the results of previous decisions and made decisions for the next round according to their strategic plans. The list of the decision options was as follows: product price, advertising costs, sales promotion costs, investment costs in manufacturing facilities, production quantity, ordering quantity of raw materials, ordering point for raw materials, R&D expense, QC expense, employment of sales personnel, employment of factory workers, layoff of sales personnel layoff of factory workers, average wages, loan, and loan repayments. All the data were numerical.

Strategy Analysis

Interpretation of Factor Scores

The game began with the same opening parameters and ended at the eighth round. We ignored decisions for the first and final round in order to avoid the influence of both preset parameters and rushed, end-of-game decisions. We thus obtained multidimensional data for 8 teams × 18 decision options × 6 rounds. Since the hypothesis was that the decision making of each team would be based on the strategic plan set up at the beginning of the game, and that a relatively small number of latent factors would reflect this strategy (Ichikawa et al. 1981), we employed the Normal Vairmax Method of Factor Anaylsis for analyzing the actual strategies implied by the decision values.

Factor Analysis of the decision values produced between the second and the seventh rounds generated the corresponding factor loading matrix for each round. Table 2 shows the factor loading matrix of round 2 as an example. We tried to interpret the factor loading matrix by making statements such as the following:

— Factor 1: Boosting production by taking out loans and depending on borrowed capital; loan repayments causing financial difficulties
— Factor 2: Increasing sales promotion with lower investment in manufacturing facilities and improving corporation identity in the market

Table 2. Factor loading matrix of Round 2

	Factor no.				
	1	*2*	*3*	*4*	*5*
Factor values	5.7	4.2	1.7	0.9	0.8
Loan	0.95	0.26	−0.03	0.03	0.14
Loan repayments	0.95	0.26	−0.03	0.03	0.14
Production quantity	0.95	−0.01	−0.05	0.16	0.12
Employment of factory workers	0.92	−0.20	−0.11	0.32	0.01
Employment of sales personnel	0.81	0.35	−0.11	0.00	0.02
QC expense	0.69	0.22	0.31	0.06	−0.06
Investment costs in manufacturing facilities	−0.23	−0.89	0.11	0.31	0.13
Advertising costs	0.26	0.89	0.17	0.27	−0.05
Sales promotion costs	−0.02	0.87	0.03	−0.18	−0.30
Product price	0.01	0.63	0.30	−0.36	−0.59
R&D expense	−0.09	0.09	0.99	−0.03	−0.08
Average wages	0.28	−0.12	−0.01	0.94	0.15
Ordering point of raw materials	0.49	−0.17	−0.21	0.00	0.83
Ordering quantity of raw materials	−0.08	−0.42	0.45	0.34	0.68
Layoff of factory workers	(No differences)				
Layoff of sales personnel	(No differences)				

— Factor 3: Producing high quality goods with higher investment in R&D and QC
— Factor 4: Employing skilled personnel at high wage levels
— Factor 5: Maintaining a narrow profit margin

Table 3 shows the interpretations of all factor loading matrices for all of the rounds. We can see that some strategies were in conflict with others. This could have been caused by wrong decisions by the team.

Business Strategy Analysis

At this point we had statements interpreting all of the factors. The next step was the calculation of factor scores for each team, based on the factor loading matrices. In this paper, we will present the factor scores in a figure instead of a score table. Figure 1 shows the factor scores for round 2. The numbered factors correspond to the factors described above and to those in Table 2.

The interpreting factor statements and the factor scores for each team are helpful in unveiling the latent strategies of each team. Some strategies found in the game were as follows:

— Team E: Investing capital aggressively in both production and sales promotion; using borrowed capital
— Team F: Lowering all expenses and selling lower priced goods
— Team H: Influencing consumer preference with brand image

We could not make out clearly the operating strategies for the other teams. Table 4 shows all the other strategies we found from the factor scores.

Table 3. Interpretations of latent factors

Round no.	Factor no.	Interpreted factors (key words)
2	1	Production by borrowed capital, financial difficulties
	2	Sales promotion, corporation identity in the market
	3	High quality products, R&D, QC, brand image
	4	High level wages
	5	Lower priced products, narrow profit margin
3	1	High quality products, aggressive sales promotion
	2	Mass production, lower priced products
	3	No loans, capital independence
	4	Investments in manufacturing facilities
	5	High average wages
	6	Small quantity production, sales promotion
4	1	Cost reduction, sales promotion
	2	Higher pricing, sales promotion, good profits
	3	Production oriented
	4	Mass production
	5	Sales oriented
	6	Financial difficulties
5	1	Streamline management, sales promotion
	2	Share oriented
	3	Financial difficulties
	4	Inventory adjustments
	5	Higher pricing
	6	Investments in manufacturing facilities
6	1	Minimum utilized production
	2	Cost reduction
	3	Investments in all costs
	4	Minimum utilized sales promotion
	5	Highest priced products
7	1	Higher pricing, mass production
	2	Financial difficulties
	3	Risky streamline management
	4	Corporate identity, long-term planning
	5	Big business oriented

Consistency of Decision Making

Now we were ready to compare the strategies put down at the beginning of the game with those extracted from actual decision values using qualitative analysis. First, we focused on Team H to study its consistency.

— Round 2: Influencing consumer preference with brand image
— Round 3: Producing high quality goods and selling them with intense sales promotion
— Round 4: Conducting an advertising campaign and increasing advertising costs
— Round 5: Selling high priced goods by sales promotion
— Round 6: Strengthening the sales capability by reducing production costs
— Round 7: Strengthening the production of high priced goods

Factor Score

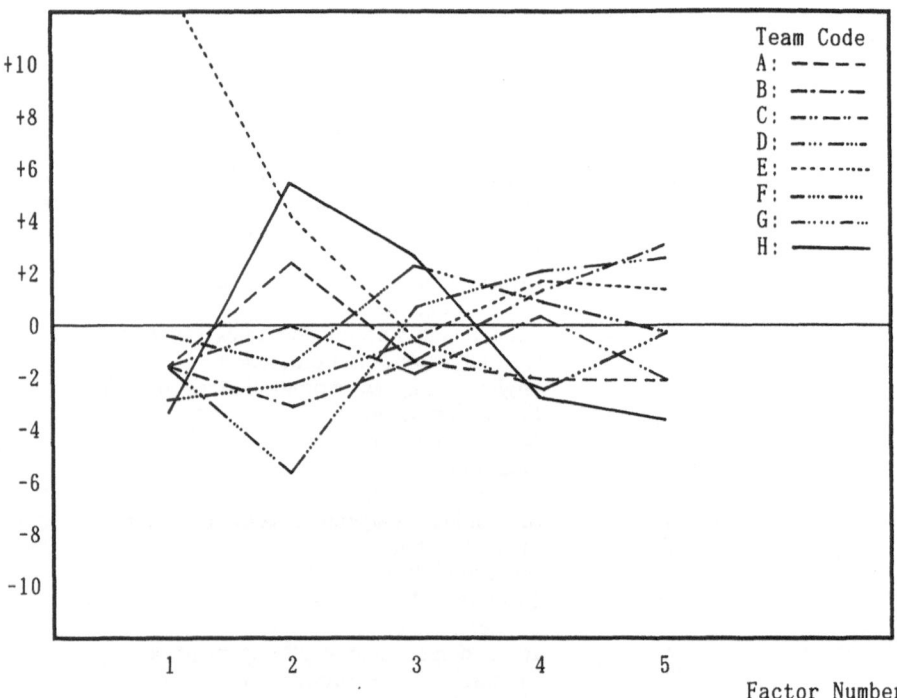

Fig. 1. Factor scores of round 2

Because the key words of the strategies of Team H were "brand image oriented," "first class products," and "small business is beautiful" (Table 1), it can be seen that the decisions made between rounds 2 and 7 were in accordance with the planned business objectives. Team H showed the best team performance in strategic decision making. Next we focused on Team E to see if it was as consistent as Team H.

— Round 2: Investing borrowed capital in both production and sales promotion
— Round 3: Facing financial difficulties and selling goods at the lowest price
— Round 4: Avoiding bankruptcy with streamline management
— Round 5: Being in debt and repaying loan interests
— Round 6: Reducing all costs and even selling goods at a sacrifice
— Round 7: Going into bankruptcy intentionally

Because their key words were "high quality products" and "big business is beautiful," we can see that their decisions were only slightly in accord with their planned business strategy. Thus, Team E showed a not-so-good team performance in pursuit of their goals in spite of heavy capital investment. It was also found that some of the other teams' decision making was full of inconsistencies. But, on the whole, each team was trying to make decisions consistent with their own strategies.

Table 4. Extracted strategies of decision making

Round no.	Team	Analysed strategies (key words)
2	E	Borrowed capital, investments in production and sales Promotion
	F	Lower pricing
	H	Brand image
3	B	Mass production
	D	Market share oriented
	E	Financial difficulties, lowest pricing
	H	High quality products, sales promotion
4	C	Financial difficulties
	D	Highly utilized production
	E	Streamline management
	H	Advertisement oriented
5	C	Inventory adjustments, lower utilized production
	E	In debt, loan repayments
	H	Higher pricing, sales promotion
6	C	Investments in production
	E	Sacrifice sales
	H	Sales promotion, lowering costs
7	B	Expanding production facilities
	E	Bankruptcy
	H	High priced products

This sort of evaluation of the decision-making process in business gaming should be given more emphasis in the future. We have here a new performance measurement tool for financial statements and management analysis reports.

Conclusions

Hitherto, whenever we use business gaming as a teaching learning system, we have faced difficulty in achieving our educational objectives (Kurozawa 1990). Players, usually college students, have a propensity towards seeking a big profit at the end of the game. The reason that bankruptcy is of such vital concern to all players is simply because going bankrupt means losing everything in both the game (the company) and the business course (the grade). The main objective should not be gambling, but rather rational decision making. Needless to say, there are many procedures and instructions to the students to help them avoid such profit-only-oriented decisions.

The experiment reported here and its qualitative analysis of strategies has shown that players can make strategic decisions without ending in bankruptcy. Gaming directors, usually professors, can make objective evaluations of the performance of each team by looking at their decision consistency as well as their class attitudes. Also, students themselves can easily understand the results of their own decision making because they are explained, not in figures, but with understandable statements.

In the future, we intend to gather strategic key words found through Factor Analysis to organize a database. This will require a support system for the interpretation of factor scores produced by a Factor Analysis program. A powerful personal computer having multitasking capability will be used to integrate simulation models and evaluation tools (Ichikawa 1984). Currently, manual interpretation would be a time-consuming task for gaming directors.

References

Davidson WH (1984) The amazing race. Wiley, New York
Ichikawa A (1984) A personal-computer-based business game and its uses (in Japanese). Operations Research (Japan) 29(2):96–104
Ichikawa A, Mukuda M, Inaba H, Yokoyama K (1981) On latent policies in a business gaming simulation and its educational uses (in Japanese) Research Report of Kogakuin University 51:49–60
Kurozawa T (1990) A survey of gaming/simulation for management education in 40 industrial engineering departments (in Japanese). Simulation and Gaming (Japan) 1(1):89–93
Okuno T, Yamada B (1978) Management analysis in the information (in Japanese). Tokyo University Press, Tokyo
Takeuchi K, Tsukuda Y (1990) Managerial statistics (in Japanese). Yuhikaku, Tokyo
Teach RD (1990) Profits: The false prophet in business gaming. Simulation & Gaming: An International Journal of Theory, Design, and Research 21(1):12–26

Arata Ichikawa is an Associate Professor at Ryutsu Keizai University. He was one of the organizers of the Japan Association of Simulation and Gaming, 1989, and served as financial chair of ISAGA '91. His research interests in simulation and gaming originated from participation in an US-Japan political gaming/simulation conducted by the East-West Center, University of Hawaii in 1973. He was a Fulbright visiting professor at the Systems Science Department, University of Southern California from 1983–4, doing research on decision-support systems and public administration and environment gaming systems. He was a member of the research promotion committee for the development of a small business gaming simulation under the Ministry of International Trade and Industry, Japan, 1991.

Minoru Mukuda is an Associate Professor at Bunri College. His primary research interest is in the use of network technologies, particulary intelligent network gaming environments. He has developed many personal-computer-based gaming simulations.

Hideo Inaba is an Associate Professor at Komatsu College. His research interest focuses generally on management science, particularly price modelling. He has done empirical research on the decision-making procedures of small businesses.

Management Games in the International Business Classroom

Ronald D. Klein[1]

Abstract. Management games have been a cornerstone in the Business Policy course for two decades and have spread to the areas of marketing and finance. They have not made the same impact as a pedagogy tool in international business (IB) courses, where the case study remains the primary experiential learning tool. One of the reasons IB is slow to adopt the business simulation game may be due to their perceived lack of validity as a teaching tool, that is, their inability to address the critical issues of international business. This paper attempts to assess their suitability for the classroom by examining the degree to which the games address relevant international business issues. If they are found to have an acceptable degree of validity, some of the reluctance to use the games may be alleviated.

Key words effectiveness; evaluation; experiential learning; gaming; international business; pedagogy; review; simulation

Simulation/games have been heavily adopted for use in policy, marketing, and finance courses by American business schools. They also play a major role in executive training programs. The majority of these games are domestic business simulation/games. However, some of these games have been internationalized to broaden student exposure to international business (IB) concepts, but relatively few of these games have been adopted for use in IB courses.

There may be several reasons for their lack of popularity. One reason is that they are not well known to IB faculties. In reviewing the literature, references to specific games are very few. Even teachers who have adopted a given simulation/game generally have little knowledge of the games available beyond the one being used. Interviews with the IB game authors themselves disclosed that they were generally unfamiliar with other IB games. This is probably due to the dearth of games available. Another reason is that IB faculties have not been interested in using this type of experiential tool; the case study has been the mainstay in the IB classroom for many years. Finally, and perhaps most

[1] Department of Business Administration, Abbott Turner School of Business, Columbus College, Columbus, GA US; phone 404-568-2284 (w); facsimile 404-568-2040

important for those who are familiar with simulation/games, there may be a reluctance to adopt them because they are not regarded as effective learning tools.

This paper addresses the reluctance to use these IB tools in the classroom. It attempts to provide the reader with a list of games available, and reviews their suitability, potential, and effectiveness as learning tools. Hopefully, this review will be helpful to IB faculties who are interested in adopting a game for classroom use. The games selected for review are broad-based management games such as those used at the business policy level. Games which focus on more specific areas of interest such as foreign exchange or negotiation simulations were not included.

Methodology

The games listed below were the only high-level, corporate management games devoted to the IB environment which were found in a literature search and in interviews with game authors. The games vary in terms of sophistication, complexity, and time requirements, representing a spectrum from easy to difficult that will fit most academic programs.

Business Strategy Game (BSG)
International Operations (INTOP)
Multinational Management Game (MMG)
Thunderbird International Management Simulation Game (TIMS)
World Wide Simulation Exercise (WISE)

Three international business faculties were asked to develop a set of critical issues which could be used to assess the face validity of the candidate games. The critical issues were developed from four of the five leading Principles of International Business textbooks. The game decision inputs and the report

Table 1. International business simulation overview

	INTOP	*TIMS*	*MMG*	*WISE*	*BSG*
Number of products	2	1	2	2	1
Product type	gen	spec	gen	gen	spec
Home office	L'stein	USA	USA	USA	USA/Europe
Active subsidiaries	0	0	3	3	2
Expansion options	3	16	3	3	4
Minimum decisions	24	6	24	30	42
Maximum decisions	100	200+	24	40	70
Manual length (pp)	55	118	81	47	81
Teams per industry	4–25	9	8	7	3–16
Worksheets	no	yes	yes	no	yes
Support programs	gen	gen	gen	yes	gen
First decision time (h)	8–12	12–16	8	8	6

gen, generic; spec, specific

outputs were then examined by the same group in terms of critical IB issues and attempts were made to determine how many and how well these issues are addressed by each game.

All of the information presented here comes from the student user manuals which provide background data, decision forms, report forms, and decision procedures, and from the authors who were interviewed via telephone and who provided answers to specific questions that were not addressed by the manuals.

Results

Table 1 shows the major parameters of each of the games. It describes the types and numbers of products, the starting position, expansion options, and number of decisions required. The minimum and maximum number of decisions quoted below are not exact figures. Exact figures would depend on the number of products produced, the number of markets selected, or other decisions. The numbers given are simply reasonable limits.

In general, WISE has two consumer products. INTOP has two grades of each product, which are unspecified medium-sized appliances. The TIMS product is a single 4-ft^3 refrigerator. MMG has a consumer and an industrial product. The BSG's product is footwear. Those games which start with active subsidiaries may have plants, sales offices, or corporate shells in place in foreign locations. This accounts for the large number of initial decisions required. The type of foreign operation for MMG, INTOP, and BSG is a manufacturing facility. Other options are available with TIMS and WISE. The decision times are estimates based on three- or four-person teams.

Other unique attributes should be pointed out. WISE has no student worksheets to aid in decision making, but does have two supplementary computer programs to aid in predicting cost and demand. Other authors indicated that future revisions of their games would also have decision-support programs. The worksheets for MMG are especially easy to use.

Critical Issue Review

Table 2 is a list of critical issues and a game rating for each of the candidates by issue. The scale used is from 0 to 3, where zero indicates that the issue is not addressed at all and a three indicates that the issue is heavily addressed. Note that the games vary considerable in this respect providing an opportunity for adopters to select games which match a unique set of needs.

Table 2 shows that TIMS rates very high in IB attributes; it offers opportunity to operate behind tariff walls, includes political unrest, has a degree of nationalism built into its marketing functions, and uses a controlled random function for inflation, economic index, political unrest, and foreign exchange rates. WISE permits inter-company actions such as contracting, loans, and collusion; it is very well suited to executive training because of the decision-

Table 2. Critical issue comparison

	INTOP	*TIMS*	*MMG*	*WISE*	*BSG*
Comparative advantage	1	2	2	2	2
Direct foreign investment	3	3	3	3	3
Differential inflation	3	3	1	2	0
Tariff effects	2	3	1	1	3
Socio-cultural forces	2	0	2	1	0
Technology transfer	0	0	0	0	0
Transfer pricing	2	3	2	3	0
Consolidated operations	2	3	3	3	3
Imperfect competition	3	2	1	3	1
Nationalism	1	2	2	0	0
Exchange gains/losses	1	3	0	2	0
Hedging	1	3	0	2	0
Exchange rates	0	3	2	2	2
Foreign sourcing	1	3	1	2	2
Taxes	1	2	2	3	1
License agreements	2	0	0	0	0
Political stability	0	3	0	0	0
Totals	25	38	22	29	17

support programs. INTOP also permits collusion and team interchange; teams are even allowed to sell factories to each other. MMG and BSG are the "smallest" of the simulations. MMG is easily the most popular with more than 100 adoptions; it is easy to use in the classroom both for the instructor and the student. The game is currently being used as a competitive vehicle for an annual national competition among business schools hosted by Georgia Southern. BSG is the newest of the games.

Narrative Game Review

The following section provides the reader with a detailed description of each of the simulations.

BSG allows players to build three additional factories in its basic leisure footwear industry—one in Texas and one each in Asia and Europe. Different quality grades are available for the firm's two products and the simulation's overall rate of economic activity is tied to the Standard and Poor 500 Index. Exchange rates for deutsche marks and yen against the US dollar are obtained from the *Wall Street Journal*.

Using Liechtenstein as its corporate headquarters for tax haven benefits, INTOP allows companies to operate in the United States, Brazil, and the EC. These operations may be accomplished through product exportation, local production, or the cross-licensing of a company's products. While no decision-support tools are provided in the game's manual, the authors encourage players to create spreadsheet programs to simplify decision making. The game is rich, however, in other ancillary materials: the management consulting firm of Arthur DeBig is available for special services, the *Gazette* of the fictitious World Federation of Appliance Manufacturers can provide industry gossip and updated information.

MMG is also a two-product game, which allows operations in two foreign areas plus the United States. Each decision round encompasses a year's worth of company activities and the real world's exchange rates for ringgits and deutsche marks are employed for its geographic operations in Malaysia/Asia and Germany/Europe, respectively. The player manual is especially rich in describing and employing real-world data to delineate the characteristics of each market area available in the simulation.

TIMS allows the production and/or sales of a small, apartment-sized refrigerator in 16 countries in the Western world and the Far East. The decision period is in quarters and decision-making operations are described very thoroughly in the Manager Manual. Computer-based, decision-support materials provide production costs, cash flow projections, sales forecasts by country, and so on.

WISE is a simulation in which two generic products can be sold or manufactured in the United States, and one country each within the EC and South America. Companies can sell their products to each other within the simulation on a quarterly basis and these goods can be shipped by either air or surface transport. Decision-support spreadsheets are available for forecasting demand for products by country as well as forecasting the company's sales value and for costing out other sourcing options.

Discussion

As described above, the games differ significantly in their degree of complexity and difficulty. A simple sum of the Table 2 columns indicates the relative degree to which critical issues are addressed and is a subjective measure of their complexity and difficultly. The values range from 17 to 38. The lower level games would be suitable for undergraduate courses. The higher level games would be more suitable for graduate courses or executive training programs.

In all of the games, the instructor may control several variables. The play of the game can be controlled to minimize the carry over among students from term to term by adjusting the product demand function and economic index. MBG and TIMS grade student performance and the grade weights may be adjusted so that market share may be very important during one semester or quarter and net profit be very important the next term. The number and types of decisions may be varied in some games by limiting students to specific options. Other options for control include changing inflation rates, tariffs, or taxes. INTOP is exceedingly flexible in that every game parameter may be set by the instructor.

Conclusions

The games in general address a large number of critical issues giving the student good exposure to IB concepts. This exposure should allay some of the misgivings potential users have in adopting such simulations for classroom use. Although there is no consensus that games teach concepts, there exists a large measure of face validity which supports such a theory. In addition, the games provide a stimulating and competitive environment for learning.

The business simulations reviewed here represent several levels of difficulty. Potential adoptees may wish to consider the amount of class time available especially at the beginning of the term when choosing a particular game. The other major consideration is the level of the course in which a particular game will be used. The more difficult games are definitely suited to the graduate level or at the least senior level of an IB undergraduate program.

Acknowledgements

The author thanks Robert A. Fleck Jr. of the Abbott Turner School of Business, Columbus College, and Joseph Wolfe of the College of Business Administration, University of Tulsa for their help in identifying critical issues and assessing their contributions to the selected simulations.

References

Ball DA, McCulloch Jr. WH (1988) International business: Introduction and essentials. Plano, TX

Czinkota MR (1988) International business. Dryden, New York

Daniels JD, Radebaugh LH (1989) International business: Environment and operations. Addison-Wesely, Reading, MA

Edge AB, Keys B, Remus W (1991) Multimanagement management game: Student's manual. Business Publications, Dallas, TX

Grosse R, Kujawa D (1988) International business: Theory and managerial applications. Irwin, Homewood, IL

Hoskins WR (1989) The Thunderbird international management simulation. Manager manual. American graduate school of business. Thunderbird Software, Phoenix, AR

Ingo W, Murray T (1988) Handbook of international business. Wiley, New York

Thompson AA, Steppenbeck BJ (1990) The business strategy game: A global industry simulation. Homewood, IL

Thorelli HB, Graves RL (1989) INTOP international operations simulation: Player's manual. The Free Press, MacMillan, New York

WISE: World wide simulation exercise for management development. Barker, Temple, Sloan (1988) Lexington, MA (33 Hayden Street, Lexington, Massachusetts)

Ronald D. Klein is a professor of management at Columbus College and is currently serving as interim dean of the Abbott Turner School of Business. He served as a research project manager and consultant for 18 years with the University of Michigan and Stanford Research Institutes, specializing in experimental design and statistical analysis. Since entering academia, his research interests have focused on business simulations and classroom pedagogy. Dr. Klein has published over 40 articles. He received his doctorate in International Business and MBA from Georgia State University.

Business and Engineering Gaming in the Ukraine

Victor I. Rybalski[1]

Abstract. A number of business simulations have been developed at the Civil Engineering Institute in Kiev. The games are described and discussed in terms of typical situations in the Ukraine and neighboring republics. A trilogy of games, PRESIDENT I, II and III, allows participants to run a country—its economy, its politics, and its social order. Two other games, SPUSK and KROSS deal with aspects of complex construction projects, such as control. Also mentioned is the use of these games in the Ukraine and beyond.

Key words business simulation; construction projects; economics; engineering; management; politics; social order

Business games form an integral part of a wide class of methods used for active training in higher educational establishments and technical institutes. At the same time, business games can be used not only for training processes, but also for research, the development of innovation, the solving of planning or production problems, and for the certification of staff.

A number of management games have been developed at the Kiev Civil Engineering Institute. These games have been widely used in recent years for training engineers, working out national business solutions, designing automated systems, and training managers and engineers in the use of effective management methods. This paper describes some of these simulations.

PRESIDENT-I (Economics)

In PRESIDENT-I, the aim of each participant is to take on the role of president, who then must develop the internal economy of a country as well as establish trade and diplomatic contacts with neighboring countries. Responsibilities also include the development of certain managerial skills, such as

---

[1] Civil Engineering Institute, Vozdukhoflotsky pr. 31, Kiev-37, Ukraine; phones 276-53-30, 276-40-25, 272-94-98, 272-95-95, 272-95-00; telex 131280 SACC SU

leadership ability, resourcefulness, creative conflict resolution, compassion, and cooperation rather than confrontation. Characteristically, the president must possess good decision-making skills and the ability to be flexible in order to deal effectively with unexpected events such as inflation, labor disputes, work stoppages, and the threat of war.

The basic version of the game allows four participants, each acting as a president of one of four countries. The presidents take positions around the common game field—a geographical map. The aim of each participant is to gain as many special money units as possible. These money units are used to evaluate the total winnings gained by the participants at the end of the game.

Before the game starts, each president is advanced a sum of money which is to be returned at the end of the game. The president can use this money for the following purposes:

— Production of resources in his own country
— Purchase and shipping costs of foreign goods and resources
— Purchase of various modes of transportation
— Construction of additional manufacturing plants
— Construction of tourist hotels

The presidents start the game by initiating negotiations leading to trade agreements. The agreements reflect fixed dates for the delivery of goods and products at agreed prices, and penalties in case of delivery delays. Consideration should be given to the modes available for transporting the imported resources. In addition, negotiations can be initiated, setting up various business ventures for the future.

In each round, the president draws a card from the "chance situations" deck. The card may denote such events as an accident in one of the plants, an influenza epidemic, a major repair of a tourist hotel, a natural disaster, or a labor strike. Another card deck is used to randomly introduce other items, such as price increases, availability of advanced technology and equipment, refurbishing particular building sites, and any inclement weather that would disrupt deliveries.

Presidents make their own decisions or collaborate with each other if necessary. Game parameters may be set up that would eliminate disaster scenarios, require a penalty payment for a delay in product delivery, operate import and export goods, set up a discretionary clause that would cancel agreements in force (with corresponding losses), or make available the purchase of resources from other suppliers at lower prices. When new business projects are forecast, it becomes possible to investigate how cost-effective the refurbishing of existing structures would be as opposed to building new ones. If conflict is anticipated, the purchase of merchant and military ships and aircraft is authorized. These ships and aircraft would act as deterrents in the event of harbor blockades or any other disruption that might cause a discordant note in the agreements.

When a country experiences deficit spending, it may attempt to resolve the situation by one of two avenues: (1) negotiate an "aid package" from another country, with terms amenable to both parties, or (2) utilize a loan from the

International Bank at a rather high interest rate. If the country defaults on the loan, then the lender will assume ownership of all goods and property (transportation means, buildings, etc.) of the insolvent country. In order to recoup the loss, the lender will "sell off" all goods and property thereby incurring a profit.

The game is considered to be over when one of the countries becomes bankrupt, or the predetermined number of rounds has been exhausted. The winner is determined by the amount of money accrued. In so doing, the money spent on construction of buildings and projects and the purchase of transportation (except that for military purposes) are all taken into account.

Upon completion of the game, the participants, using critical analysis, independently assess the outcome and the methods used in achieving the results. This allows performance-based rating of the various managerial skills and abilities exhibited by each participant during the game, such as cooperation, resourcefulness, sympathy, and so on.

PRESIDENT-II (Politics)

This version of the game concentrates on political dimensions. It is currently under development and will run on a personal computer.

PRESIDENT-III (Law and Order)

In contrast to PRESIDENT-I (Economics) and -II (Politics), this version of the game pays special attention to the problems of law and order. In PRESIDENT-III, each participant assumes the role of a world leader (president) of an imaginary state. The president has to see to the business of running the country and to ensure that it continues to evolve successfully as a nation. The president must combine all this with the country's global responsibility toward interacting with other nations. In addition, the president must preserve internal harmony by championing the cause of law and order. This may be done by trying to eliminate smuggling and illegal drug trading, in addition to preventing crime in other areas, reducing the threat of terrorism, attending to problems of emigration and immigration, responding in a timely fashion to the threat of strikes, mediating conflicts between nations, and dealing with the desires of regions whose goal is to achieve autonomy.

Game participants, the presidents, take their places on four sides of a table where the game field is situated. In the course of playing the game, none of the presidents should be privy to the decisions of the other presidents, nor the amounts of their monetary holdings. The developmental level of each state is evaluated on the basis of its respective position on a state matrix. The implication being that, during the game process, certain indicators will be identified, such as the average human life span, mortality rate of children, services offered to the population, and the moral condition of the society. As the game progresses, any level on the state matrix can be reached—depending largely on the president's actions. The lower portion of this matrix indicates that the country is in a state of depression, and the upper that it is thriving economically.

In the final analysis, each president attempts to collect as much game money as possible. This amount is used to evaluate the total score when all game rounds are completed. At the same time, an evaluation is made of the country's rate of growth while under each president's regime. The winnings depend both on a level achieved on the state matrix and on a strategy chosen by the president during each round. In addition, there exists some specified cases of adding on gains or fines regardless of the state matrix. Actions taken by presidents during each round consist primarily of showing colored cards:

— A red card indicates intense activities
— A green card symbolizes moderate activities and the allocation of time for leisure and entertainment
— A black card indicates drawing excessive, illegal income
— A yellow card symbolizes passivity

After the presidents have shown their strategies (cards), the supervisor informs them about their gains, and casts the die separately for each country. The die's sides characterize the total activities of the population of a given country. After each round or once after several rounds, presidents may distribute gains among the following budget categories: defense, law and order, social problems, and customs concerns.

Next, each president obtains (through the process of random selection) information garnered from a roulette wheel or a deck of cards. The information may take the form of a single card and focus on accidental events. These may include acts of smuggling, drug trafficking, or illegal entry or exit of persons. Also possible are strikes in parts of the country or throughout the country, conflicts between nations in the regions with mixed populations and border disputes between neighboring states. In addition, acts of international terrorism are possible, including seizure of the presidents as hostages, as well as declarations of independence by regions, crime increases, and so on.

Each president then has to respond to a specific situation. Thus, the probability of unimpeded importation of drugs, illegal emigration, and immigration are less likely with a sufficiently large pool of funds allocated to border services and customs agencies. If such events did take place, the president would bear losses. Strikes occur less frequently if the mark approaches the prosperity area. Since the risk of such an event still exists, one form of insurance may be to institute funding for social problems. Concerning conflicts between nations or outbreaks of crime, the chances for such incidents increase as the mark approaches the depression area. The president may introduce a state of emergency, either throughout the country or within a region. For this purpose, however, an irrevocable loss to a large law-and-order fund is required. Collisions between different nations over disputed territories may result in a military conflict. In order not to suffer from defeat in such a conflict, a considerable defense fund is required. In the event of a region leaving the state, the president may either agree under certain conditions to a compromise, but subsequently lose a part of the gain in all the following rounds, or disagree and introduce the state of emergency, which would cause considerable loss of law and order and defense funds. Also, to be considered is a marked decrease in the position on the state matrix. The president may also make demands of other countries, to conclude various treaties with these countries (e.g., on mutual assistance, nonaggression, and cooperation in the control of illegal drug trafficking).

The game is considered over when one of the countries reaches the upper or lower limits on the state matrix, or after completion of a specified number of rounds. The winner is decided by the amount of the presidential fund accrued and by the level attained by the country on the state matrix.

Upon completion of the game, the participants analyze their strategies and actions, taking into account both their personal winnings and the levels achieved on the state matrices of their countries. It is important to evaluate the state of a country as it pertains to the leadership of a president. This analysis may serve as the basis for discussing such aspects as competence, democracy, authoritativeness, resourcefulness, responsiveness, ability to cooperate with neighbors, and in particular the tendency to compete or confront. In the course of such discussion, special attention is paid to the efficiency of the presidents' steps with respect to law and order problems.

SPUSK and its Derivatives

In the business game SPUSK, 25–30 people collectively ensure the construction of a large-scale industrial complex within a minimum time frame. A final result of the joint efforts of various construction organizations and the Kiev Engineering and Construction Institute was a transformation of the SPUSK business game into a special business game called OPTIMUM (Optimization of the Training Process-Game Modelling of the Unit Method), using unit network models. This is comparable to the systems developed for large and complicated construction projects.

Another direction in modernizing the business game SPUSK has appeared as a result of running it with a number of construction managers in Moscow. It was developed into the game for GLAVMOSSTROY, which involved a control system for the construction of a series of particularly important and complex projects using network models. After mastering the methods and the documentation used in the process of the game SPUSK, the users made substantial alterations to the project for future use, the contents of the input and output data, and the forms and routes of the information flow. Further modifications took into account control mechanisms on construction sites (e.g., the services of "external" organizations which are suppliers of structures and equipment). This version became known as KROSS (Controlling and Regulating the Supply and Condition of Construction).

Conclusion

Over 2,000 people from 300 educational institutions, scientific research institutes, and construction organizations have now (July 1991) been trained in the use of business and management games. The above games are being used successfully in most of the 300 institutes and organizations.

Victor Rybalski—President of EESAGA (Eastern European Simulation And Gaming Association) works at the Kiev Civil Engineering Institute. In September he was one of the organizers of the 18th International Seminar on Gaming-Simulation in Education and Scientific Research, held in September, 1991, with support from the Ukrainian Academy of Sciences.

Stock Market Simulation

Ivo Wenzler[1] and Marijo Polic[2]

Abstract. Eastern European countries are undergoing a transition from centrally controlled societies to market-oriented democracies. The securities exchange and how it functions much be understood as part of this transition. The STOCK MARKET SIMULATION (SMS) is an attempt to help this process. Participants in the simulation take roles of investors (shareholders), house brokers, and floor brokers. They engage in different activities, and receive, process, and send different types of information, all with the purpose of fulfilling their tasks and achieving their investment objectives. Through its extensive application, the SMS has proved to be a powerful tool to help participants understand the basic principles of a stock market and its functions.

Key words Eastern Europe; experiential learning; gaming/simulation; market economy; stock market

Problem Background

Through the end of the 1980s and on into the early 1990s there have been tremendous changes in the Eastern European business environments. Whereas in the past, different sectors of the economy were mostly government controlled and state owned, an increasing number of companies are now attempting to undergo the process of privatization. One of the major paths to present ownership of these companies is through issuing securities, which are then distributed or sold on the securities exchange market.

In several Eastern European countries, after close to 5 decades, the stock markets are being reopened to accommodate this new need. Not only were the markets closed for the last 5 decades, but their entire economies were centralized. As a result of these events, an enormous gap was created in the knowledge that is needed to function within a market-oriented versus a centrally planned economy.

---

[1] Multilogue International, Inc., 329 Lake Park Lane, Ann Arbor, Michigan 48103, US; phone 313-663-3690, fax 313-663-3623
[2] Institute for Development and International Relations, Ulica 8 maja 82, 41000 Zagreb, Yugoslavia; phone 41-444-522

More pointedly, there is a general lack of knowledge and understanding of what a stock market is and how it functions. As an increasing number of companies within Eastern Europe become privatized and are issuing securities to be traded on the market, greater numbers of people will be required to have an adequate understanding and knowledge of how to function on the securities exchange markets, and will need to acquire an understanding of the new conditions in an effective way.

In its effort to respond to this challenge, the Zagreb Business School commissioned the Stock Market Simulation (SMS), an experiential learning tool which can achieve the following objectives:

— Provide participants with the basic knowledge and understanding of a stock market and its functioning.
— Provide participants with the opportunity to experience the dynamics of participating in stock market activities.
— Provide educators with a simulated stock market environment through which participants' understanding of the stock market issues and their potential performance on a real stock market can be tested.

Schematic Presentation of the Problem

As part of SMS design process, a large problem environment schematic was developed. This was an attempt to create a visual presentation of the securities exchange problem environment, on which this exercise focuses. Once developed, the schematic was used as a basis for deciding which issues were important enough to be represented in the final exercise. A simplified version of this schematic is presented in Figure 1.

Why Gaming/Simulation?

The decision by the Zagreb Business School to use a gaming/simulation exercise as a tool for educating people on how to function within the newly opened stock markets was primarily based on the understanding that gaming/ simulations provide a safe environment for learning, and are an efficient and effective approach to experiential learning. Lectures alone were perceived as unable to provide the participants with a full understanding of a stock market and its functioning, and in providing them with the opportunity of experiencing the dynamics of participating in stock market activities.

Context of Use

The primary intent is for the simulation to be used as part of a seminar on the issues of securities exchange (stock market) activities. The seminar contains five modules of about 4 h each. Two of the modules are lectures and presenta-

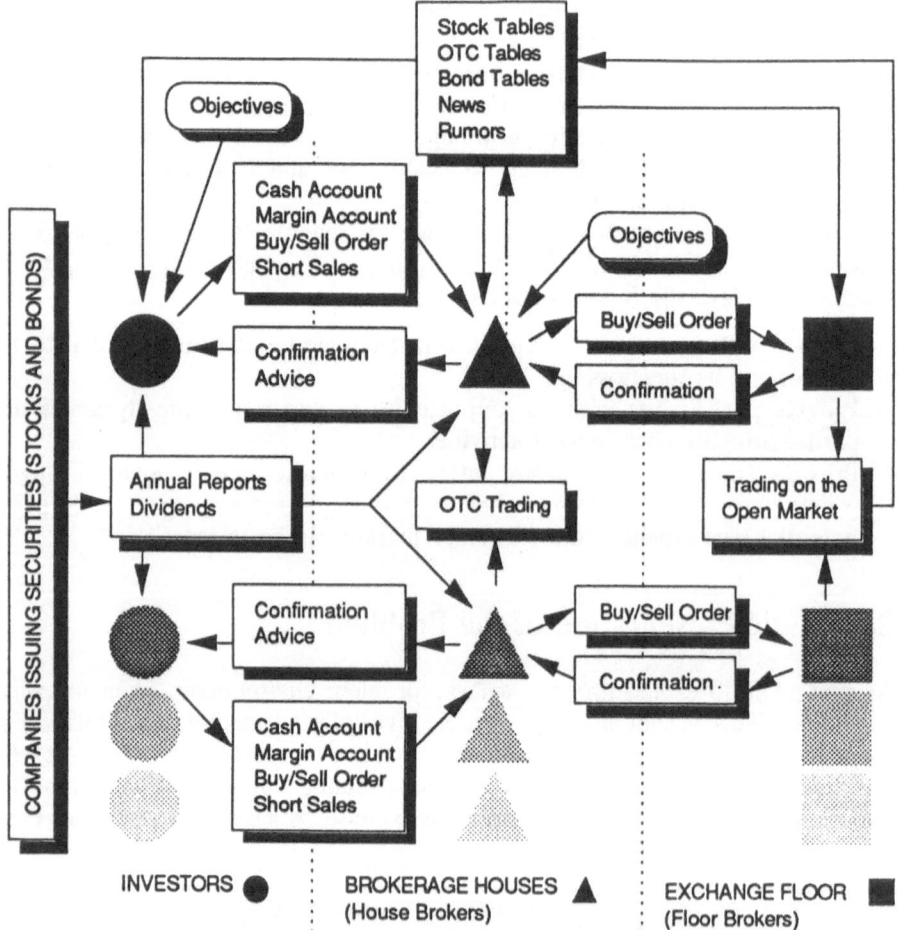

Fig. 1. Problem environment schematic

tions focusing on the issues that will be experienced during the simulation run. The simulation alone lasts 12–15 h and thus occupies the remaining three modules. The number of participants can vary, with a minimum of 30 and a maximum of 70.

Sequence of Activities

The SMS is composed of two distinct phases. Phase I activities include trading on the open market, trading on the OTC market, margin trading, and short selling. Phase II includes listed options, index options, and stock index futures. In order not to overwhelm the participants with the full complexity of the SMS at the outset, and also to let participants gradually acquire new knowledge, the simulation starts with only the open market trading activities, while other activities are added in later cycles of the run.

Each phase of the SMS consists of three modules, each one being five cycles long, with the activities of each cycle divided into two broad stages. The first stage represents the time when the securities exchange floor is closed; the second stage represents the time when it is open. All cycles are 45–60 min in length and consist of the same steps of play. Each step has a specific purpose designed to achieve the objectives of the cycle, as well as the objectives of the whole run.

Following the last cycle played, an extensive debriefing session takes place. During the debriefing the participants discuss, analyze, and evaluate their experiences. The role of the facilitator is to pull the various perspectives into a coherent package, placing an emphasis on the learning that took place during the simulation.

Roles

In the SMS there are two broad groups of roles. One group of roles is performed by participants in the simulation run (gamed roles). The second group of roles are introduced to the run either through the facilitators' activities or through different written materials during the run (simulated roles).

The gamed roles in the SMS are investors, house brokers, and floor brokers. Investors are the owners of securities issued by different companies and institutions and are expected to buy and/or sell securities through their brokerage houses, depending on their investment objectives. House brokers are employees of several brokerage houses present on the market. Their primary role is to work with investors, helping them to sell and/or buy securities. Brokerage houses are also participating on the OTC market as investors. The primary role of the floor brokers is to work on the securities exchange floor, executing buy and sell orders received from investors through their brokerage houses. They execute the orders by dealing with floor brokers who represent other brokerage houses.

The simulated roles are reporter, simulated investor, and companies and institutions whose securities are on the market. Reporters are securities exchange officers and are played by facilitators. Their role is to record all deals that happen on the floor of the exchange, make sure that these deals are properly executed, and record any price changes.

The simulated investor role is also played by one of the facilitators. The primary role of the simulated investor is to load the simulation with a constant flow of different sell and buy orders distributed among all brokerage houses, thus creating the necessary dynamics for the activities in the brokerage houses as well as on the floor of the exchange. Through repetitive loading and cycle-specific interventions on the market, the simulated investor controls the securities exchange market. This ensures that the exercise behaves as realistically as possible, and thus prevents some wild and unrealistic market fluctuations from taking place.

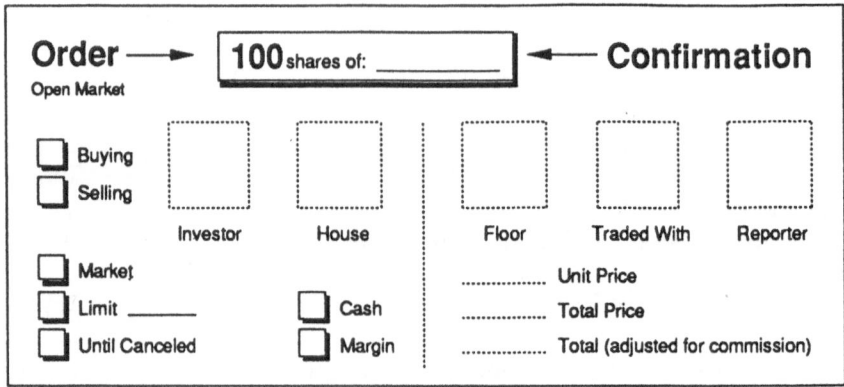

Fig. 2. Order/confirmation form

The third category of simulated roles is the companies and institutions whose securities are available on the market. They are selected and designed to cover a wide spectrum of characteristics which real companies on the real securities exchange market would have. They are simulated through information provided in the written materials introduced during the run.

Accounting System

The accounting system in the SMS is a set of procedures incorporated directly into the exercise with the purpose of recording, processing, and providing feedback on participants' decisions that take place during the run. The major elements of the accounting system are information input, accounting of transactions, accounting of finances, and information output, all of them supported by specially designed forms and software.

Information input consists of the newspaper, cycle reports, and rumors. The newspaper is issued each cycle and covers events ranging from the general state of the economy to those about specific companies and institutions present on the market. Another source of information is the cycle reports (financial reports) for the companies and institutions on the market. Rumors contain information about an event or expected action which, if true, could significantly effect the demand and supply of particular securities on the market.

Another element of the accounting system involves transactions. Several different forms and software packages are used for that purpose. The order/confirmation forms are used for ordering and confirming transactions. There are several different types of these forms depending on the type of transaction. A sample order/confirmation form is presented in Figure 2. The third element of the accounting system deals with finances, such as payments for executed transactions, dividends for the securities owned, and interest on the money deposited into a bank.

Stock, bond, and OTC Tables comprise information outputs of the activities on the market, providing participants with such items as latest price, opening price, quantity of securities traded, dividends for the cycle, and price/earnings ratio for each security on the market.

Results and Benefits

Since its development, the simulation has been run numerous times. The experiences from these runs have shown this simulation to be successful in providing participants with the basic knowledge and understanding of a stock market and its functions. During debriefing sessions, the participants emphasized that the simulation made the information presented through the lectures much more understandable and vivid. The results and comments following the runs have also shown that the simulation provided the participants with the opportunity to experience the dynamics of participating in stock market activities.

The power of the simulation became evident in one of the first runs of the exercise. During the first cycle, floor brokers were trading while sitting on the chairs they brought with them to the floor of the exchange. In the second cycle they did not sit on the chairs any more. In the third cycle they pushed the chairs aside, moved around vigorously and started to develop a system of symbols between themselves and their house brokers. All of this happened without a single hint from the facilitators. Learning about some of the most important dynamics of the securities exchange floor happened in less than 3 h. Participants were forced to make decisions in a very short time and without enough information. This experience provided them with the confidence that they are, despite all the pressures, able to function in an imperfect environment. As a result of the experiences in Eastern Europe, development of a third and fourth phase is under way. These phases will expand the activities in which participants engage by incorporating trading with gold, foreign currency, and futures.

Ivo Wenzler was born in Zagreb in 1955. His undergraduate degree is from the Faculty of Architecture at the University of Zagreb, and he worked as a construction and project manager until moving to the United States in 1984. He has a master's in urban planning, a certificate in gaming/simulation, and a doctorate in architecture, all from the University of Michigan in Ann Arbor. He lives in Ann Arbor and works as a research scientist at the University of Michigan, as well as a senior associate at Multilogue International, Inc. At both places his primary activity is designing policy exercises and large scale gaming/simulation exercises for clients all over the world.

Marijo Polic was born in Zagreb in 1944. He graduated from the Faculty of Economics at the University of Zagreb, and since then he has been working at the Institute for Development and International Relations, which was formerly known as the Institute for Developing Countries. He is primarily working on the issues of development economics, technological development and transfer of technology, and international finance. For the last few years he has been working on implementing gaming/simulation methodology, as well as other experiential learning methods, in the areas of economics and business management.

Using Spreadsheets for Complex Business Simulations

Fernando E. Arellano[1] and Richard D. Johnson[2]

Abstract. Electronic spreadsheets are used as learning tools in business courses. As instructors have become more familiar with spreadsheets, they have used them increasingly to achieve their teaching objectives. Spreadsheet templates are also being used as business-simulation complements to aid in decision making by participants. The objective of this paper is to promote the development of business simulations using electronic spreadsheets as a programming language. A brief description of the parts of a Lotus-based simulation will be given, along with a summary of advantages and limitations found in the development of a complex business simulation.

Key words business games; computer modeling; electronic spreadsheets; simulation design; simulation/gaming; training

Business instructors are increasingly using electronic spreadsheets as a teaching aid for problem solving and case analysis. Textbooks often come with spreadsheet templates as supplements offering students enhanced learning. With additional spreadsheet experience, business instructors are also developing class materials of their own. Businese students are becoming more familiar with the use of electronic spreadsheets as an analytical tool.

Electronic spreadsheet capabilities are also being improved with the release of new versions, providing programming features that were previously in the realm of traditional computer languages. Macros allow programming for automatic execution of keystroke sequences and easy data manipulation. Given that spreadsheets are now the choice programs in many business applications, they may also be superior as the programming environment where business simulations are developed.

Several papers have touted the use of electronic spreadsheets as an aid in the design and evaluation of business simulations, as well as their value in helping students make decisions (see, e.g., King et al. 1988, Schwartz et al. 1989). The

[1] Department of Economics, Colorado State University, Fort Collins, CO 80523 US; phone 303-491-6324; facsimile (303) 491-6441
[2] Department of Finance and Real Estate, Colorado State University, Fort Collins, CO 80523 US; phone 303-491-5062; facsimile 303-491-0596

BUSINESS STRATEGY GAME is a business simulation currently on the market that uses Lotus 1-2-3 as the only operating system. This paper espouses the use of electronic spreadsheets for the complete development and operation of a complex business simulation.

A business game that simulates the operations of a commercial bank within the environment of the US banking system, was developed using lotus 1-2-3 as the sole programming environment. The simulation, which was originally developed in 1987, has been used and tested with bank managers and college business students. With the advances offered in updated versions of spreadsheets, the simulation has been greatly improved.

Research Methodology

In a business simulation, the generation of financial statements includes the combined preparation of cash flow, income, and balance statements. This entails the calculation of sales revenue in terms of price and quantity sold. In most simulations, participants decide on price while quantity is the result of simulation algorithms. Factors such as economic conditions, price, promotion, and other competition variables are often important in determining quantity. Cash flow, income, and balance statement items, such as labor, materials, promotion, and depreciation, also comprise participant decisions or are parameters set by the developer of the simulation. These parameters include labor costs, depreciation rates, materials costs, and interest rates.

All of the elements involved with the simulation are easily programmed into a spreadsheet environment. Once you have developed a spreadsheet model for financial statements, the development of a full blown simulation requires construction of an assignment module linked to financial statements. Auxiliary modules, including the logging and evaluating of decisions, and the continued enhancement of the economic environment, can also be modelled and linked in the same way. In the case of the bank simulation, the complexity and detail of the financial statements resulted in the need for different files; one for the evaluation of decisions, one for the assignment module, and one for the processing of financial statements. An additional file contained a model that allows the instructor to generate the economic environment and the initial bank financial statements.

Developing the Simulation

A prototype was first developed with many of the simulation operations being performed manually. The development continued with the use of program commands and built-in functions and macros to automate some of the simulation's operations. The simulation was designed with separate files to enter and check participant decisions, to make competitive assignments, to generate the financial statements, and to create an environment for the administrator to model different economic scenarios. Separate files were used to overcome size limitations.

The Decision Model

This model contains the decisions of all participating firms. Decisions are entered manually by the administrator, or automatically from each firm's file by using a "file combine" command. Once the decisions are loaded, a subroutine checks each decision using "if" conditions. A report including the number of firms for which unallowed decisions were detected is generated and printed if desired. Some incorrect entries are automatically corrected by the program.

This model allows the instructor to control the number of decisions that can be made by the participants, with default values for those that are excluded. This option allows for a gradual introduction of the simulation to the participants, and for a better understanding of the effect of some variables when variability in others is eliminated.

The Assignment Model

The firm's decisions as well as information on the economic environment are loaded into this file from the decision file, and the results are extracted and later loaded into the file that generates the financial statements. This model contains the algorithms for allocation. The interaction between economic environment conditions and the firm's decisions determines the allocation of sales to each firm, which in this case is based on loans and deposits.

The spreadsheet format allows for easy, quick, and efficient testing of the assignment model. Changes in parameters and decisions are easily incorporated, and the testing requires only the use of the familiar recalculation procedure. One of the characteristics of a spreadsheet is that the formulas and programming code (macros) are in the environment where the reports or results are displayed. This permits immediate testing of changes made in formulas and in decisions. The spreadsheet format also allows partitioning of complex formulas into simple ones to observe the response of the algorithms to different levels of decisions.

The Financial Statement Model

The financial statement model is a straightforward application of an electronic spreadsheet. It builds the cash flow, income statement, and balance sheet for each team. Footnotes, or detailed reports, are used to provide additional explanation. Parameters such as interest rates, tax rates, personnel salary, depreciation rate, and others are explicitly shown in cells, allowing for easy visualization and modification.

To conserve memory, individual team financial statements are processed sequentially. After each firm is processed, the new report is saved in a file, and then copied into the firm's disk for use in the following period. Printing of each firm's report is optional and takes place after the report is generated. This model also generates the instructor's report, which includes a summary of the firm's decisions, the assignments, main financial statements variables, and a performance ranking.

The Economic Environment and Firm Initial Statements Model

The simulation contains an economic environment and a set of initial financial statements, which can be changed. For example, one can design an inflationary environment by entering the quarterly inflation rate and setting a desired quarter variability through the use of the random number generator. A table contains the firm's balance structure in percentage terms. Finally, individual cells hold the simulation parameters, which allows for easy change of parameters. The administrator can choose not to model the economic environment, in which case, a default economic environment is already built into the simulation.

The Participant Model

The participant model, which allows teams to forecast future performance, was derived from the financial statement model: it is contained in a similar file, with the difference that the participants must estimate deposit and loan levels and certain current economic conditions. Since the basic model is given, participants can concentrate on making decisions and estimating their effect on sales rather than on model building. Performing what-if analyses is efficient and accurate.

Advantages of Using a Spreadsheet Environment

Using a spreadsheet as a simulation environment has three major advantages from both the point of view of program development and program operation. First, use of built-in commands and functions facilitates easier construction of routine financial reports. The programming that is required is far less complicated in a spreadsheet environment. Second, administrators and participants are commonly familiar with the operation of the spreadsheet program. Therefore, it requires less training and education for both administrators and participants. Some instructors are unwilling to use simulations because of the start-up costs of learning the program. The initial costs are clearly lower when the spreadsheet program is already being used. Third, reports and data from the simulation are formatted in standard spreadsheet format. This makes it easier for participants to analyze performance and forecast future performance. Data is available for graphical presentation if the administrator requires a report to shareholders for each team. The administrator can also use the data to help demonstrate performance to the participants.

Limitations of Using a Spreadsheet Environment

There are a few significant limitations in developing simulations in a spreadsheet environment. First, special care has to be taken to protect parts of the code. With later versions of some spreadsheet programs, capabilities exist to store macros in separate files and to protect sensitive algorithms. The possibility also exists of compiling the allocational part of the program to make it inaccessible. Some limitations on file size can present problems if the simulation is large and complex. This limitation requires some additional work in setting up separate files and some additional programming to move and combine files.

Conclusions

The characteristics of electronic spreadsheets make them a suitable programming alternative in the development of business simulations. Given the widespread use of spreadsheets, it is anticipated that more business simulations will be developed in the future using this programming tool.

References

BUSINESS STRATEGY GAME. Thompson A, Stapepenbeck G (1990) Richard D, Irwin (Homewood, IL 60430, US)

King B, King A, Crookall D (1988) Using spreadsheets for games calculations and model development. Simulation/Games for Learning 18(1):95–101

Schwartz R, Teach R, Letson S (1989) Teaching forecasting, cash budgeting, and inventory model building using SBTOOLS. In: Wingender J, Wheatley W (eds) Developments in business simulations and experiential exercises, vol. 17, p 213

Fernando E. Arellano, from Peru, is a PhD graduate student and instructor in the Department of Economics at Colorado State University. He has coauthored BANMAN, a bank management simulation, and has also developed MANECSIM, a simulation for use in managerial economics courses.

Richard D. Johnson is a PhD in Finance. He is an associate professor in the Department of Finance and Real Estate at Colorado State University. He has coauthored BANMAN and three books about commercial banking. He also conducts seminars about specialized topics in banking.

A Holistic Approach to Using a Marketing Strategy Simulation

Olivia de Bergerac[1]

Abstract. MARKSTRAT is a complex marketing strategy game which invites five teams to compete with a portfolio of products in the consumer electronic markets. However, like most of the traditional business games, MARKSTRAT is characterized by three unfortunate features: a mono-disciplinary approach, a tendency for cultural blindness, and an ethnocentric vision. In order to address these issues, I will investigate three different points of view: the designers, the lecturer, and the students. From the congruence and discrepancy of these different perspectives, I will recommend a multi-disciplinary and an integrative approach for using MARKSTRAT in an MBA program. Based on a strategy which combines the business simulation with the "simulated environmental laboratory" created by the exercise, the new approach extends the use of MARKSTRAT from a marketing strategy simulation to (1) a management game, (2) an intercultural communication exercise, and (3) a global strategy game.

Key words business game; computerized simulation; global management game; intercultural communication exercise; marketing strategy

Originally conceived as a research environment for strategic studies in marketing, MARKSTRAT has become instead a popular classroom game in graduate and management-education programs. The MARKSTRAT simulation is today used by several hundred business schools and corporations in many countries. The purpose of this paper is to explore the use and the potential use of MARKSTRAT in the curriculum of an MBA program. In order to do so, I will investigate three different points of view. Starting from the designers who created the game, I will move on to the lecturer who uses it, and then to the students who have been through the experiential learning process. From the congruence and discrepancy of these three different perspectives, I will recommend a multidisciplinary and integrative approach for using MARKSTRAT in an MBA program. This takes advantage of the following facets of the simulation: (1) as a management game, (2) as an intercultural simulation, and (3) as a global strategy game.

[1] Australian Graduate School of Management, University of NWS, NSW 2033 Australia; phones 2-931 9200 (w) 2-665 0712 (h); facsimiles 2-662 2451 (w), 2-664 2018 (h)

MARKSTRAT: A Marketing Strategy Game

MARKSTRAT's inventors, Jean-Claude Larréché (INSEAD) and Hubert Gatignon (Wharton School), describe its raison d'être, definition, and story in the following manner: Competence in marketing is related to taking the right course of action. One may have read all existing articles on marketing and know all the theories and still be a poor marketing practitioner if this knowledge is not properly translated into action. Marketing strategy knowledge is not an end in itself but a means to an end. Marketing strategy being an action-oriented discipline, an essential part of the learning process is actually to perform tasks in a real environment. Simulations emphasize the application of concepts in an action-oriented approach; students have to make decisions (as opposed to recommendations) and live with their consequences.

In addition to marketing-mix considerations, the simulation allows students to make decisions on product portfolios, market segmentation, and product-positioning issues. It requires a segmentation of markets, an evolving product line with suppressions and introductions, and hence the development of non-marketing functions, such as research and development.

Five firms (teams) in a MARKSTRAT industry begin with different relative competitive positions, but each firm has two brands in the marketplace. Throughout the course of the game, firms are free to introduce new brands into the market, withdraw old ones, and modify existing brands. In addition, firms must make decisions related to production, advertising, sales, pricing, and product-positioning for each of their brands in each period.

How MARKSTRAT is Used in the AGSM MBA Program

David Midgley, professor of Marketing Strategy Australian Graduate School of Management (AGSM), justifying the use of MARKSTRAT, emphasizes the point that the simulation helps to bridge the gap between textbook learning and the operational circumstances that hold in real life:

> While lectures are included to revise key strategic frameworks, these are then applied and reinforced by case studies (48% of the final grade) and by MARKSTRAT simulation (52% of the final grade). Indeed a significant proportion of the subject is devoted to MARKSTRAT since the exercise provides an excellent vehicle to illustrate longer-term strategies, and it also allows you to practice your strategic skills in a realistic setting.

The format chosen is an interrupted or discrete game time. This mode has the advantage of not upsetting delicate MBA class scheduling. It takes place during the first class sessions and then during students' free time (evenings and weekends) over the 10 weeks of the MBA term. Students are free to form teams according to their affinities.

At the beginning of the simulation exercise, the teams are assigned to a firm and receive a report of their firm for period zero. The teams are able from the company report information and the information in the manual to evaluate the relative market strength of their respective firms compared to the competition. Each team is asked to submit two reports: (1) an initial report detailing the team's definition of its corporation, its analysis of the starting product portfolio, and the marketing strategies proposed to bring successful performance over the course of the game, and (2) a final report detailing what the team has learned from the game—particularly with respect to the concepts explored in the text, lectures, and cases. In addition, each team will make a final 10 min presentation on their corporate performance over the game.

Student Feedback on MARKSTRAT Experience

Data were collected from two main sources. The AGSM students are asked at the end of each term to fill in evaluation forms with specific questions on the performance of the lecturer and open questions on the course. The quantitative data consists of these 71 students' comments over 4 years from 1987 to 1990. From these data, three main issues stand out:

1. Can a lecturer be a facilitator?
2. Is MARKSTRAT a win-lose game?
3. What is the group dynamic learning experience?

The majority of students enjoy and value the learning-by-playing process, meanwhile a minority have some doubts about the learning experience because it is a game. A lot of students find it hard to accept the fact that the lecturer is just a facilitator.

The normal classroom of lectures is sender-centered communication, in which knowledge is dispensed in a linear fashion from the teacher (sender) to passive receivers (students). With experiential learning, communication is changed to being primarily receiver-centered. The role of lecturers is therefore altered from being that of an expert to that of facilitator (Freeman and Dumas 1989). The contradiction is that students recognize the value of the learning process but cannot identify the lecturer's input. In the receiver-centered model, a low lecturer profile is an indication of a well-designed learning experience.

Some students questioned the competitive atmosphere created. It is recognized that MARKSTRAT generates a very competitive atmosphere in the MBA program. There are many anecdotes about teams cheating, stealing competitors' reports and sabotaging competitors' answers. Such actions lead to the following questions: Does MARKSTRAT, like recreational games (e.g., DUNGEONS AND DRAGONS) teach aggressive and competitive values? Is MARKSTRAT a win-lose game of destructive competition? However, students emphasize the fact that MARKSTRAT is a very powerful group dynamic learning process.

MARKSTRAT'S LEARNING PROCESS

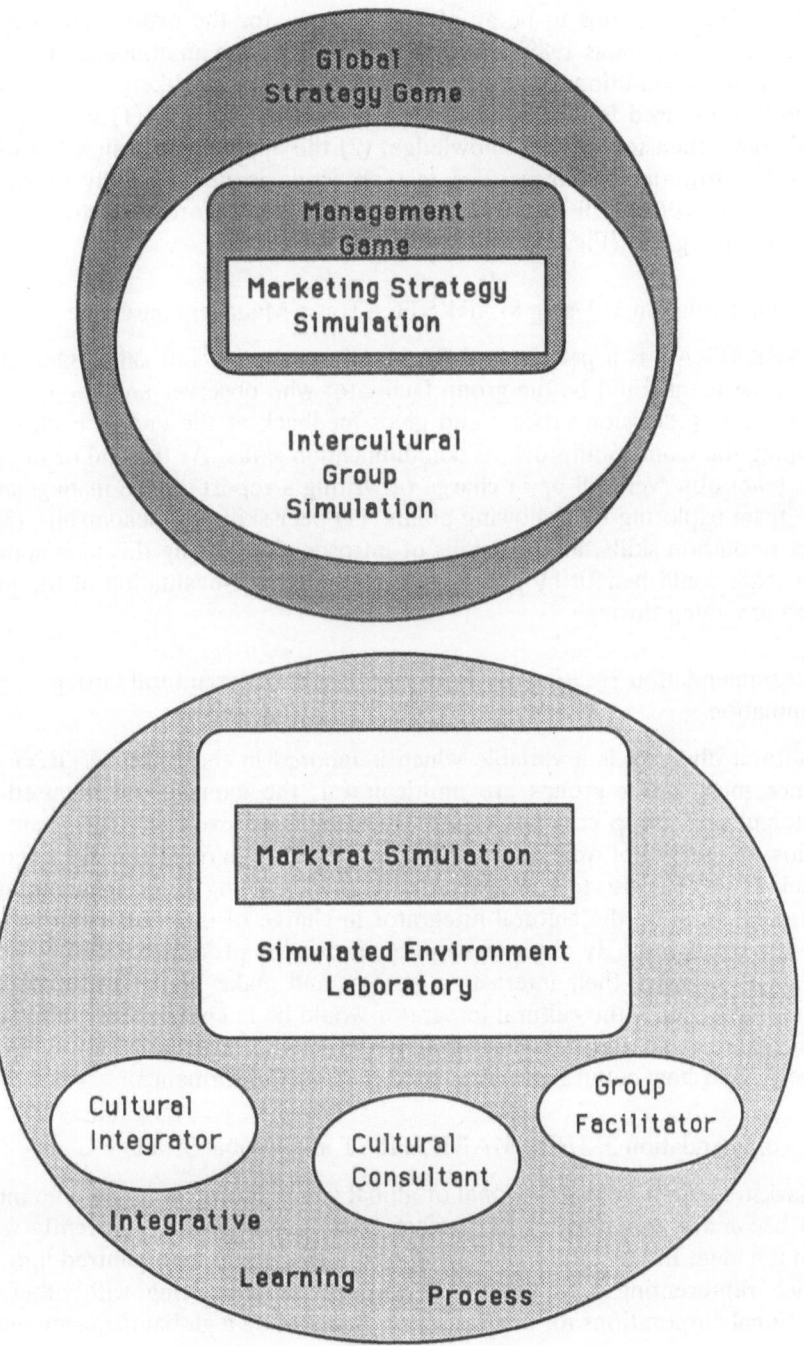

Fig. 1. MARKSTRAT model

Recommendations

MARKSTRAT seems to be an effective means for the expression of feelings (Freeman and Dumas 1989) amongst learners. In such an atmosphere of highly participative conditions, attitude change appears to be likely. The game can therefore be used for the achievement of three objectives: (1) the acquisition and comprehension of new knowledge, (2) the application of new knowledge, and (3) attitudinal change. The last objective is not yet fully explored in MARKSTRAT. I will therefore recommend a new, more "holistic" approach to using the game (Fig. 1).

Recommendation 1: Using MARKSTRAT as a Management Game

MARKSTRAT is a perfect tool for learning management skills. One student from the team could be the group facilitator who observes and keeps a record of the group-decision process and gives feedback at the end of each session, helping the team to improve its communication skills. At the end of the game, the team observer will be in charge of writing a report on the management of the team exploring the following points: (1) peer skills, (2) leadership, (3) conflict resolution skills, and (4) skills of introspection. Using this new approach, the team could benefit by providing a continual self-evaluation of the process they are going through.

Recommendation 2: Using MARKSTRAT as an Intercultural Group Simulation

Cultural diversity is a variable which is ignored in the MARKSTRAT game. Since most MBA groups are multicultural, the game could be used as an intercultural group communication and simulation exercise. The game would allow the team not only to understand cultural similarities and differences and their cause, but also to know how to act when working in foreign cultures. One student could be the cultural integrator in charge of the cross-cultural process of the simulation. By observing and analyzing the process, s/he could help the team to improve their interaction process and make better decisions. At the end of the game, the cultural integrator would be in charge of writing a report. The game could allow students to understand what managing cultural diversity means and how cultural synergy could increase performance.

Recommendation 3: Using MARKSTRAT as a Global Strategy Game

Markstrat is not an international or global game. However it could be modified to become a global game. A project development team is currently working on the idea. In this future global game, students could be organized into teams, each representing a multinational corporation competing with other multi-national corporations for market share. It would be a global management game with international trade and cross-cultural communication problems, with firms selling all around the world. The game would mix students for a cross-cultural learning-experience. In this way, cultural differences are no longer ignored, but start to be valued as a corporate asset and managed as a competitive advantage by capitalizing on cultural diversity to understand the foreign market better. The learning objective will then go beyond marketing strategy concepts. Successful managers of the next decade will need to be cultural integrators.

A holistic approach to using MARKSTRAT could be a powerful tool for developing the truly global manager, culturally sensitive in applying both theory and practice. I expressly use the word global instead of international. We can be international managers by acknowledging cultural differences; we are global managers once we start to value cultural diversity and manage it in a synergistic way. The only limit we would like to point out is that MARKSTRAT is excellent in terms of teaching about the real world of marketing strategy, yet it appears to skirt the moral or ethical issue of what the product actually is. In other words, it demonstrates prevalent attitudes of consumerism, in the sense that it does not matter what you are selling, as long as you sell it. Environmental issues, for example, are not addressed.

References

Cook VJ (1987) Introduction to strategic studies in MARKSTRAT. Journal of Business Research 15(6):467–468

Cook VJ, Page JR (1987) Assessing marketing risk. Journal of Business Research 15(6):519–530

Crookall D, Saunders D (1989) Towards an integration of communication and simulation. In: Crookall D, Saunders D (eds) Communication and simulation: From two fields to one theme. Multilingual Matters, Clevedon, Aron

Dodgson H (1987) Management learning in MARKSTRAT: The ICL experience. Journal of Business Research 15:481–489

Freeman JM, Dumas P (1989) Business games: From business schools to business firms. In: Crookall D, Saunders D (eds) Communication and simulation: From two fields to one theme. Multilingual Matters Ltd, Clevedon, Aron

Glazer R, Winer RS, Steckel JH (1987) Group process and decision performance in a simulated marketing environment. Journal of Business Research 15(6):545–557

Harris P, Moran R (1987) Managing cultural differences. Gulf Publishing, Houston

Ibe M, Sato N (1989) Educating Japanese leaders for a global age: The role of the international education centre. Management Development in Japan 8

Kinnear TC, Klammer SK (1987) Management perspectives on MARKSTRAT: The GE experience and beyond. Journal of Business Research 15(6):491–501

Larréché JC (1987) On simulations in business education and research. Journal of Business Research 15:559–571

Larréché JC, Gatignon H (1977) MARKSTRAT strategy game; Participant's Manual. Scientific Press, Palo Alto, CA

Olivia de Bergerac has a PhD in French Literature from the Université de Nice, France and an MBA from the University of NSW, Australia. She has designed and run language in-house training programs in France, the US, the UK and Australia for large companies (Schlumberger, Mitsui and Dow Chemical). She is currently doing research for the Australian Graduate School of Management.

Simulation Gaming and the Improvement of Quality

James M. Freeman[1]

Abstract. Improvements in quality performance in the Far East are slowly being mirrored in the West. The role of analytic methods, including charts, in raising quality standards is justly attracting wider recognition. SQCCHART is a new simulation/game, written to support training in the effective use of statistical quality control charts.

Key words   charts; quality control; simulation; training

Quality is of global significance in world trade. Western companies have been far slower than their counterparts in the Far East to exploit the potential of formal quality assurance techniques. Japanese industry, in particular, is thought to be three decades ahead of that of the US and EC countries. As a result, quality costs have tended to be significantly higher in the West (Schmidt and Jackson 1982) whilst, ironically, overall quality standards have been much lower. It was recently estimated (Dilworth 1989) that the average cost of failing to control quality in US companies was 25% of sales.

The SQCCHART package, recently developed at UMIST, is a simulation/game for supporting specialist training in quality management. In particular, SQCCHART demonstrates the most commonly applied statistical quality control (SQC) chart schemes in conformance quality testing (Schroeder 1989). The package is so designed that trainees can discover, firsthand, the technical behaviour of chosen schemes in a realistic, on-line setting. A gaming facility built into SQCCHART helps with related skill assessment.

The software is described in outline in the last section of the paper. Beforehand, two introductory sections summarise respectively (1) the nature of quality and quality control (QC) in business and (2) the role of statistics in the maintainance of quality standards.

---

[1] School of Management, UMIST, PO Box 88, Manchester M60 1QD, UK

Quality and Quality Control

The quality of a product is usually judged in terms of its fitness for use, a "quality product" being defined as one that meets the needs of the marketplace (Chan 1990). Quality, together with price and availability, is a key determinant of customers' potential purchasing behaviour.

As consumers become more affluent and discerning, so management's capability to control and improve the quality of its products becomes more critical to its continued success. To achieve its quality goals, a company traditionally needs to address three major areas of concern (Dilworth 1989):

1. *Quality of design*—with respect to a product's meeting the minimum appropriate standard for use
2. *Quality of conformance*—in terms of a products's conforming to design specifications
3. *Quality of performance or service*—as regards training and service support, in order that a product can be considered satisfactory within reasonable expectations

Statistical quality control charts focus particularly on the second area in the list.

Quality control (QC) is the generic term used to describe all activities a company undertakes to ensure a product performs to the customer's satisfaction. Such activities typically involve many different groups within the company, including engineering, production, and marketing staff. This multidisciplinary approach to the subject highlights the pervasive nature of QC, or Total Quality Control (TQC) as some prefer to name it.

Statistical Quality Control

Statistical quality control (SQC) is the major means of determining whether goods or services perform to specification. (Alternatives such as customer interviews and product testing can be expensive.) With SQC, sampling is usually used in preference to 100% inspection because of the much lower costs involved. In some situations, such as destructive testing, sampling is, of course, the only practical option.

Sampling, carried out as the product is being made (process control sampling), is typically differentiated from that undertaken after production is complete (acceptance sampling). The first is normally the responsibility of the supplier while the second tends to be the concern of the customer.

The sampling procedure itself can be by attributes or variables. With sampling by attributes, items are classified as either good or defective and lots of items as acceptable or unacceptable. The degree to which items are found to be defective is not an issue with this approach. It is, however, when sampling is by variables. With this option, measurements are taken of a characteristic, such as a dimension or weight, to establish how far these differ from a prescribed quality standard.

SQCCHART, in its present form, deals only with control charts for variables. Variables charts are more expensive to operate than attributes charts, but have the advantage of providing more information about the processes they are used on.

Control Charts

In every production process, a certain amount of natural variation always exists. This occurs regardless of how well the process was designed or implemented, or how adequately it is being maintained. The variation is uncontrolled and results from numerous small causes. When it is low, the process is said to be in statistical control.

Processes that are out of control are operating in the presence of assignable causes of variation, relating to the machines, operators, or materials used (machines can become improperly adjusted, operators make mistakes, etc.). Normally, production processes operate in control most of the time. If, however, a process slips out of control, a significant portion of output may not conform to specification.

A useful way of monitoring the performance of a production process is by means of the control chart, an example of which is shown in Fig. 1 (Jerry 1989). Each observation here relates to a sample of processed items. It may be an individual reading (when the sample size is one) or, for example, the average of a number of measurements. The center line (CL) represents a target or standard value for the characteristic, against which sample values are compared.

The UCL and LCL lines represent the upper and lower control limits respectively. These are chosen by statistical convention, so that assuming the process is in control, the probability of a sample point falling outside the limits is quite small. If a point goes above the UCL or below the LCL, this is taken as evidence that the process has gone out of control. In this case, a search is made for an assignable cause and appropriate corrective action is taken.

Benefits

Benefits arising from the effective use of control charts (Chan 1990) include good record keeping, on-line control, reduction of scrap, and the provision of diagnostic information and data on process capability. Training in this area can, therefore, prove extremely productive, as can the development of appropriate action-oriented learning materials.

SQCCHART

SQCCHART is a stand-alone, menu-driven package which runs on an IBM PC or compatible. It functions at two basic levels: demonstration and testing. These both depend heavily on the package's capability for simulating a standard repertoire of control charts. Charts used for monitoring process

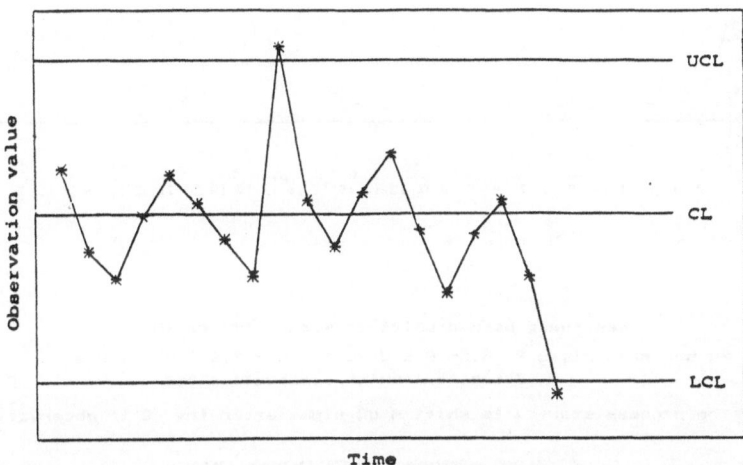

Fig. 1. Typical control chart. *UCL*, upper control limit; *CL*, center line; *LCL*, lower control limit. From Jerry (1989) with permission

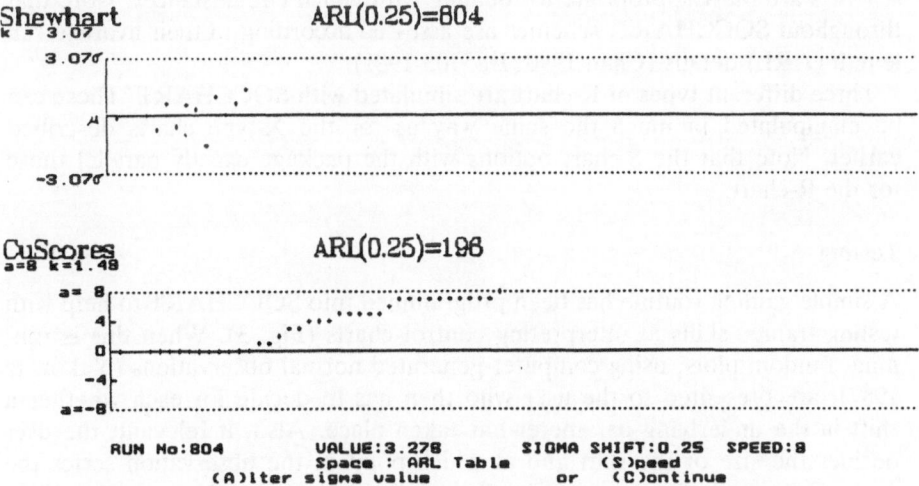

Fig. 2. Comparative plots with SQCCHART. *ARL*, average run length

variability (so-called R- and S-charts) are handled separately from corresponding $\bar{X}$-charts, which measure the central tendency of a series of samples. For $\bar{X}$-type charts in particular, six options are available with the package (Shewhart 1931, Page 1954, Munford 1980, Lucas and Crosier 1982).

Charts from this list can be screened individually or in pairs (Fig. 2). By experimentation, such as introducing different shifts in the process mean as charts are being screened, the sensitivity of schemes to changes can be observed. In this way, users are quickly able to build up an instinct as to which

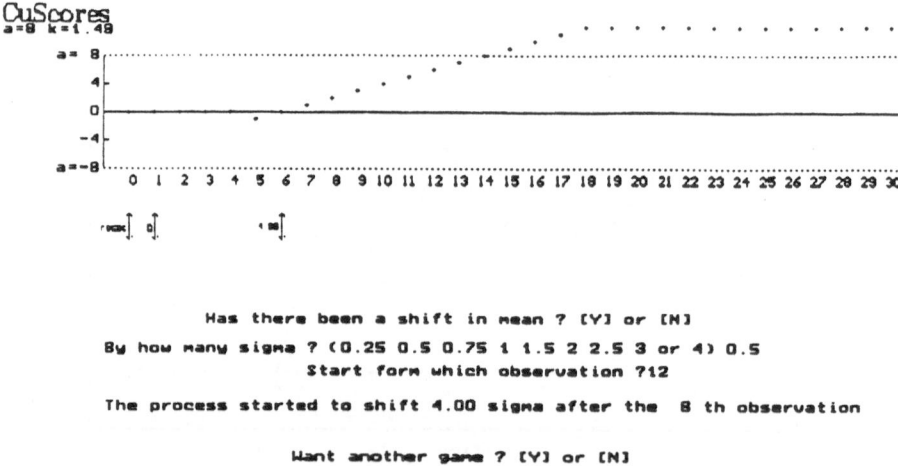

Fig. 3. Testing skill at interpreting a control chart

schemes are most appropriate for dealing with which circumstances. Note that throughout SQCCHART, schemes are assessed according to their average run length (ARL) details (Chan 1990, BS5703 1981).

Three different types of R-chart are simulated with SQCCHART. These can be manipulated in much the same way as for the $\bar{X}$-type charts described earlier. Note that the S-chart options with the package exactly parallel those for the R-chart.

Testing

A simple gaming routine has been programmed into SQCCHART to help with testing trainee skills at interpreting control charts (Fig. 3). When this is running, random plots, using computer-generated normal observations (Yakowitz 1987), are presented to the user who then has to decide for each whether a shift in the underlying parameter has taken place. Also, if relevant, the user decides the size of the shift and at which point in the observation series the change occurred. Performance in the test is scored at each stage, providing trainees with useful feedback on their skill at reading problem charts.

The ISAGA Session

The SQCCHART prototype was demonstrated at the conference to a mixed group of mainly Japanese and American delegates. A laptop with an LCD screen was specially brought into the lecture room for this purpose. Despite hardware limitations, the package worked very efficiently, one university expressing an interest in acquiring it.

References

BS5703 (1981) British Standard Institute, London

Chan WC (1990) SQCCHART, a computer-based training package relating to the use of statistical quality control charts. MSc dissertation, University of Manchester

Dilworth JB (1989) Production and operations management. McGraw-Hill, New York

Jerry B (1989) Principles of quality control. Wiley, New York

Lucas JM, Crosier RB (1982) Fast initial response for CUSUM quality control schemes: Give your CUSUM a head start. Technometrics 24:199–205

Munford AG (1980) A control chart based on cumulative scores. Applied Statistics 29:252–258

Page ES (1954) Continuous inspection schemes. Biometrika 41:100–115

Schmidt JW, Jackson JF (1982) Measuring the cost of product quality. In: "Effective quality cost analysis for increased profit and productivity/SP-512". Society of Automotive Engineers, Warrenton, PA

Schroeder RG (1989) Operations management. McGraw-Hill, New York

Shewhart WA (1931) Economic control of quality of manufactured product. Van Norstrand, New York

Jim Freeman lectures in Quantitative Methods at the University of Manchester Institute of Science and Technology. The author of many articles on management simulation, he specializes in producing computer-based packages—particularly business games—for outside organizations. He is currently working on a series of packages for use in quality management.

Section 5
Assembled Abstracts

Building More Meaning into the Organizational Modeling Process

Margaret M. Beranek and Connie E. Wells[1]

Abstract. The documentation that is generated during systems analysis and design is the basis for the development of new information systems. The contents of that documentation are determined, to a large extent, by the methodology and models used by the systems analyst/designer. The systems analyst/designer will define system characteristics as identified by these models, and exclude any organizational or system characteristics not identified by these models. In the past several years much attention has been given to the inclusion of social, cultural, and psychological characteristics of the users and user environment into the Information System modeling process. The organizational modeling process needs to be expanded to include these aspects in our system models if we are to consciously address them during the SA&D process and to therefore enrich the organizational model. We present social, political, behavioral, and cultural factors that have a profound impact on the organization and that need to be included in the modeling process and present techniques for integrating these factors in the modeling process.

---

[1] Computer Information Systems Department, Georgia State University, Atlanta, GA 30303 USA; phone (404) 651-3880; facsimile (404) 651-2804; e-mail: CISMMB@GSUVM1

Organizational Modeling: Rationales, Issues, and Methods

Minder Chen[1] *and Kunihiko Higa*[2]

Abstract. The increasing complexity and dynamics of modern organizations have made it difficult for people in them to understand and manage them. But modeling an organization is a difficult process, if not an impossible one. We raise several issues and

---

[1] Department of Decision Sciences and MIS, School of Business Administration, George Mason University, 4400 University Drive, Fairfax, VA 22030 US; phone 703-993-1788; facsimile 703-993-1809; e-mail Bitnet: mchen@gmuvax
[2] Information Technology for Management, College of Management, Georgia Institute of Technology, Atlanta, Georgia 30332 US; phone 404-894-4365; facsimile 404-894-6030; e-mail Bitnet: khiga@gtri01

assumptions about organizational modeling in this paper. The potential usages of organizational models are suggested. The large amount of information and complex relations among objects in the organization dictate the use of information technologies to support organizational modeling. An object-oriented multi-dimensional organizational modeling method is presented. Future research directions in organizational modeling are explored.

Using a Simulation Tool for Defining Global Problems

Dorien J. de Tombe[1]

Abstract. Global problems are often very complex. Some of the complex interdisciplinary global problems, such as the sudden changes in Eastern Europe, the Gulf war, and AIDS, are very hard to handle. Even defining the problem is difficult. Managers and experts often lack the knowledge and training as to how to handle these kinds of problems. Regular education provides little opportunity to learn how to handle these kind of problems, as it focuses on well-defined problems in strictly divided domains in a constant context. Policy makers need special training for analysis of complex interdisciplinary problems imbedded in a dynamic context. This training can be done using a free-form game as a didactical context in which cases can be analyzed. Cases that are not defined and well-structured give the opportunity to deal with all the aspects of a real life problem. Defining complex interdisciplinary problems is teamwork. A system dynamic modelling tool can give a graphical representation of the mental model of the problem. With this computer simulation tool a scenario of the problem can be made in which one can simulate the effect of interventions on the problem. The graphical model can function as a shared language to help experts in different fields concentrate on the same aspects of the problem during the discussion.

[1] Faculty of Social Sciences, Utrecht University, PO Box 3286, 1001 AB, Amsterdam, The Netherlands; phone 20-692-7526; e-mail detombe @lri.jur.uva.nl

A Proposal to Develop an International Business Simulation Laboratory

Joanne Velan Dunn[1]

Abstract. This paper will outline the rationale and plan presented at ISAGA 91 for the development of an International Business Simulation Laboratory. A survey assessed the interests of participants in cooperatively participating in an international consortium to develop a prototype resource model for use in schools and businesses throughout the world. Participants were asked to identify effective simulation/games and relevant

[1] Department of Instructional Communications, Community College of Allegheny County, Allegheny Campus, 808 Ridge Ave. M-523, Pittsburgh, PA 15212-6097 US; phone 412-237-2628 (w) 412-922-5159 (h), facsimile 412-922-5159

materials from their own country to share with other members. An outline for a cooperative pilot project was discussed as well as possible methods for funding, types of resources, and content areas.

Law, Contract, and Power

Kaoru Endo[1]

Abstract. The development of computer networks is affecting various aspects of society, both nationally and internationally. In such an age, the rules for the formation of social structure should be reviewed radically; for example, how can the aggregation of egocentric actions by individuals generate macro social structures from a chaotic state? Although this has long been a common theme in sociology and its importance is now increasing because of the ecological implications, up to now no clear theoretical answer has been established. In this paper, I present some models to solve this problem. The models have the following features: in the initial state, there is no given social consciousness because of the original nature of human beings. There are no given laws, ethics, morality, and so on. Each individual does not have complete information about the whole world, and there is uncertainty about the natural world. Each individual produces some kinds of goods necessary for his life and acts for self-preservation on rules set by the individual based on his experiences and limited observations. Depending on the results, the rules can be changed to interact with others. The models showed that repetition of interactions among individuals resulted in formation of certain recognizable patterns which could be interpreted as formation of natural law, power, and social structure. Based on these findings, I believe that the models offer some insights into social change and social relationships.

[1] Graduate School of Science and Engineering, Center for Humanities and Social Sciences, Tokyo Institute of Technology, 2-12-1 O-okayama, Meguro-ku, Tokyo, 152 Japan; phone 03-3726-1111 (ext. 2268)

Design of a Computer Game for Managing an Apparel Retail Store

Maiko Fukuda[1], *Katsuaki Kuroda*[2], *and Masanobu Miki*[2]

Abstract. The purpose of this study is to improve the ability of enterprise administrators of the apparel retail trade to solve problems. We have used a computer-based management game that emphasizes the ability to utilize information. As a prior step, a management game to improve this ability was designed.

[1] Department of Computer Instruction, Teikoku Women's Junior College 6-21-57 Tohda Moriguchi, Osaka, 570 Japan; phone 6-902-0791; facsimile 6-902-8894
[2] Department of Clothing and Fashion, Faculty of Domestic Science, Teikoku Women's College

A Micro World for Kanji Learning

Toshihiro Hayashi and Yoneo Yano[1]

Abstract. Kanji consist of part structures and these often have smaller part structures too. Kanji part structures represent (a) the meaning, (b) the pronunciation, and (c) various meanings by combining with other part structures. We have developed an environment for learning kanji and understanding these features using a computer, targeted at foreigners who are learning Japanese. The environment, a "micro world" for kanji learning, allows students to construct various kanji using kanji part structures and new kanji part structures by combining kanji part structures too.

[1] Department of Information Science and Intelligent Systems, Tokushima University, 2-1 Minamijyosanjima, Tokushima, 770 Japan; phone 886-23-2311; facsimile 886-23-2761; e-mail rupoxy@tokugw.is.tokushima-u.ac.jp, yano@tokugw. is. tokushima-u.ac.jp

Game Paradigm CAI

Takayuki Iida[1] *and Akira Kikuyama*[2]

Abstract. We have developed a prototype CAI course software which has an environmental interface based on the game paradigm for children who have never played with video games and do not know how to operate a computer. The course involves drill-type mathematics combined with fantastic pictures as with role-playing games. Learners walk around in the dungeon of the mathematical course and fight with problem monsters.

[1] Konami Co., Ltd, 3-25 Kanda-Jinbocho, ChiyodAku, Tokyo, 101 Japan; phones 03-3262-9126 (w) 0473-77-7062 (h); facsimile 03-3262-9115
[2] Address as [1]; phone 03-3221-7160 (w); facsimile 03-3261-6211

Residents' Attitude Shifts in an Environmental Dispute: A Case Study of a Golf Course Location Dispute

Yoshiki Kago and Sachihiko Harashina[1]

Abstract. Since the concept of sustainable growth was advocated, the balance between development and environmental protection has been a most important problem. Nowadays the general public take great interest in environmental problems, and there are often protests against development, a trend which will become even stronger from now on. This paper, looks at one of the golf course location disputes that have arisen in recent years in Japan as a case study of an environmental dispute. We carried out a field

[1] Department of Social Engineering, Tokyo Institute of Technology, 2-12-1 O-okayama, Meguro-ku, Tokyo, 152 Japan; phones 03-3726-1111 ext. 3195; facsimile 03-3729-1131

survey, including many personal interviews with the local residents, and a questionnaire sheet survey. Focusing Analysis on the attitudes of the local residents towards the plan, we found an attitude shift of local residents in the dispute. At the outset, proponents and opponents were almost equal in numbers, and therefore there was serious conflict. However, 5 months later, opponents were in the majority, and the number of unconcerned people had decreased to 9%. Analysis of the cause of this attitude shift showed that the information provided during the dispute played a very important role:

1. The major cause of the attitude shift was the availability of reliable information.
2. Lack of information results in chaos, with the dispute becoming more serious.
3. For better agreement, reliable information and face-to-face discussions among the local residents are needed.

Future Status of Airborne Infectious Diseases: Evolution and Eradication

Masayuki Kakehashi and Fumitaka Yoshinaga[1]

Abstract. We predicted the future status of airborne infectious diseases in a human population using a mathematical model of infectious diseases. The predictions were derived from the viewpoints of evolution and eradication. We predicted that, from an evolutionary viewpoint, pathogens evolve to be less virulent as the host life span increases. The condition for successful eradication was derived mathematically: susceptible host density in the absence of the pathogen should be reduced less than that in the presence of the pathogen. The predictions were considered to be valid for a wide range of infectious diseases. The extent to which the history of man and pathogens can be verified by the predictions was also discussed.

[1] Department of Public Health, Hiroshima University School of Medicine, Kasumi, Minami-ku, Hiroshima, 734 Japan; phone 082-251-1111 (ext. 2214); facsimile 082-255-8482

Development of Game-Wise CAI and the Prerequisites

Ikuo Kitagaki[1]

Abstract. When answering multiple-choice questions, students frequently use "elimination" instead of "positive selection" by looking for the right answer as a means of arriving at the correct choice. Since discrimination between these two methods is important to proper academic evaluation, the author proposes a CAI system permitting such discrimination to be conducted based upon game theory. In other words, an element of "gaming" is included within the framework of the question-answer process.

[1] Division of Research and Development, The Institute of Vocational Training, 4-1-1 Hashimotodai, Sagamihara, 229 Japan; phones 0427-61-2111 (ext. 769) (w) 0484-43-7110(h); facsimile 0427-61-9946

The author reviews the results of this system in light of both game statistical decision theory, and discusses the prerequisites to construct this framework. In the framework, if a student answers correctly, whether he/she gains points is dependent upon whether the answering was done by "elimination" or "positive selection."

Heat Trap: Methodological Considerations for a Policy Exercise on Greenhouse Gas Emissions

Jan H.G. Klabbers[1]

Abstract. In social systems an almost limitless number of ill-structured situations exist that cannot be dealt with in isolation. They hardly qualify as problems. Problem solving per se as encountered in mathematics and physics does not provide an appropriate context for dealing with issues such as environmental planning, health care planning, urban planning, and so on. It is against this background that gaming and simulation will be discussed. Gaming provides a language for combining the social/human domain with the physical, technological, and economic domains. Three types of learning environments will be distinguished for integrating these knowledge domains. The methodology presented here provides a framework for dealing with complicated issues such as sustainable development of social systems. This will be illustrated through specifications for a policy exercise on greenhouse gas emissions.

[1] Department of Public Administration, Erasmus University Rotterdam, PO Box 1738, 3000 DR Rotterdam, The Netherlands; phone 010-4082397; facsimile 010-4527842; Center for Policy Analysis and Advice Nijmegen (CBA/N), St. Canisiussingel 26, 6511 TJ Nijmegen, The Netherlands; phone 080-240222 (w) phone 08811-62455 (h); facsimile 080-238790

History Education and the Use of Simulation/Games

K. Yoshihiro Kuriyama[1]

Abstract. History as a discipline has been made a sine qua non of precollege and university lower division schooling in order to foster proper citizenship in prospective new members of a society. In performing this task history has helped develop ethnocentric biases in young minds. The use of person-computer simulation may help students overcome the pitfalls of history education and acquire a sense of global interdependence.

[1] Department of History, Indiana State University, Terre Haute, Indiana 47809 US; phones 812-237-2714, 812-237-2710

Modeling and Managing a Top Management Game by an Expert System Tool

Toshiro Kurozawa[1]

Abstract. In university education, the effectiveness of a top management game decreases if the same model is repeated every year. Instructors must modify the game programs directly to change rules and parameters, a time-consuming and problematic process. In this paper an expert system structure is proposed which supports modeling and managing of a top management game so as to avoid such programming efforts. It is composed of an expert system shell, game programs, and some blackboards, and is executable by NEC-PCs. The production-system inference engine modifies the market size and the share of each firm on a market rules-basis. It also provides diagnostics for the players on a financial diagnostic rules-basis. Instructors can easily develop and operate many games with different rules every year using the original editor of the expert system. In addition, suitable diagnostics for the players can improve the quality of education.

[1] Department of Industrial and Systems Engineering, Faculty of Engineering, Setsunan University, 17-8 Ikeda-nakamachi, Neyagawa, Osaka, 572 Japan; phone 0720-26-5101 (w); facsimile 0720-26-5100; e-mail c62042g@ccsun01.center.osaka-u.ac.jp

Differences that Make a Difference: Intercultural Communication, Simulation, and the Debriefing Process

Linda Costigan Lederman[1]

Abstract. Communication is a fundamental life process through which humans learn to make sense of their worlds and relate themselves to other people. Many of the dynamics associated with that process are culture-bound such as the explicit and implicit rules surrounding language choice and usage. To the extent that much of human communication is culture-bound, teaching about communication involves teaching about culture and its effects upon the process of communication. There is a natural isomorphy between experiential learning and the study of communication in general, and intercultural communication in particular. Experiential learning in the classroom incorporates those real life processes into the educational setting in order for them to be used and scrutinized. The heart of these sorts of learning experiences is the post-experience analytic process, generally referred to as the debriefing session. This essay focuses on the debriefing process as it accompanies one form of experiential learning, simulations and games. It provides an analysis of the debriefing process and effective strategies for its use, with special attention to debriefing in intercultural communication learning contexts.

[1] Department of Communication, Rugters University, New Brunswick, NJ US 08903; phones (908) 932-8285 (w) (609) 921-2911 (h); facsimile 908-932-6916; e-mail lederman @ zodiac

Capturing Organizational Knowledge Through Modeling

Yihwa Irene Liou[1] and Kunihiko Higa[2]

Abstract. Acquiring organizational knowledge from managers and subsequently representing that knowledge in forms such as rules are vital to the building of an organization model. It would be possible to integrate expert systems based on such an organization model with existing organizational information systems. This paper introduces the Structured Object Model (SOM) as a means to acquire and represent organizational knowledge to build expert systems. The benefits of SOM and the application of SOM to the expert system development process are discussed. The paper concludes with the limitations of this method and future research directions.

[1] Information and Quantitative Sciences Department, Merrick School of Business, University of Baltimore, 1420 North Charles Street, Baltimore, Maryland 21201 US; phone 301-625-3420; facsimile 301-752-2821; e-mail Bitnet: earvlio@ube, Internet: earvlio@ube.ub.umd.edu
[2] Information Technology for Management, College of Management, Georgia Institute of Technology, Atlanta, Georgia 30332 US; phone 404-894-4365; facsimile 404-894-6030; e-mail Bitnet: khiga@gtri01

Modeling and Diagnosing a Misconception by Hypothesis-Based Reasoning for ITS

Noboru Matsuda[1] and Toshio Okamoto[2]

Abstract. This paper describes a framework to infer a student's misconception from observed errors during problem solving processes. A human teacher can generate hypotheses about reasons for an error by observing a student's problem solving process. He or she is also able to identify the student's misconception during the verifying process of these hypotheses. Furthermore, by using these hypotheses he can generate new tasks to evaluate the student's understanding level. In this way, appropriate instructions based on the student's knowledge structure can be provided. To accomplish such a behavior within an ITS, we have defined a domain model and applied hypothesis-based reasoning to diagnose the student model. When the system finds an error in a student's problem solving process, it attempts to generate hypotheses which explain that error in terms of the domain model.

[1] Department of Center for Computer Assisted Instruction, Faculty of Engineering, Kanazawa Institute of Technology, 7-1 Ohgigaoka, nonoichi, Ishikawa, 921 Japan; phone 0762(48)1100; facsimile 0762(48)6189; e-mail mazda@cai.kanazawa-it.ac.jp
[2] Department of Educational Information Science, Faculty of Education, Tokyo Gakugei University, 4-1-1 Nukuikita, Koganei, Tokyo, 184 Japan; phone 0423(25)2111; facsimile 0423(22)9898; e-mail C05713@sinet.ad.jp

On Problem Solving and Decision Making in MUSAS, a Musical Arrangement System

Tatsuya Mikami[1] and Kazuo Inoue[2]

Abstract. MUSAS is a system for automatic music arrangement. In our research, the definition of musical arrangement is "to get a four-part melody through selection of appropriate chords from the given monotonic melody," and the problem of musical arrangement includes decision making using information combined with fuzziness and uncertainty. It is necessary to observe and analyze the intelligence of human thinking processes in order to build intelligent systems with higher performance. Therefore, we regard the realization of MUSAS not only as a development of expert systems, but also as a simulation of human thinking processes with intelligence through the problem of musical arrangement. MUSAS uses three stages to process the musical arrangement, the melody interpretation stage, the chord selection stage, and the harmony generation stage. In these stages, there are some sub-knowledge-based systems to solve each problem independently. But, in each stage, plural solutions obtained by some sub-systems have to be reduced to one solution for one problem. Therefore, MUSAS consists of a cooperative-distributed problem solver with strata. The prototype system was implemented for Japanese nursery songs, and, we are now developing a second system for jazz standards. In this paper, the problem of the musical arrangement, mainly focusing on the problem of chord selection, is described, and the results of the simulation of chord selection through sub-systems are shown.

[1] Advanced Software Technology and Mechatronics Research Institute of Kyoto (ASTEM RI), 17 Chudoji Minami-machi, Shimogyo-ku, Kyoto, 600 Japan; phone 75-315-8652; facsimile 75-315-2898; e-mail mikami@astem.or.jp
[2] Department of Computer Science and System Engineering, Faculty of Science and Engineering, Ritsumeikan University, Tojiij Kita-machi, Kita-ku, Kyoto, 603 Japan; phone 75-465-1111; facsimile 75-465-8239

Simulation of Polarization from Each Group Decision

Mieko Nakamura[1]

Abstract. Group polarization is a tendency in which the mean of group decisions is shown to be more extreme than the mean of individual decisions. Though this is a stable tendency, it says very little about each group decision. In this paper, we simulated group polarization in terms of each group decision. We assumed that each group has a focal point to which every group member concentrates his/her attention. The focal point refers to a tacit understanding, concerning one, whose decision is considered to be right for the group decision. To calculate the focal point and to predict each group decision, we used the Meta-Contrast Ratio (MCR). Accumulating the predicted group decisions, we simulated group polarization.

[1] Faculty of Economics, Ryutsu Keizai University, Hirahata 120, Ryugasaki, Ibaraki, 301 Japan; phones 297-64-0001 (w) 298-73-2589 (h); facsimile 297-64-0011

Gaming/Simulation for Research into Road Pricing

Toshinori Nemoto[1]

Abstract. A gaming/simulation model has been developed in order to test hypotheses on the decision-making process concerning the introduction of road pricing. Although economic theory holds that road pricing brings net benefits to cities with traffic problems, in most cases it has not been accepted by the public. This implies that the real decision-making process is ruled not only by the "efficiency" that the economic model is based on, but also by other criteria. This paper reviews the political controversies in several cities and clarifies the dynamics where a social decision, either for or against road pricing, was made. It is thus hypothesized that the real issue is how distribution effects result from the policy package, including uses of the tolls collected. A gaming/simulation model on road pricing is then introduced. In order to make the outcomes of simulations understandable, the model focuses on the economic motivation of the players and predicts their behavior with relation to gain or loss.

[1] Faculty of Economics, Fukuoka University, Nanakuma Jonan-ku Fukuoka, 814-01 Japan; phone 092-871-6631; facsimile 092-864-2904

A Simulation Based on a Self-Referential Model of Organizational Intelligence

Toshizumi Ohta[1]

Abstract. Interactions between managers and subordinates were analyzed, within the context of a self-referential system, using data derived from an actual opinion survey. As a result, three patterns were obtained that can be understood as stable reaction patterns. Each of these patterns can be considered as a loop having its own identity, in which the manager's behavior and reaction of his subordinates are made stable. We find here an example of a self-referential system endowed with its own identity in a group or an organization. The existence of these three loops suggests various problems to be resolved, and they are essential for an understanding of any self-referential system. However, there are also issues regarding non-constant aspects about these loops. The question to be asked in this connection is the following: if there arises any confusion in the group at some point of time, how will the confusion be cleared away? Will it be settled by being shifted or passed on to another loop, or by the effect of some compensatory actions? In order to examine transitions between these loops and production of new loops, a simulation model of interactions between managers and subordinates can be effective in discussing a new problem solving capacity of organizations or organizational intelligence.

[1] Department of Knowledge-Based Information Engineering, The Toyohashi University of Technology, 1-1 Tempaku, Toyohashi, 441 Japan; phones 0532-47-0111 (w) 0532-47-0536 (h); facsimile 0532-47-5301; e-mail ohta@miel.tutkie.tut.ac.jp

Business Games in Managerial Training/Development and the Transition to a Free Market Economy

Eduard Rădăceanu[1]

Abstract. Some considerations concerning the general situation of the Romanian economy and society are presented to better understand the major changes in the approach to management training and development there. This paper examines several business-games developed by the author that have been used successfully over a long period of time in management training/development programs. The paper then evaluates efforts and directions for adaptation or even complete change in Romanian society. Finally, the paper emphasizes the utility of active, participative methods of management teaching/learning, such as business games, complex case studies, and simulation models, as well as how these methods may accelerate the implementation of economic reform.

[1] Institutul Român de Management, şos. Odăi no.20, Bucureşti 71601, Romania; (home) str. Aleea Circului nr. 2, sc.1, ap.95, Bucureşti, Romania; phones 335250/13 (w) 100882 (h); facsimile 334902

A Simplified Simulation Model for Country Risk Evaluation

Kanji Sato[1]

Abstract. Country risk evaluation (CRE) models have been used to aid decision makers to determine whether to invest in particular countries or not. Recently, CRE models have started evolving into more sophisticated and rather complex models. However, there is doubt about the necessity of their complexity. In this paper, a new simplified CRE model is proposed and simulation results are compared with an existing CRE model. The proposed model is simple enough to be used by novice users and easily expandable. Analysis of the simulation result also suggests that the proposed CRE model may perform as well as a complex model. Lastly, some findings from this study are discussed as future research directions.

[1] Information and Computer Science, College of Business Administration, Soka University, 1-236 Tangi, Hachioji, Tokyo, Japan; phone 0426-91-2211; facsimile 0426-28-0582

Analysis and Simulation of Credential Competition

Hiroyuki Shiraishi[1]

Abstract. In most advanced nations, educational background (i.e., credentials) has come to be all important. This report sketches a model of credential competition in Japan and shows the negative influences caused by credentialism. Individual workers' traits, talents, and skills are not directly observable, so employers use credentials as one of the most important signals when screaning. In our model, workers with higher credential produce more output. Higher credentials result in higher wages, because of the added production and also the higher estimate of individual ability. This all provides incentives for workers to pretend their credentials are higher than they are in reality. Private returns for additional credentials then exceed the additional output. Furthermore, individual workers are spurred on by knowing that they share the output of workers of greater ability in a group of workers with higher credentials. In the model, there are four different classes of workers. The utility of workers of class n (U_n) depends upon the goods they consume (G) and the credential group to which they belong (E). The core of this simulation can be written:

$$U_n = G - E - 3/8 \, (E - n)^2 \quad n = 1 \ldots 4$$

where 3/8 is the fraction representing the ordeal of moving to an upper credential group. The conclusion of this simulation is that everyone except workers of one class are working in a group higher than the optimum.

---

[1] Faculty of Economics, Graduate School of Tokyo University, 7-3-1 Hongo, Bunkyo-ku, Tokyo, 113 Japan; phone 03-3812-2111

Strategy Formation in Universities: Changing Strategic Decision Processes of Loosely Coupled Systems Through Information Technologies

Shigehisa Tsuchiya[1]

Abstract. It has been widely accepted among organizational researchers and practicing administrators that "loose-coupling" and "garbage-can models" are accurate descriptions of organization and decision making in universities and colleges. They all agree that strategy formation is difficult in loosely coupled organizations. According to the dominant theory, organizations in times of crisis should tighten up their couplings and adopt centralized analytical strategies. However, since the characteristic feature of university decision making is ambiguity, if they become tightly coupled, universities will lose sensitivity and flexibility to the environment and fail to adapt themselves to environmental changes. Through intensive case studies of a private college, I have developed and verified new hypotheses: information technologies can improve strategic

---

[1] Chiba Institute of Technology, 17-1 Tsudanuma 2-Chome, Narashino, Chiba, 275 Japan; phones 0474-78-0215 (w) 03-3262-3230 (h); facsimile 0474-78-0259; e-mail NIFTY-Serve NBD03024

decision processes at universities and enable them to formulate and implement effective strategies without tightening the couplings. Unfortunately, university administrators in Japan have no concern for utilizing information technology in management. I want to make ISAGA'91 an occasion to change their attitude.

Concept Formation Model of the Shape of Two-Dimensional Multimodal Functions and Its Application to Optimization

Mitsuru Tsukamoto[1], Katsuari Kamei[2], and Kazuo Inoue[3]

Abstract. This paper describes a method of concept formation of human beings and its application to optimization. Human beings are able to grasp general shapes with only a small amount of information by using heuristics and fuzziness. We use the words "TOP," "SIDE," and "BOTTOM" of the mountain as basic concepts for representing a two-dimensional multimodal shape. The basic concepts for each point on a surface are defined by "IF-THEN" rules. First, a maximal point search experiment is carried out and the concept formation proceeding is observed. Secondly, the heuristical algorithm of human subjects is extracted. Next, the algorithm is represented using some simple rules. Finally, the human subjects' shape-grasping process is simulated on a computer. The results are comparable to concepts of shape formed by human subjects, so the similarity between them can be confirmed. The proposed method is useful in making a machine-oriented concept formation of shape such as a topographic map.

[1] Department of Computer Science and Systems Engineering, Faculty of Science and Engineering, Ritsumeikan University, Tojiin Kita, Kita-ku, Kyoto, 603 Japan; phone 75-465-1111 ext. 3792; facsimile 75-465-8239; e-mail inouelab@cs.ritsumei.ac.jp
[2] Address as[1] ext. 3663; e-mail kamei@cs.ritsumei.ac.jp
[3] Address as[1] ext. 3664; e-mail inoue@cs.ritsumei.ac.jp

A Classroom Simulator for Computer Language Education

K. Tsushima, H. Kaga[1], and K. Fujii[2]

Abstract. A training system called "Classroom Simulator" for unexperienced teachers and tutors of the BASIC language tutorial has been developed. The system can simulate the personal computer of individual students using history files from an actual classroom. The teacher can learn how to correct individual student's mistakes.and so on in a tutorial situation.

[1] Faculty of Engineering, Osaka Electro-Communication University, 18-8 Hatsucho, Neyagawa, Japan; phone 0720-24-1131
[2] Dept. of General Education, Osaka University, Matikaneyama, Toyonaka, Japan; phone 06-844-1151

One Step Beyond: Problems with Traditional Game Evaluation

Bart van Linder[1]

Abstract. Game designers claim to deal with complex problems. However, they sometimes treat complex problems as if they were simple problems, thereby destroying all kinds of rich resources, like local variation, which are needed to cope with complex problems. In this paper I argue that this approach is partly due to the traditional methodology of evaluating and designing games. New, less traditional, ways have to be opened up. As a first step, three new criteria are introduced for the evaluation of an externally structured game called SWITCHER.

[1] University of Amsterdam, Center for Innovation and Cooperative Technology (CICT), Grote Bickersstraat 72, 1013 KS Amsterdam; phone 20-525-1233; facsimile 20-525-1211; e-mail Bart_van_Linder@ooc.uva.nl; Counterpart Business Consultants, Carnegielaan 4-14, 2517 KH Den Haag, The Netherlands; phone 70-356-0828; facsimile 70-360-7599

Automating Model Formulation for Decision Support

Ajay S. Vinze and Arun Sen[1]

Abstract. Many still believe that model formulation is a by-product of selecting a proper problem or tool representation technique. The recent trend is to divorce model formulation from model execution and study it separately. By automating the model formulation process, we create an internal representation of the formulated model. In this paper, we discuss an architecture that has been developed to support the model formulation process. Model formulation in our context is defined as the process that helps transform a problem description into a solvable algebraic notation. The problems being focused on are transportation problems in the production planning domain. As a first step, a brief overview to a cognitive model developed by the authors for the model formulation process is discussed. Design features extracted from this cognitive model for facilitating the automation of the model formulation process are presented. The architecture is based on the blackboard paradigm and attempts to capture the nuances of the cognitive model. A prototype has been developed using this architecture, which bears out the viability of the approach.

[1] Department of Business Analysis and Research, College of Business Administration, Texas A&M University College Station, TX 77843-4217; phone (409) 845-1616; facsimile (409) 845-5653; e-mail VINZE@TAMCBA.bitnet, SEN@TAMCBA.bitnet

A Simulation Approach to Process Modeling in Information Systems Analysis and Design

Rosemary H. Wild[1] *and Kenneth A. Griggs*[2]

Abstract. Typical modeling techniques for information system analysis and design treat key system requirement parameters as static. In addition, system dynamics reflected in time-path behavior, such as queues and bottlenecks, are not captured in traditional information system process models. A more realistic approach to information system analysis and design, which would allow decision makers to make more informed choices on information system design alternatives, might be to include the dynamic aspects of a system and to model those components for which uncertainty exists in a probabilistic fashion. In this paper we propose a paradigm for integrating conventional process modeling in systems analysis and design with simulation modeling and analysis techniques. Simulation analysis enhances the modeling process by allowing systems analysts to experiment with and analyze alternative system designs. In addition, by including the distributional characteristics and, thus, the variability of key system parameters in the model, sensitivity analysis may be performed and the robustness of alternative system designs can be explored. Our proposed methodology for information system analysis and design is illustrated with an example of an order entry information system.

[1] Decision Science Department, University of Hawaii, 2404 Maile Way, Honolulu, Hawaii 96822; phone (808) 956-7714; facsimile (808) 956-3261; e-mail wild@uhccvx.uhcc.edu
[2] Address as [1]; phone (808) 956-7494; facsimile (808) 956-3261; e-mail griggs@uhccvx.uhcc.edu